Edited by
Hans-Joachim Leimkühler

Managing CO_2 Emissions in the Chemical Industry

Further Reading

Aresta, M. (Ed.)

Carbon Dioxide as Chemical Feedstock

2010
ISBN: 978-3-527-32475-0

Anastas, P. T. (Ed.)

Handbook of Green Chemistry

12 volume set
ISBN: 978-3-527-31404-1

Battarbee, R., Binney, H. (Eds.)

Natural Climate Variability and Global Warming
A Holocene Perspective

2008
ISBN: 978-1-4051-5905-0

Coley, D.

Energy and Climate Change
Creating a Sustainable Future

2008
ISBN: 978-0-470-85312-2

Deublein, D., Steinhauser, A.

Biogas from Waste and Renewable Resources
An Introduction
2nd, revised and expanded edition

2010
ISBN: 978-3-527-32798-0

Centi, G., Trifiró, F., Perathoner, S., Cavani, F. (Eds.)

Sustainable Industrial Chemistry

2009
ISBN: 978-3-527-31552-9

Edited by Hans-Joachim Leimkühler

Managing CO_2 Emissions in the Chemical Industry

WILEY-VCH Verlag GmbH & Co. KGaA

The Editor

Dr. Hans-Joachim Leimkühler
Bayer Technology Services GmbH
Process Design
51368 Leverkusen
Germany

Library of Congress Card No.: applied for

British Library Cataloguing-in-Publication Data
A catalogue record for this book is available from the British Library.

Bibliographic information published by the Deutsche Nationalbibliothek
The Deutsche Nationalbibliothek lists this publication in the Deutsche Nationalbibliografie; detailed bibliographic data are available on the Internet at <http://dnb.d-nb.de>.

Typesetting Toppan Best-set Premedia Ltd., Hong Kong
Printing and Binding Strauss GmbH, Mörlenbach
Cover Design Formgeber, Eppelheim

Printed in the Federal Republic of Germany
Printed on acid-free paper

ISBN: 978-3-527-32659-4

Contents

Managing CO_2 Emissions in the Chemical Industry. Edited by Leimkühler

ISBN: 978-3-527-32659-4

Preface

Dear Reader,

The aim of this book is to produce an integrated overview of the challenges facing companies operating in the chemical industry on account of climate change and the need for energy efficiency. Yet the two topics–climate change and energy efficiency–are not dealt with separately or simply side by side. The interdependencies that exist between the reduction of greenhouse gas emissions and the cutting of energy consumption in production plants are simply too great.

For the anthology, it has been possible to win a group of scientists with a broad theoretical and application-oriented horizon. This ensures not only methodical penetration of the complex material, it also allows a very precise and detailed description of the technical measures necessary for achieving the environmental targets.

Although German authors took a leading role in many of the chapters, very profound and expert contributions have also been made by scientists from Denmark, UK, Portugal and Brazil.

This reflects the global background of climate protection and energy efficiency, and underlines the need to share and exchange knowledge and experience at a global level, now more than ever.

The scope of the book is, however, broader than normal. CO_2 reduction and energy savings are not things that just 'happen' by themselves. They have to be prepared, organized and implemented, in other words, they have to be made effective via a management approach. A number of chapters deal explicitly with these important role model functions and managerial tasks.

But what is a book on climate change and energy without a vision and a challenge? The last two chapters are devoted to these topics. The articles on 'Carbon Capture and Storage' and 'CO_2-Neutral Production – Fact or Fiction' describe trends and take an initial look at the possibilities for their technical implementation. It shows how fascinating and challenging climate protection and energy supply will be for the chemical industry in the coming decades.

The book is therefore targeted not only at the practitioner but also at the broad community of people interested in being kept expertly and graphically informed about the way to Low Carbon Production.

Managing CO_2 Emissions in the Chemical Industry. Edited by Leimkühler

ISBN: 978-3-527-32659-4

Progress towards climate protection and energy efficiency is possible and necessary. It is my hope and also my firm conviction that this anthology will stimulate ideas, examples and fresh impetus in this direction.

Dr. Wolfgang Große Entrup

Senior Vice President
Head of Group Area Environment & Sustainability
Bayer AG

List of Contributors

Carlos Augusto Arentz Pereira
Petroleo Brasileiro S.A.
Petrobras
Av. Almirante Barroso 81
Centro
Rio de Janeiro RJ 20031-004
Brazil

Thomas Böhland
Evonik Degussa GmbH
Weißfrauenstr. 9
60287 Frankfurt/M.
Germany

Benjamin Brehmer
Evonik Degussa GmbH
Creavis Technologies & Innovation
Paul-Baumann-Str. 1
45764 Marl
Germany

Ana Isabel Cerqueira de Sousa Gouveia Carvalho
Instituto Técnico
Lisboa
Portugal

Alan Eastwood
KBC Process Technology Ltd
KBC House
42-50 Hersham Road
Walton on Thames
Surrey KT12 1RZ
UK

Rafiqul Gani
Danmarks Tekniske Universitet
Institut for Kemiteknik
Computer Aided Process Engineering Center
Soltofts Plads
Bygning 227
2800 Lyngby
Denmark

Roger Grundy
Breckland Ltd
Beech House
Steep Turnpike
Matlock, Derbyshire DE4 3DP
UK

Birgit Himmelreich
Bayer Technology Services GmbH
51368 Leverkusen
Germany

Managing CO_2 Emissions in the Chemical Industry. Edited by Leimkühler

ISBN: 978-3-527-32659-4

Andreas Jupke
Bayer Technology Services GmbH
51368 Leverkusen
Germany

Jörn Korte
Bayer Technology Services GmbH
51368 Leverkusen
Germany

Christos Lecou
Westfälische Wilhelms-Universität
Münster
Leonardo Campus 1
48159 Münster
Germany

Hans-Joachim Leimkühler
Bayer Technology Services GmbH
Process Design Geb. E41
51368 Leverkusen
Germany

Henrique A. Matos
Instituto Superior Técnico
Lisboa
Portugal

Zoran Milosevic
KBC Process Technology Ltd
KBC House
42-50 Hersham Road
Walton on Thames
Surrey KT12 1RZ
UK

Susanne Mütze-Niewöhner
RWTH Aachen
Lehrstuhl und Institut für
Arbeitswissenschaft
Human Resource Management
Bergdriesch 27
52062 Aachen
Germany

Stefan Nordhoff
Evonik Degussa GmbH
Creavis Technologies & Innovation
Paul-Baumann-Str. 1
45764 Marl
Germany

Markus Röwenstrunk
RWTH Aachen
Lehrstuhl und Institut für
Arbeitswissenschaft
Human Resource Management
Bergdriesch 27
52062 Aachen
Germany

Yvonne Schiemann
Evonik Degussa GmbH
Creavis Technologies & Innovation
Paul-Baumann-Str. 1
45764 Marl
Germany

Frank Schwendig
RWE Power AG
Dpt. PCR-N / CCS and New
Technologies
Huyssenallee 2
45128 Essen
Germany

Nathan Steeghs
EcoSecurities
1st Floor
40/41 Park End Street
Oxford OX1 1JD
UK

Thomas Tacke
Evonik Degussa GmbH
Creavis Technologies & Innovation
Paul-Baumann-Str. 1
45764 Marl
Germany

Nancy Wayna
B.A. Medien und
Kommunikationswissenschaft
Hausdorfstrasse 343
53129 Bonn
Germany

Martin Wolf
Bayer Technology Services GmbH
51368 Leverkusen
Germany

Trends in Energy and CO_2 Reduction in the Chemical Process Industry

Hans-Joachim Leimkühler

1
Climate Change

There are two main reasons why the chemical process industry should be motivated to reduce energy consumption and CO_2 emissions: rising concerns in companies, the public and scientific community about climate change or global warming, and the increasing fraction of energy in manufacturing costs.

'Climate change' [1] in this context, means a change of climate, which is attributed directly or indirectly to human activity that alters the composition of the global atmosphere and which is in addition to natural climate variability observed over comparable time periods.

'Global warming' [2] is the increase in the average temperature of the Earth's near-surface air and oceans since the mid-twentieth century and its projected continuation. Global surface temperature increased $0.74 \pm 0.18\,°C$ during the previous century [3] (see Figure 1).

The impacts of global warming are described in the Fourth Assessment Report [5, 6] of the Intergovernmental Panel on Climate Change (IPCC). An excerpt:

- Dry regions are projected to get drier, while wet regions are projected to get wetter: 'By mid-century, annual average river runoff and water availability are projected to increase by 10–40% at high latitudes and in some wet tropical areas, and decrease by 10–30% over some dry regions at mid-latitudes and in the dry tropics ...'.
- Drought-affected areas will become larger.
- Heavy precipitation events are very likely to become more common and will increase flood risk.
- Water supplies stored in glaciers and snow cover will be reduced over the course of the century.
- The resilience of many ecosystems is likely to be exceeded this century by a combination of climate change and other stressors.

Managing CO_2 Emissions in the Chemical Industry. Edited by Leimkühler

ISBN: 978-3-527-32659-4

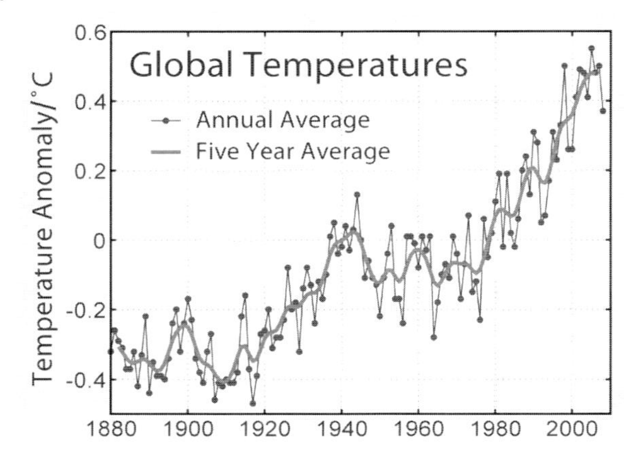

Figure 1 Development of global temperature [4].

- Carbon removal by terrestrial ecosystems is likely to peak before mid-century and then weaken or reverse. This would amplify climate change.
- Globally, the potential food production will increase for temperature rises of 1–3 °C, but decrease for higher temperature ranges.
- Coasts will be exposed to increasing risks such as coastal erosion due to climate change and sea-level rise.
- Increases in sea-surface temperature of about 1–3 °C are projected to result in more frequent coral bleaching events and widespread mortality unless there is thermal adaptation or acclimatisation by corals.
- Many millions more people are projected to be flooded every year due to sea-level rise by the 2080s.

There are many reasons to take action against climate change, the main reason being that climate change has become an important issue discussed by the public, in politics, society and industry.

The IPCC concludes that increased greenhouse gas (GHG) concentrations resulting from human activity, such as burning fossil fuels and deforestation, were responsible for most of the observed temperature increase since the middle of the twentieth century. The main GHGs in the Earth's atmosphere are water vapor, carbon dioxide, methane, nitrous oxide, and ozone. One example of that could be the concentration of CO_2 in the atmosphere, which increased from 320 ppmv in 1960 to nearly 390 ppmv by 2008 (measured at Mouna Loa, Hawaii) [7] (see Figure 2).

The IPCC also concludes that variations in natural phenomena such as solar radiation and volcanoes produced most of the warming from pre-industrial times to 1950 and had a small cooling effect afterward [8, 9].

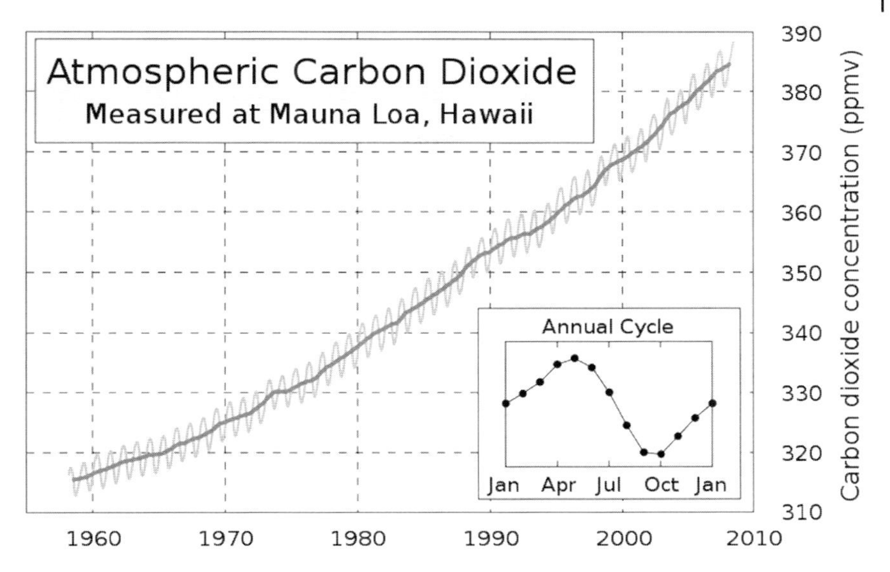

Figure 2 Development of the atmospheric CO_2 concentration [7].

These basic conclusions have been endorsed by more than 45 scientific societies and academies of science, including all of the national academies of science of the major industrialized countries [10]. A small number of scientists dispute the consensus view.

The GHG emissions are normally expressed in CO_2 equivalents because GHGs differ in their warming influence on the global climate system, due to their different radiative properties and lifetimes in the atmosphere. These warming influences may be expressed through a common measure based on the radiative forcing of CO_2. The CO_2-equivalent (CO_2e) emission is the amount of CO_2 emission that would cause the same time-integrated radiative forcing, over a given time horizon, as an emitted amount of a long-lived GHG or a mixture of GHGs. The CO_2e emission is obtained by multiplying the emission of a GHG by its global warming potential (GWP) for the given time horizon. For a mix of GHGs it is obtained by summing the equivalent CO_2 emissions of each gas. Equivalent CO_2 emission is a standard and useful measure for comparing emissions of different GHGs, but does not imply the same climate change responses. Figure 3 shows that CO_2 is the most important anthropogenic GHG (data from [5]).

Between 1970 and 2004, annual CO_2 emissions have increased by about 80%, from 21 to 38 gigatonnes (Gt), and represented 77% of total anthropogenic GHG emissions in 2004. Industry directly emitted 19.4% of the GHG in 2004 and is responsible for a part of the 25.9% of GHG emissions caused by the energy supply.

Reduction of GHG emissions is a global task to minimize the effect of climate change. As shown in Section 5 there is a strong political and societal commitment

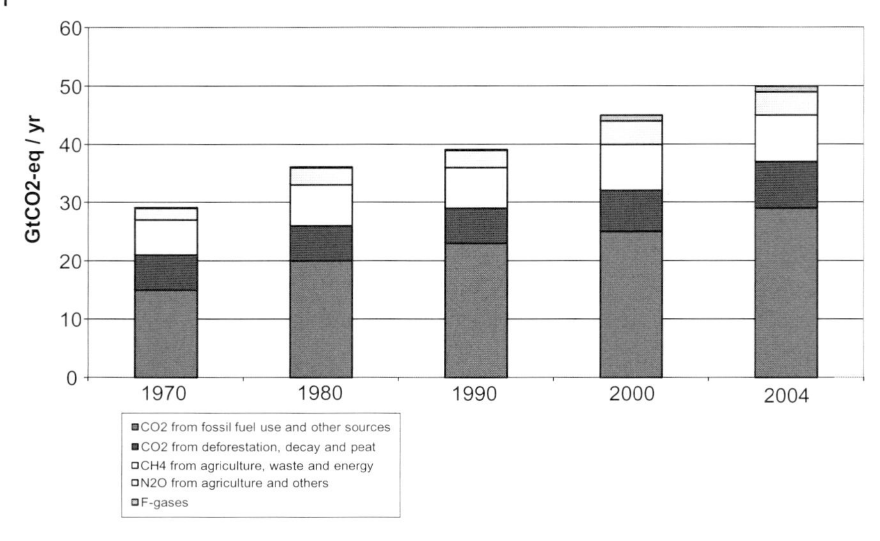

Figure 3 Global annual emissions of anthropogenic GHGs from 1970 to 2004 (after [5]).

to GHG reduction. In this book we want to discuss ways of reducing GHG, especially CO_2, from the perspective of the chemical process industry.

2
Overview of the Chemical Process Industry

By referring to the chemical process industry, we mean the industry that comprises the companies converting raw materials into different products [11]; it can be classified into the following three categories:

- The basic chemicals industry, which manufactures polymers, bulk petrochemicals and intermediates, other derivatives and basic industrials, inorganic chemicals, and fertilizers.
- The life sciences industry which includes differentiated chemical and biological substances, pharmaceuticals, diagnostics, animal health products, vitamins, and crop protection chemicals.
- The specialty chemicals or fine chemicals industry, which produces electronic chemicals, industrial gases, adhesives and sealants as well as coatings, industrial and institutional cleaning chemicals, and catalysts.

The chemical process industry is involved in the GHG topic and in climate change in two different ways:

1) On the one hand, due to the energy-intensive nature of its manufacturing processes, the chemical process industry is directly responsible for the burning

of fossil fuels and resulting CO_2 emissions. Furthermore, the chemical industry processes fossil fuels like petroleum or its derivatives as raw materials for the manufacture of different products. Therefore, it is part of the problem.

2) On the other hand, the chemical process industry delivers materials and products (like insulation materials, fuel additives and many more) that help to enhance the energy efficiency in many different fields and thereby reduces the emission of CO_2. For this reason, the chemical process industry is also part of the solution.

The International Council of Chemical Associations (ICCA) estimates that in 2005 the CO_2e emissions linked to the chemical industry amounted to about 3.3 Gt CO_2e +/− 25% [12]. The majority of these emissions, 2.1 Gt CO_2e, resulted from the production of chemicals from feedstock and fuels delivered to the chemical industry. An additional 1.2 Gt CO_2e of emissions arose during the extraction phase of the feedstock and fuel material, and during the disposal phase of the end products. Figure 4 illustrates the numbers.

The ICCA study [12] also states that gross savings by the use of a chemical product vary from 6.9 to 8.5 GtCO_2e depending on the scope and assumptions used. This translates into a gross savings ratio of 2.1 : 1 to 2.6 : 1. The gross savings ratio is the amount of CO_2e saved through the use of a chemical product measured against the amount of CO_2e emitted. In other words, for every GtCO_2e emitted by the chemical industry in 2005, CO_2e savings of 2.1 to 2.6 Gt were achieved, thanks to the products and technologies provided to other industries or users.

The ICCA study was reviewed by the Öko-Institut (in [12]). Öko-Institut challenges the savings claimed by the chemical industry and points out that 'it is

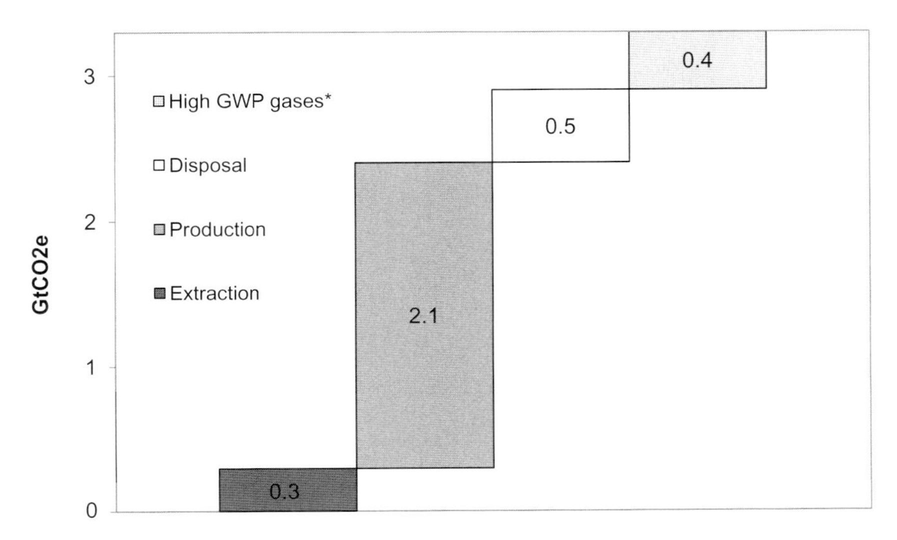

Figure 4 Total life cycle CO_2e emissions linked to the chemical industry (2005) (after [12])* Gases with high Global Warming Potential like SF_6 (see [12] for details).

problematic to impute specific shares of products' potential of avoidance to individual players as the life cycle of many products important for the protection of the climate crosses many sectors of industry and consumers and is dependent on various political ancillary conditions. (...) Hence, it is impossible in an economic market system that producers still may impute the ecological benefits proprietarily, especially if beyond that financial implications may be deduced (...). In the end, only the consumers who have bought the product and who are their owners may claim this ecological benefit.'

Nevertheless it is the goal of the chemical industry to further reduce the specific emissions of GHG per ton of product. A study by McKinsey ([13], cited in [12]) leads to the result that the 2.1 Gt CO_2e emissions linked to the chemical production in 2005 can be nearly equally assigned to direct energy emissions (fuel consumption required by the process to run, 0.6 Gt CO_2e), indirect energy emissions (energy generated off-site, 0.8 Gt CO_2e) and process emissions (mainly N_2O, CO_2 and chlorofluorohydrocarbons, 0.7 Gt CO_2e). In other words, two-thirds of the GHG emission of the chemical production is caused by its energy consumption. The direct energy emissions for different fuels and regions are presented in Figure 10, in Section 3.2. Energy efficiency is therefore an important lever for GHG reduction and a way to achieve the GHG reduction goals of the industry.

At the same time energy has gradually become an important cost factor. Energy costs amount to a major fraction of the manufacturing costs of a product. Enhancing the energy efficiency of the production means therefore not only reduction of CO_2 emissions but also reduction of costs. This is an additional reason why the issue of CO_2 emissions should be addressed by companies who want to remain competitive.

3 Energy Consumption, CO_2 Emissions and Energy Efficiency

3.1 Energy Consumption and CO_2 Emissions in General

Energy consumption increases with economic development in most countries. From 1990 to 2006 the total primary energy consumption rose from 370 to 500 Exajoule (EJ, 1 EJ = 10^{18} J) or 35% [14], as shown in Figure 5.

The biggest rates of increase can be observed in Asia and Middle East (plus 110% each), whereas the rates of increase are the lowest in Eurasia (minus 25%) and Europe (plus 13%). This reflects the economic development of these regions, but on the other hand, it is also a signal for an enhanced energy intensity of the industry.

The energy consumption (500 EJ in 2006) has been satisfied by a relatively stable mix of energy sources since 1990. Fossil fuels like petroleum, coal and natural gas and electricity from nuclear and hydropower are the main energy sources. Other renewable energy sources, here named as 'others' in the graphic, still only play a

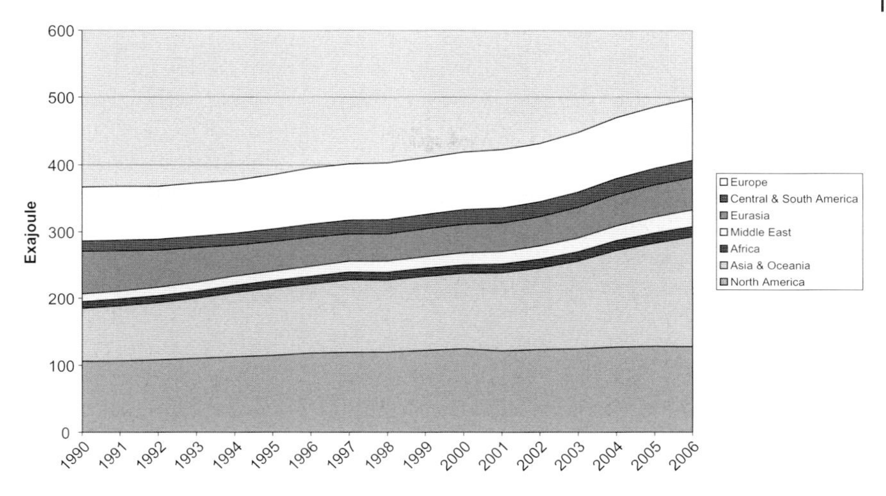

Figure 5 Primary energy consumption by region (data from [14]).

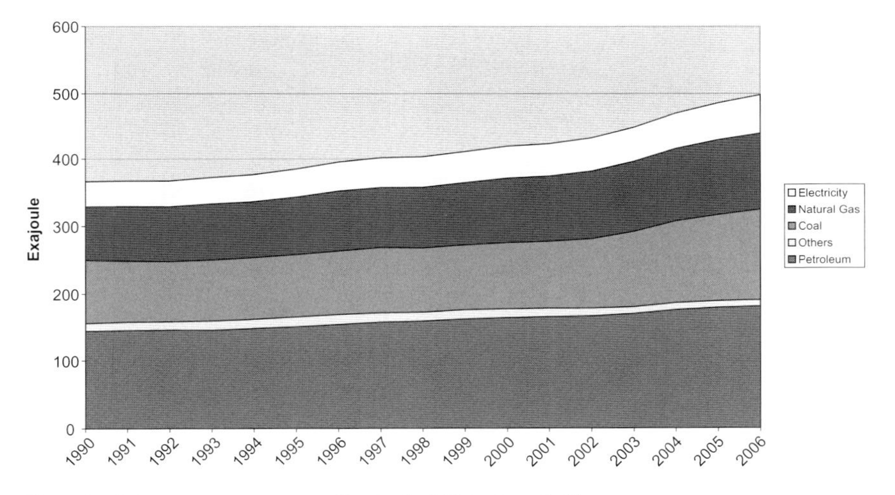

Figure 6 Primary energy consumption by fuel (data from [14]).

negligible role (except hydroelectric power generation). Figure 6 shows the development.

The highest rates of increase can be seen for electricity (plus 57%), whereas the growth rates for natural gas and coal are smaller (plus 43%). Petroleum with plus 26% has grown subproportionally. Nevertheless the fraction of different energy sources of the total consumption did not change significantly, as can be seen from Table 1.

Petroleum remains the main energy source with 37–38%, coal (25–27%) and natural gas (21–23%) are the other important elements of our energy supply.

Table 1 Energy mix 1990 and 2006 (consumption of energy sources in % of the total consumption) (data from [14]).

	1990	2006
Petroleum	38.17	36.38
Natural Gas	21.09	22.86
Electricity	9.94	11.84
Coal	24.96	26.99
Others	3.13	1.93

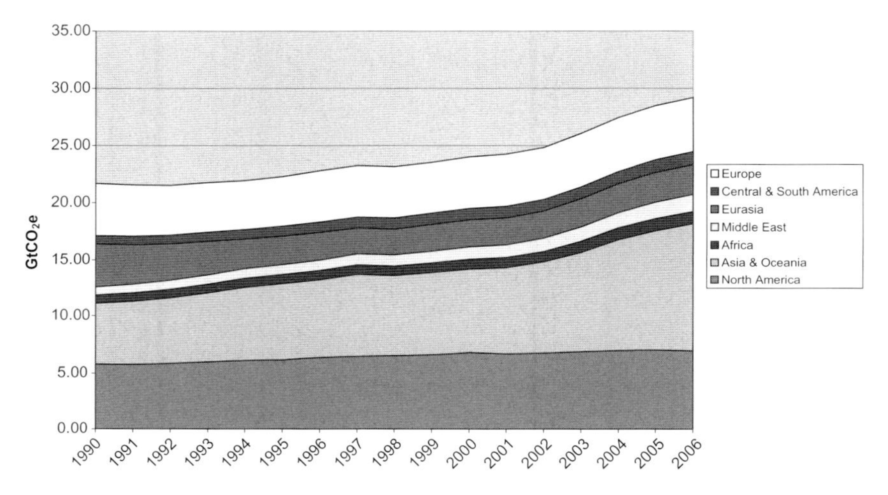

Figure 7 CO_2e emissions from the consumption of energy by region (data from [14]).

The emission of GHG, expressed as CO_2e, is of course closely connected to the energy consumption and the mix of energy sources. Since renewable energy sources still are small on a global scale, the GHG emission from fossil fuels has risen in the same range as the energy consumption.

In Figure 7 we see the total emissions from the consumption of energy shown for the regions.

We measure the emissions in gigatonnes of CO_2 equivalents (GtCO_2e). Development from 1990 until 2006 is more or less parallel to the development of the energy consumption. Starting from 22 GtCO_2e in 1990 the emissions rose to 29 GtCO_2e in 1990, the same 35% increase we could observe for the energy consumption. The main increases come from Asia (plus 112%) and Middle East (106%), whereas Eurasia (minus 32%) and Europe (plus 3%) show a slower and more climate-friendly development.

Looking at the different energy sources in Figure 8 we see that petroleum and coal are the main contributors to GHG emissions with similar absolute CO_2e amounts. Since coal emits more CO_2 per energy unit, its fraction is relatively high in comparison to its contribution to the total energy supply. Today coal is the main source for CO_2 emissions (12 GtCO_2e).

Natural gas with its higher hydrogen content emits fewer GHGs per energy unit and is therefore the more climate friendly fuel.

In Table 2 we show the contributions of the fuels to the total GHG emission in 1990 and 2006.

Coal and petroleum each carry about 40% of the CO_2 load, whereas natural gas with 20% has the smaller part of the emissions. There is no noticeable difference between the relations in 1990 and 2004. Looking at Tables 1 and 2 a remarkable difference for coal between its fraction of energy consumption (27% in 2006) and its fraction of GHG emissions (42% in 2006) can be noticed. These numbers reveal that the higher carbon fraction in coal compared to petroleum or natural gas results in a high specific CO_2 emission per energy content.

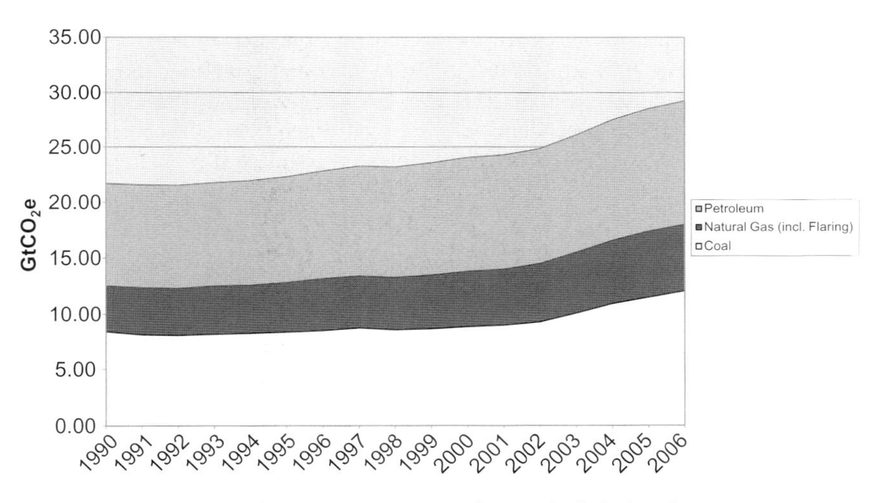

Figure 8 CO_2e emissions from the consumption of energy by fuel (data from [14]).

Table 2 GHG emissions (emissions from energy sources in % of the total emission) (data from [14]).

	1990	2006
Petroleum	42.00	38.43
Natural gas (incl. flaring)	19.22	20.25
Coal	38.78	41.32

For the consequences of GHG emissions, especially climate change, the absolute amount of emissions is the decisive factor. But in order to document the economic, scientific and industrial development we want to add a graphic describing the GHG intensity of the national economies. In Figure 9 we see the world CO_2 emissions from the consumption and flaring of fossil fuels per gross domestic product for the regions (in metric tons of CO_2 per 1000 US$ GDP, using market exchange rates of the year 2000). Data before 1994 are not available globally, therefore we show the period from 1994 to 2006.

Most regions have carbon intensities of the economy between 0.5 and 1 ton of CO_2 per 1000 US$ of GDP. The Middle East and Africa lie between 1.4 and 1.9 tons of CO_2 per 1000 US$ of GDP and the Eurasian countries show a development from 7 tons of CO_2 per 1000 US$ of GDP in 1994 to a value of 4.5 in 2006. Partially the flaring of natural gas is responsible for these high values, but the carbon intensity of these regions is nevertheless very high in comparison to the developed countries.

The development of the carbon intensity is shown in Table 3. Here we quote the values in tons of CO_2 per 1000 US$ of GDP for the years 1994 and 2006 and their percentage development.

The most developed countries (Europe and North America) reduced their carbon intensity by around 20% since 1994, whereas Asia still is getting more carbon intensive. The petroleum and natural gas producing regions (Eurasia, Africa) could reduce the carbon dependence of their GDP since 1994. The specific global CO_2 consumption increased by 8% since 1994 and lies now at 0.77 t CO_2 / 1000$ GDP.

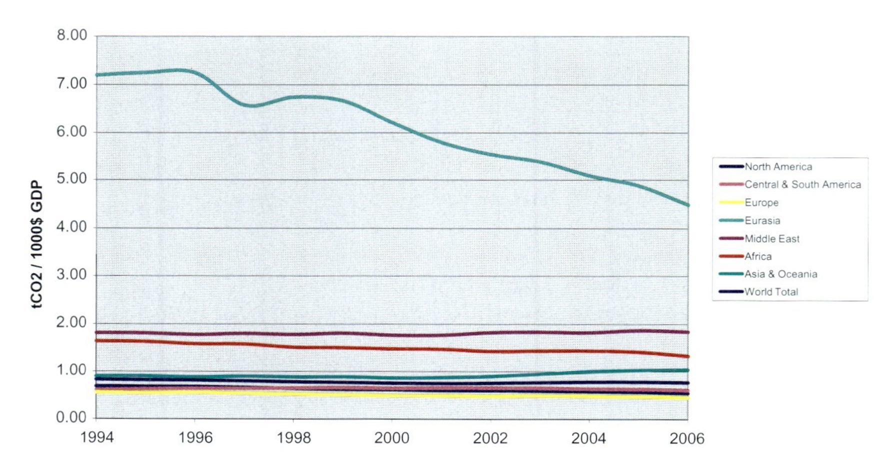

Figure 9 World carbon dioxide emissions from the consumption and flaring of fossil fuels per thousand dollars of gross domestic product using market exchange rates (metric tons of carbon dioxide per thousand (2000) US$) (data from [14]).

Table 3 CO_2 emissions from the consumption and flaring of fossil fuels per gross domestic product of the regions in 1994 and 2006 (in metric tons of CO_2 per 1000 US$ GDP, using market exchange rates of the year 2000) (data from [14]).

	1994	2006	Development from 1994 to 2006 in %
North America	0.68	0.54	−21
Central and South America	0.61	0.61	−1
Europe	0.55	0.45	−18
Eurasia	7.20	4.49	−38
Middle East	1.81	1.84	+2
Africa	1.64	1.33	−19
Asia and Oceania	0.90	1.04	+15
World total	0.83	0.77	−8

3.2 Energy Consumption and CO_2 Emissions in the Chemical Industry

The chemical process industry is an energy intensive industry. The ICCA study [12] shows that the fuel consumption linked to the chemical industry amounts to 9 Exajoule in 2005. Here the fuel consumption for energy generation is considered, but not petroleum, coal or natural gas used as raw materials. The global energy consumption in 2005 was about 485 EJ (see Figure 5 in Section 3.1), that means the chemical industry consumes about 1.9% of the total global energy.

Figure 10 shows the breakdown of the fuel consumption for energy generation linked to the chemical industry by regions and fuels.

We see that on a global scale ('total') natural gas is the main energy source for the chemical industry and covers more than half of the consumption (58%). The significance of coal and oil is similar. Both cover less than a quarter of the energy requirements of the chemical industry in the world (21% each).

The picture changes if we look at the relevance of the energy sources in the regions. In Asia coal and oil are by far the most important energy sources, whereas gas dominates in Europe, America and Middle East. This asymmetric distribution of the energy sources is one of the reasons for the lower carbon intensity of the economies in Europe and America in comparison with Asia (see Section 3.1).

There is a very close correlation between the fuel consumption for the generation of energy used for the chemical industry in Figure 10 and the GHG emissions linked to the chemical industry in Figure 11.

On a global scale natural gas as the main energy source is also the main source of CO_2 emissions. Because of the higher hydrogen content of natural gas in comparison to coal and petroleum the fraction is lower than for the energy consumption (47% of the total CO_2 emission instead of 58% of the total energy consumption).

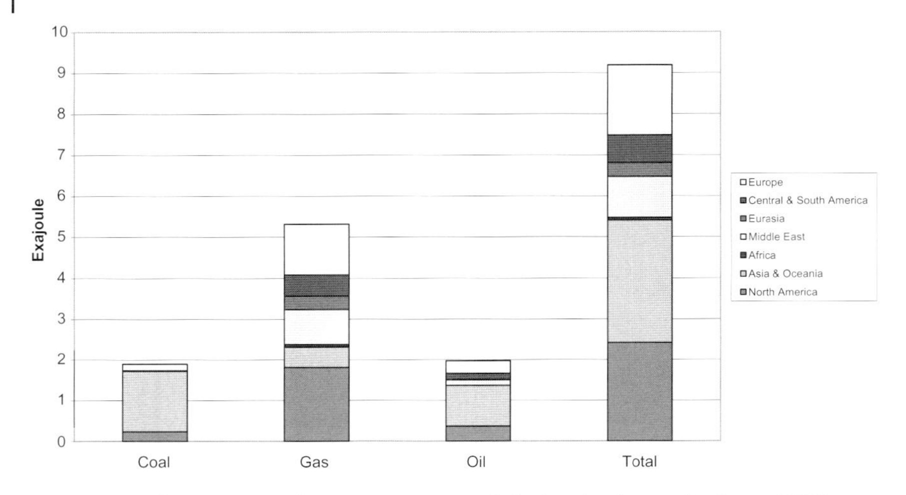

Figure 10 Fuel consumption for energy generation linked to the chemical industry (2005) (data from [12]).

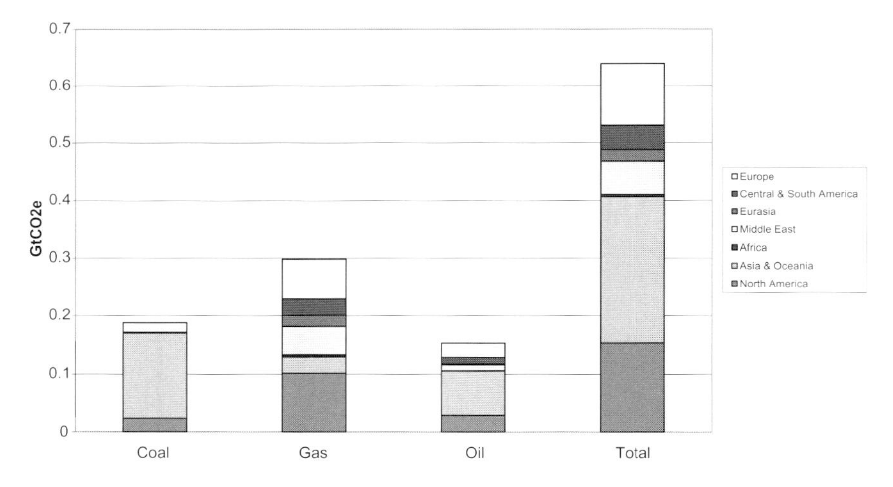

Figure 11 GHG emissions linked to the chemical industry (2005) (data from [12]).

Coal has the highest C/H relation and therefore emits relatively much GHG (29%) in comparison to its 21% contribution to the energy supply.

By the analysis of data from the perspective of regions it follows that countries with a higher fraction of energy generation from natural gas contribute specifically less to CO_2 emissions than countries which base their energy generation on coal or petroleum. That leads for instance to the result that Asia's chemical industry (with a high energy generation from coal) emits 40% of the global GHG emissions although the energy consumption is only 33% of the global value.

These discussions show that the energy mix is one of the most important levers for the reduction of the climate impact of the industry. The increased use of natural gas for energy generation instead of coal or petroleum leads to the reduction of GHG emissions, although a shift from coal or oil to natural gas will not influence the energy costs dramatically. Costs will only be reduced by achieving enhanced energy efficiency.

3.3 Energy Prices

We all noticed in recent years that energy prices, and by this energy costs, have risen dramatically. As an example in Figure 12 we see the petroleum price development over the last 20 years. While the petroleum price was relatively stable during the 1990s at a level around 20$ per barrel we observe a strong increase in the last 10 years. In 2008 the petroleum price rocketed, achieving its peak at 140$ per barrel. Although it came down during the economic crisis in 2009 the prices of petroleum and other energies sources will probably continue to rise in the long term.

When we talk about energy prices we have to bear in mind two different effects that have impact on the chemical industry: on the one hand, chemical industry consumes energy for operating its plants. Fuels like petroleum, coal or gas are needed for the energy generation. On the other hand most of the raw materials, which the chemical industry transforms into products, are based on the same compounds that are used for energy generation: mainly petroleum and gas, to a lesser extent coal. For instance polymers, agrochemicals or even pharmaceuticals are mostly based on petroleum as raw material.

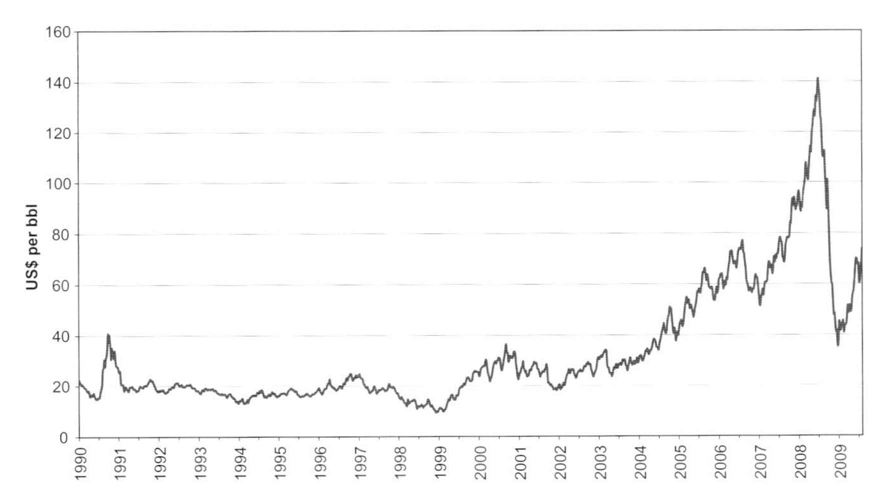

Figure 12 Development of petroleum price (Weekly Europe Brent Spot Price FOB) (data from [12]).

As a consequence, rising fuel prices affect the chemical industry in two ways: by increased production costs and by increased manufacturing costs. Dow emphasizes for instance [15] that overall energy costs (fuel and raw materials) at Dow in 2005 exceed 50% of total costs.

The question for the chemical industry now is: How to react to rising fuel prices in order to stay competitive? Two possible levers are: less raw material input by yield increase and less energy consumption by higher energy efficiency. In this book we mainly focus on enhançement of the energy efficiency in the chemical process industry.

3.4
Energy Efficiency in the Chemical Industry

Energy efficiency has become one of the key topics not only in the chemical industry. There are three main reasons for that development:

1) As mentioned in the last section energy prices rose significantly during the last ten years. In parallel with the energy prices the energy costs increased as a percentage of the production costs. Therefore there was and there is a high pressure on the industry, and especially the energy-intensive chemical industry, to work on the reduction of the specific energy consumption.

2) Secondly climate change became an ever more important issue in public discussion. The targets given by politics and public and the self-commitments of the industry led to an increased level of attention to energy efficiency and by this to additional efforts in the chemical process industry.

3) Last but not least many employees and managers of the chemical companies feel the need for responsible management of resources. They influence and define the culture and the behavior of companies. This leads to company initiatives for CO_2 and energy reduction that will be described in Section 6.

To quantify energy efficiency we follow the CEFIC, the European Chemical Industry Council, which defines energy intensity as the energy input per unit of chemicals production [16]. Since 1990 this energy efficiency has increased significantly. Figure 13 shows the development since then for the European chemical and pharmaceutical industry (for EU countries).

While production has increased by 3.3% per year on an average, the absolute value of the energy consumption has remained more or less constant. That means for the energy efficiency as defined above, that it has decreased by 4.6% every year for the last 16 years or by 50% in total. This result is an impressive proof that the chemical industry has understood the need for action in the field of responsible use of natural resources.

But also in the USA energy efficiency has increased in recent decades. The American Chemical Council (ACC) states that since 1974 the business of chemistry in the USA has reduced the energy consumed per unit of output by half [17].

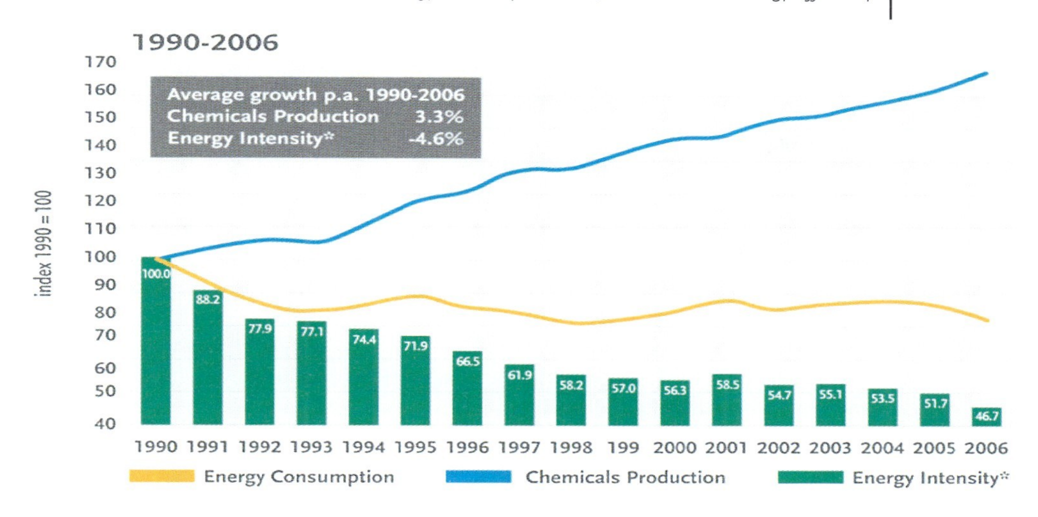

Sources: Cefic Chemdata International and Eurostat
*Energy intensity is measured by energy input per unit of chemicals production
** Including pharmaceuticals

Figure 13 Energy intensity in the EU chemicals industry [16].

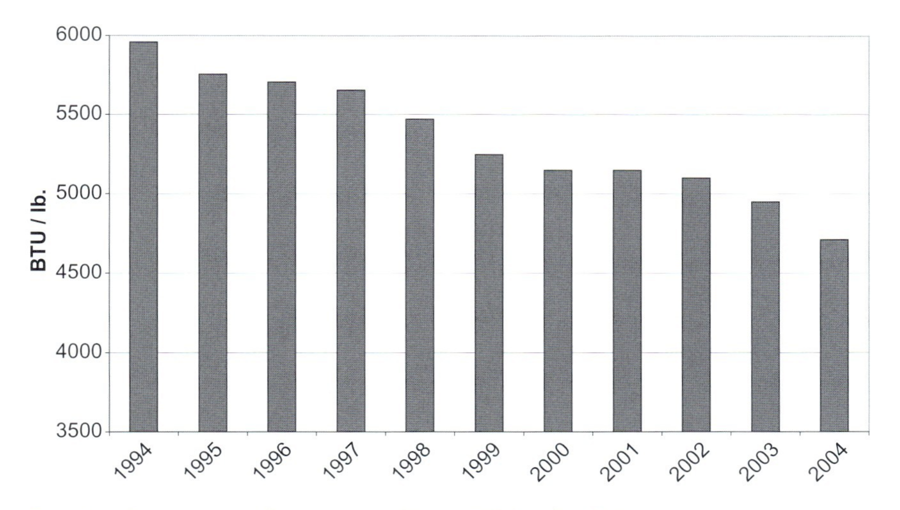

Figure 14 Dow's energy efficiency from 1994 until 2004 (after [18]).

As an example of a chemical company we can look at Dow, who published [18] that in the ten years from 1994 to 2004 the energy efficiency was enhanced by 21% from 5958 BTU lb^{-1} to 4711 BTU lb^{-1}. The development is shown as an example in Figure 14.

Since then the energy consumption could be reduced until 2007 by another 8% to 3811 $BTU\,lb^{-1}$ [19].

As seen in the example of Dow many companies could reduce their energy consumption in recent decades and at the same time increase their production capacity. This development is still ongoing and we will see a further enhancement of the energy efficiency in the chemical industry.

4
Political Framework and Trends

The beginning of the political activity against climate change was the UN Conference on Environment and Development in Rio de Janeiro in June 1992, also called The Earth Summit [20]. Three international treaties were the main result on this conference: the United Nations Framework Convention on Climate Change (UNFCCC), the Convention on Biological Diversity and the Convention to Combat Desertification. These conventions all involve matters strongly affected by climate change.

The starting point for the International Climate Policy is the UNFCCC. The Convention sets an ultimate objective of stabilizing GHG concentrations 'at a level that would prevent dangerous anthropogenic (human induced) interference with the climate system.' It states that 'such a level should be achieved within a timeframe sufficient to allow ecosystems to adapt naturally to climate change, to ensure that food production is not threatened, and to enable economic development to proceed in a sustainable manner' [20].

The Convention is a 'framework' document and is to be amended or augmented over time so that efforts to deal with global warming and climate change can be focused and made more effective. The first addition to the treaty, the Kyoto Protocol, was adopted in 1997 (see Section 5.1 and Chapter 2).

The Kyoto Protocol committed the industrial nations involved to cutting greenhouse gases as a whole by 5.2% between 1990 and 2008. The Kyoto Protocol expires in 2012. Despite the protocol, GHG emissions rose globally by around 30% between 1990 and 2005 (see Figure 3). The increase is caused mainly by the rapidly emerging countries Brazil, India and, in particular, China. In 2008, China topped the list of CO_2-emitting countries in absolute emission amounts, having about a quarter share in global CO_2 emissions (24%), followed by the USA (21%), the EU-15 (12%), India (8%) and the Russian Federation (6%) [21]. Values per capita show a different order. The top five regions in CO_2 emission per person are: USA (19.4 t CO_2 per person), Russia (11.8 t CO_2 per person), EU-15 (8.6 t CO_2 per person), China (5.1 t CO_2 per person) and India (1.8 t CO_2 per person) [21].

There are approaches to use the per capita values as target for future CO_2 emissions. The German Chancellor Merkel presupposed that the industrialized countries cut their share of energy consumption as far as possible, thus reducing per capita emissions of CO_2. The emerging economies, on the other hand, need to grow if they are to reduce poverty. The downside is, of course, that their emissions of CO_2 will continue to rise in the years to come. In the final analysis the per capita emissions in emerging economies will meet those of industrialized countries. If the agreement is to be just, the emerging economies should not be permitted to emit more CO_2 per capita than the industrialized countries [22].

5
Kyoto Process and National Programs

5.1
Kyoto Protocol

The Kyoto protocol and its instruments will be discussed in detail in Chapter 2.

The Kyoto Protocol [23] is an international agreement linked to the United Nations Framework Convention on Climate Change [24]. The major feature of the Kyoto Protocol is that it sets binding targets for 37 industrialized countries and the European Community for reducing GHG(emissions. This amounts to an average of five per cent against 1990 levels over the five-year period 2008–2012.

The major distinction between the Protocol and the Convention is that while the Convention encouraged industrialized countries to stabilize GHG emissions, the Protocol commits them to do so.

Recognizing that developed countries are principally responsible for the current high levels of GHG emissions in the atmosphere as a result of more than 150 years of industrial activity, the Kyoto protocol places a heavier burden on developed nations under the principle of 'common but differentiated responsibilities.'

Under the treaty, countries must meet their targets primarily through national measures. However, the Kyoto Protocol offers them an additional means of meeting their targets by way of three market-based mechanisms.

1) emissions trading (ET)
2) clean development mechanism (CDM)
3) joint implementation (JI).

The mechanisms help stimulate green investment and help parties meet their emission targets in a cost-effective way.

5.2
Flexible Mechanisms [24]

Emissions Trading Parties with commitments under the Kyoto Protocol (Annex B Parties) have accepted targets for limiting or reducing emissions. These targets are expressed as levels of allowed emissions, or 'assigned amounts,' over the period of commitment between 2008 and 2012. Emissions trading, as set out in the Kyoto Protocol, allows countries that have emission units to spare (i.e., emissions permitted to an individual country, that are higher then the level of current emissions) to sell this excess capacity to countries that are over their targets.

Clean Development Mechanism (CDM) Under the CDM, countries with emission caps (Annex I countries) assist non-Annex I parties, which do not have emission caps, to implement project activities to reduce GHG emissions (or remove CO_2 by sinks). Credits are issued on the basis of emission reductions achieved through the project activities. These credits which are known as certified emissions reductions (CERs, in t CO_2e) can be bought and sold. This injects financial incentives into the system. Purchasing these CERs can help companies to meet their emission reduction targets, as set under the EU Emissions Trading Scheme.

Joint Implementation (JI) Under the JI regime, countries with emission caps (Annex I countries) assist other Annex I parties to implement project activities to reduce GHG emissions (or remove CO_2 by sinks). Credits are issued on the basis of emission reductions achieved through the project activities (also called emission reduction units, ERUs).

In the European Union (EU) the link between the flexible mechanisms CDM and JI and the European Union Emission Trading Scheme (EU ETS) is the 'Linking Directive' [25]. The directive regulates that JI/CDM credits can be used by operators to fulfill their obligations under the EU ETS, because it allows the conversion of CERs and ERUs into the EU ETS. That implies the recognition of JI/CDM credits as equivalent to allowances from an environmental and economic point of view.

Out of the Kyoto Protocol is the so called *Offset Market*. Verified emission reductions (VERs) are frequently used for voluntarily balancing of GHG emissions, demonstrating in this manner a person's or company's responsibility and awareness of climate change issues and contributing to reasonable investments by offsetting carbon emissions. The voluntary carbon offset market allows companies, public bodies and individuals to purchase credits generated from projects that either prevent or reduce an amount of carbon entering the atmosphere, or that capture carbon from the atmosphere. Prices for VERs are lower than for CERs or ERUs.

It is important to consider that market mechanisms can only work if there is a supply and a demand of a product. In our case the products are the CO_2 emission certificates. The supply of certificates comes from the EU ETS trading or the flexible mechanisms CDM/JI. Finally the demand is the need of producing companies to emit a certain amount of CO_2 and the cap which was set on the political layer. Therefore the Linking Directive (see above) which combines supply and demand is necessary to let the market work.

What do the flexible mechanisms look like from the perspective of the chemical industry? Let us have a look at CDM trades. There we can notice some incentives: companies can use the CDM or JI mechanism to create certificates and sell them to companies which need them. If investments in developing countries are planned, then it should be proved whether a CDM project can be registered. On the other hand, companies can reduce costs by buying CERs or ERUs, if for instance a capacity increase is planned and the necessary CO_2 certificates are not available. Furthermore companies can use offsets for reputation. And of course there are new jobs and business opportunities in sectors like finance, insurance and consultants. The dynamics of the market can also be seen by the development of trading platforms. There is a growing number of market places creating CO_2 prices. Nevertheless most of the trades are direct trades between the partners.

In 2009 about 1800 CDM projects were registered [26]; the main host countries are China (35%), India (25%), Brazil (9%) and Mexico (7%). The investor countries with the highest number of registered projects are UK (29%), Switzerland (21%), Japan (11%) and the Netherlands (11%). About 4200 CDM projects are in the pipeline.

The market volume of the flexible mechanisms is enormous: In 2008 the volume of the emissions trading amounts to 3300 $MtCO_2e$ or 93 billion US$, where the EU ETS is by far the biggest market. The CDM projects had a total value in 2008 of 390 $MtCO_2e$ or 6.5 billion US$. A smaller role plays JI with 20 $MtCO_2e$ or 200 million US$ and the voluntary market with 50 $MtCO_2e$ or 400 million US$ (all values from [26]).

The Kyoto protocol and its mechanisms therefore are not only important with respect to their effect on climate change and CO_2 emissions, but they also have a strong economic impact.

5.3 Post-Kyoto Negotiations

As mentioned in Section 4 the Kyoto protocol will expire in 2012. Therefore at the conference of signatory states to the UN Framework Convention on Climate Change in Bali in 2007, an official start was made to negotiations to draft a new climate protocol which follows the Kyoto Protocol and to adopt the new protocol at the conference of signatory states in Copenhagen in 2009. The key negotiation blocks are:

- Mitigation, which means an agreement of all countries to reduce GHG emissions in order to limit global warming to 2 °C compared with the level before industrialization.
- Adaptation, which means precaution to the effects of climate change.
- Financing, which means models for poor developing countries, funds or finance mechanisms for avoiding further forest clearance and expansion of R&D to develop climate-friendly technologies and products.
- Technology means technology transfer, which is needed to combat climate change in all countries of the world.

5.3.1 Mitigation Policy

The long term vision is agreed by relevant parties like the decision of the G8 in Heiligendamm 2007 or in Toyako 2008: Halving the global 1990 level of GHG emissions by 2050. But more of the interest is the discussion for intermediate targets for 2020. There the USA position is fixed by President Obama's commitment to seek legislation to reduce GHG emissions by 14% below 2005 levels by 2020, and approximately 83% below 2005 levels by 2050. This means: coming back to 1990 levels in 2020. The Waxman–Markey Bill (see Section 5.4) would go beyond this value.

Japan committed itself under the Kyoto Protocol to cutting emissions by 6% from 1990 levels in the 2008–2012 period, but has struggled to meet that goal. The 2020 target is equivalent to a cut of 8% below 1990 emission level.

The EU Climate Policy has the strong climate target of a 20% reduction in EU GHG emissions compared with 1990 levels, and if other developed nations agree to take similar actions a reduction of 30% was promised. The target also includes an increase in the use of renewable energy to 20% of all energy consumed and a 20% increase in energy efficiency.

5.3.2 Adaptation

Adaptation to climate change is vital in order to reduce the impacts of climate change that are happening now and increase resilience to future impacts. An important element is the Nairobi work programme on impacts, vulnerability and adaptation to climate change. The objective of the Nairobi work programme (NWP) (2005–2010) is to help all countries improve their understanding and assessment of the impacts of climate change and to make informed decisions on practical adaptation actions and measures [27].

5.3.3 Financing

The contribution of countries to climate change and their capacity to prevent and cope with its consequences vary enormously. The Convention and the Protocol therefore foresee financial assistance from parties with more resources to those less endowed and more vulnerable. Developed country parties shall provide financial resources to assist developing country parties implement the Convention. To facilitate this, the Convention established a financial mechanism to provide funds to developing countries.

In the UNFCCC technical paper on 'Investment and financial flows to address climate change' [28] the costs of investment and financial flows needed are estimated. For mitigation the UNFCCC indicates a total net additional investment in 2030 of 200–210 billion US$ with a proportion of 50% needed in developing countries. For adaptation in 2030 the total net additional investment sums to 50–170 billion US$ with a proportion of >50% needed in developing countries.

5.3.4 Technology

It is the goal that the developed countries take all practicable steps to facilitate the transfer of environmentally sound technologies and know-how to developing countries to enable them to combat climate change [28].

In the Marrakesh conference in 2001 the parties reached an agreement to work together in technology transfer activities considering five main themes:

- Technology needs and needs assessments: identifying and analyzing technology needs.
- Technology information: establish an efficient information system in support of technology transfer and to improve technical, economic, environmental and regulatory information.
- Enabling environments: improve the effectiveness of the transfer of environmentally sound technologies by identifying and analyzing ways of facilitating the removal of barriers at each stage of the process.
- Capacity building: improve scientific and technical capabilities and to enable them to develop environmentally sound technologies.
- Mechanisms for technology transfer: enhance the coordination of stakeholders in different countries and regions, engage them in technology partnerships and facilitate the development of projects and programs.

Since then different groups work on the implementation of these mechanisms. The Subsidiary Body for Implementation (SBI) has the task to assist the Conference of the Parties in the assessment and review of the effective implementation of the Convention. The Subsidiary Body for Scientific and Technological Advice (SBSTA) provides the Conference of the Parties with advice on scientific, technological and methodological matters. The Expert Group on Technology Transfer (EGTT) has the target to enhance the transfer of environmentally sound technologies and Know-How and to make recommendations to this end to the SBSTA and the SBI [20].

5.4 National Programs

Countries and regions established national programs and set targets for the reduction of CO_2e emissions. Some of the national programs are presented here in order to give an overview about the activities taken in important regions of the world. We limit the overview here to the four main emitters of GHG as described in Section 4: China, the USA, the EU and India.

5.4.1 United States of America [29]

In June 2009 the House of Representatives approved the American Clean Energy and Security Act (ACESA), an energy bill that would establish a variant of a cap-and-trade plan for greenhouse gases to address climate change. The bill is also known as the Waxman-Markey Bill. It was approved by the House of Representatives, but until today (August 2009) has not yet been approved by the Senate.

Implementation of the cap would reduce emissions from covered sources to 17% below 2005 levels by 2020 and 83% below 2005 levels by 2050. Total U.S. greenhouse emissions would be reduced, as a result, by 15% below 2005 levels by 2020. This is slightly more than President Obama's goal of reducing emissions to

1990 levels by 2020 (approximately 14% below 2005 levels). However, implementing the ACESA would result in significant additional emissions reductions beyond those generated by the pollution cap.

The bill contains substantial complementary requirements including emissions performance standards for uncapped sources and emission reductions from forest preservation overseas. When these are taken into account, GHG emissions would be reduced by 28% below 2005 levels by 2020 and 75% below 2005 levels by 2050. When additional reductions from requirements related to international offsets – used for compliance with the federal cap and trade regime – are also factored in, potential emission reductions from the bill would be even greater. They could reach up to 33% below 2005 levels by 2020 and up to 81% below 2005 levels by 2050, depending on the quantity of offsets used.

5.4.2 **Japan** [30]

Under the Kyoto Protocol Japan committed itself to cutting emissions by 6% from 1990 levels in the 2008–2012 period. Absolute CO_2 emissions have risen by nearly 15% from 1990 to 2006 [14]. The actual 2020 target is equivalent to a cut of 8% below 1990 emission level. Nevertheless climate policy may change due to the recent elections in Japan. In September 2009 the new Prime Minister announced plans for a commitment to reduce national emissions by 25% by 2020, relative to 1990, and to implement a cap-and-trade system within the Japanese economy.

5.4.3 **European Union** [31]

In 2008 the European Union (EU) agreed on a far-reaching package that will help transform Europe into a low-carbon economy and increase its energy security.

The EU is committed to reducing its overall emissions to at least 20% below 1990 levels by 2020, and is ready to scale up this reduction to as much as 30% under a new global climate change agreement when other developed countries make comparable efforts. It has also set itself the target of increasing the share of renewables in energy use to 20% by 2020.

The 'Climate action and renewable energy package' sets out the contribution expected from each member state to meeting these targets and proposes a series of measures to help achieve them.

Central to the strategy is a strengthening and expansion of the Emission Trading System (EU ETS), the EU's key tool for cutting emissions cost-effectively. Emissions from the sectors covered by the system will be cut by 21% by 2020 compared with levels in 2005. A single EU-wide cap on ETS emissions will be set, and free allocation of emission allowances will be progressively replaced by auctioning of allowances by 2020.

Emissions from sectors not included in the EU ETS – such as transport, housing, agriculture and waste – will be cut by 10% from 2005 levels by 2020. Each Member State will contribute to this effort according to its relative wealth, with national emission targets ranging from −20% for richer Member States to +20% for poorer ones.

The national renewable energy targets proposed for each Member State will contribute to achieving emissions reductions and will also decrease the EU's dependence on foreign sources of energy. These include a minimum 10% share for biofuels in petrol and diesel by 2020. The package also sets out sustainability criteria that biofuels will have to meet to ensure they deliver real environmental benefits.

The package also seeks to promote the development and safe use of carbon capture and storage (CCS), a suite of technologies that allows the carbon dioxide emitted by industrial processes to be captured and stored underground where it cannot contribute to global warming. Revised guidelines on state aid for environmental protection will enable governments to support CCS demonstration plants (see also in Chapter 11 of this book).

5.4.4 China

China has established various policies and instruments in order to reduce emissions. The Chinese government mainly focused on renewable energy and energy efficiency.

In January 2006 targets were set by the Renewable Energy Law to generate 16% of China's primary energy and 20% of the electricity from renewable energies such as hydro, wind, solar and biomass.

China has set the target to increase its energy efficiency to minimize the environmental impacts connected to fulfilling the further economic development needs of its population. The 11th Five-Year Plan includes a program to reduce energy intensity by 20% below 2005. This implies a reduction of 4% annually. From 2003 to 2005 the energy intensity increased each year and energy growth surpassed economic growth, but from 2006 on this process has been reversed.

On the international level the Chinese government points out that developed and developing countries must work together and contribute to tackling climate change with a special focus of industrialized countries on their historical responsibility.

5.4.5 India

In June 2008 the Government of India released its first National Action Plan on Climate Change (NAPCC [32]) which comprises eight so-called National Missions. The plan sketches existing and upcoming policies and programs with a special focus on climate mitigation and adaptation. Among other the National Missions on energy efficiency, solar energy and sustainable agriculture will have significant impacts on India's development.

6 Company Initiatives

Beside the treaties and agreements linked to the Rio process and the UNFCCC Convention, many companies have set up initiatives for the reduction of their GHG emissions. The reasons are manifold: responsibility for environment and climate, a certain pressure for action from the public and economic benefits for instance from energy conscious behavior and energy efficient production. Some case studies from the chemical process industry shall exemplify company initiatives. They are taken from the website of the World Business Council for Sustainable Development (WBCSD)[33].

6.1 Novartis [33]

Novartis sees itself as a leader in tackling global environmental problems. The company takes advantage of opportunities to achieve its ambitious energy efficiency and GHG emission reduction targets. Cost-saving opportunities and the increase of overall business efficiency go hand in hand.

Early in the climate discussion, Novartis made a voluntary commitment to reduce GHG emissions globally in line with the objectives of the Kyoto Protocol – that is, a reduction of 5% by 2012 compared with 1990 levels.

In 2007, the company exceeded its annual 2.5% energy efficiency target with an improvement of 7.5%. A new target was established that calls for a further 10% improvement by 2010, based on 2006 figures. In 2007, for the first time, Novartis was able to reduce GHG emissions in absolute terms for both on-site emissions and emissions from purchased energy – despite the continued growth of the company.

Novartis' continued success in energy efficiency is founded on total commitment from top management and highly motivated employees, who fully support energy efficiency projects and carbon dioxide mitigation programs.

A wide variety of measures is helping Novartis to reach its energy and emissions reduction targets. These range from raising energy awareness among employees, to installing best practice technology.

The Novartis Energy Excellence Awards, now in their fifth year, recognize and reward employees who encourage energy efficiency and take initiatives toward implementing renewable or alternative energy technologies. The awards underscore the fact that energy efficiency is good business.

In 2007, the award scheme included 46 projects with potential for more than US$ 40 million in annual net savings by 2012. More than one-third of the projects have a payback time of less than a year – and more than half will repay their initial investment within two years.

To sum it up, the implementation of Novartis' energy and climate strategy results in improved performance with substantially lower environmental impact and enhanced efficiency.

6.2 Roche [33]

Since 1996, the global healthcare company Roche has doubled its energy efficiency, saving money as well as reducing the intensity of its environmental impact.

In 2005, it set itself a new goal of reducing energy consumption by a further 10% over the next five years, on a per employee basis.

But energy managers had already attacked the 'low hanging fruit' of low-cost, no-cost and quick to payback energy efficiency opportunities. In competing for limited capital, they were beginning to come up against investment hurdles that made it difficult for them to demonstrate the feasibility of energy conservation investments.

The problem they faced was that simple payback and return on investment calculations tend to underestimate the cost savings from energy efficiency investments and fail to take into account other benefits such as lower emissions, reduced exposure to energy price fluctuations, increased staff comfort and better public relations.

So Roche changed the rules about the way it assesses the net present value of energy conservation measures. Net present value is a common tool for financial feasibility analysis that allows the future income (or savings) flows from different investment alternatives to be rolled up into a measure of their 'present value' at the beginning of the project, incorporating the cost of capital invested. If a project has a positive net present value, it is profitable.

Roche set out four simple changes to the standard net present value calculation in order to capture the true balance of costs and benefits of energy efficiency investments:

- Lower discount rates – Energy efficiency investments are much less risky than normal pharmaceutical investments, therefore the high discount rate used in most net present value calculations within the company are reduced significantly when considering energy conservation.

- Higher energy prices – Future energy costs will keep rising. Actual and realistic future energy price escalations are included when calculating future energy savings.

- Multiple benefits from energy efficiency investments – Financial measures for benefits such as increased comfort, productivity, environmental benefits, better public relations, as well as utility rebates from energy providers and government grants are included in the calculation.

- Full life-cycle analysis – Design alternatives must be compared for the impacts of energy and all other costs (investment, maintenance, etc.) over the expected life of the asset.

This methodology allows Roche to rigorously compare different design alternatives and select the most profitable contender, which will also be the most energy efficient because of the strong emphasis on the future costs.

The new methodology for assessing feasibility was backed up by a Board level commitment that if the net present value of an energy efficiency proposal is positive then the project has to be done, unless there is a more profitable alternative currently available.

In the two years since this methodology was introduced, Roche has managed to reduce energy use per employee by 8%, despite growing the business and incorporating new enterprises.

6.3
Dow [33]

For the chemical company Dow, energy means both power and raw materials – like natural gas liquids and naphtha, which feed its chemical processes. In 2007 the company went through nearly 850 000 barrels of oil equivalents each day, and had a fuel bill of US$ 27 billion, accounting for nearly half of the company's operating costs.

The business case for energy efficiency at Dow is therefore simple: Saving energy makes the company money.

Between 1990 and 1994 Dow reduced the energy it used to make each pound of its product by 20%. It went on to set itself a series of 'stretch' targets: committing to reduce energy intensity by a further 20% per pound of product between 1995 and 2005, and by another 25% between 2005 and 2015.

To achieve these very public and ambitious goals, the company developed a comprehensive energy management system reaching from the boardroom into every site, plant and business unit. Key elements of this management system were leadership, targets, measurement and improvement:

- Leadership – Dow created a Strategy Board of very senior leaders, to drive energy efficiency within an overall approach to climate change and energy policy.
- Targets – Dow's Energy Efficiency and Conservation Management System makes energy efficiency and conservation a normal part of everyday work processes by ensuring that energy efficiency objectives are pursued as business targets at the corporate, business unit, site and individual level.
- Measurement – Every one of Dow's 120 sites and 685 plants and facilities reports on its energy intensity. Businesses use this to benchmark their plants, estimate potential energy savings and develop long-range improvement plans.
- Improvement – Dow uses the Six Sigma process, (problem measurement, rigorous analysis, focused improvement and controls to ensure that problems stay fixed to realize and sustain energy efficiency improvements) as its corporate-wide methodology to accelerate improvement in quality and productivity.

By improving the efficiency of power generation and chemical processes, installing and operating the most energy efficient equipment and recovering energy from waste, Dow managed to exceed its 2005 goal, cutting energy intensity by 22%.

The keys to this success were applying the focused, consistent and structured management system, pursuing energy efficiency opportunities with the same aggressive implementation as other business goals, and working with partners such as the USA Government's 'Save Energy Now' program to take advantage of external expertise.

However the ambitious 2015 energy intensity target of 25% reduction, on top of the 38% already enacted, may require more than 'more of the same'. In some processes, the company is already pushing up against thermodynamic limits. It recognizes that it will need to create game-changing innovations if it is to achieve its goals.

Dow has committed to investing a significant portion of its research and development into developing energy alternatives, less CO_2 intensive raw material sources, and other breakthrough solutions to contribute to the necessary slow, stop and reverse of non-renewable energy use and climate change.

6.4 **DuPont** [34]

The global manufacturing company DuPont has established several energy and climate targets:

- Reduce absolute GHG emissions 65% from 1990 levels by 2010.
- Hold energy consumption flat on an absolute basis from 1990–2010.
- Obtain 10% of total energy needs from renewable energy sources by 2010.

To help reach these targets in a cost-effective manner, DuPont participates in several GHG emissions trading markets, including the Chicago Climate Exchange® and the UK Emissions Trading Scheme. Participation in these markets enables the company to meet several business goals, including the following:

- Capture financial value from the company's GHG emissions reductions. Emissions markets have helped DuPont 'monetize' and receive a cash flow from the environmental benefits of its GHG reductions. For instance, through several activities DuPont reduced its N_2O emissions from adipic acid manufacturing facilities in the United Kingdom. These initiatives enabled the company to sell 10 000 metric tons of vintage 2002 GHG emissions allowances to the energy trading and marketing firm MIECO, Inc. This sale helped defray the costs of the N_2O emissions reduction investment and will encourage additional measures to reduce GHG emissions.

- Clarify the value of investing in sustainable business practices such as procuring green power. By placing a value on GHG emissions, markets signal to corporate managers the costs and future business risks of unmitigated GHG emissions. Consequently, they signal the value of avoided emissions and therefore help managers understand the financial value of investing in green power.

- Provide a competitive advantage. Participating in emissions markets positions DuPont with tools, information, and strategies to 'get ahead of the curve' and obtain a competitive advantage as emissions markets develop.

6.5
Bayer [35]

Bayer is a global enterprise with core competencies in the fields of health care, nutrition and high-tech materials. The products and services are designed to benefit people and improve their quality of life. At the same time Bayer wants to create value through innovation, growth and high earning power.

In 2007 Bayer launched the 'Bayer Climate Program' in order to meet its responsibilities in matters of climate change. Ambitious production targets and major investment in the development of climate-friendly products and processes are the hallmarks of the Bayer Climate Program – an action plan that pools the company's technological expertise. 1 billion € is available for this project over the period to 2010 alone. Bayer has been actively involved in climate protection for many years. Between 1990 and 2007, the Group reduced its carbon dioxide emissions by over 37%. Bayer Material- Science, the most energy-intensive of the subgroups, is set to lower its specific GHG emissions per metric ton of sold product by a quarter by 2020. Over the same period, Bayer CropScience and Bayer HealthCare will reduce their greenhouse gases by 15% and 5% respectively in absolute terms. The well-established 'Ecological Evaluation of New Investments' procedure (for capital expenditure >10 million €) will be supplemented with climate evaluation criteria. Bayer products are already helping to save energy and conserve resources in many different ways. Examples include products for insulating buildings and refrigerators and for constructing lightweight automotive components.

The Bayer Climate Program has already initiated several lighthouse projects, which provide groundbreaking examples of initiatives aimed at tackling the consequences of climate change and supporting climate protection: Bayer EcoCommercial Building, stress-tolerant plants, biofuels and the Bayer Climate Check for production processes and investment projects. The Bayer Climate Check will be explained more in detail in Chapter 4 of this book.

Acknowledgment

The author thanks Dr. Detlef Schmitz from Bayer AG for fruitful discussions.

References

1 UNFCCC (1992) Text of the Convention, Article 1, http://unfccc.int/resource/docs/convkp/conveng.pdf (accessed 1 February 2010).
2 Global Warming (2010) http://en.wikipedia.org/wiki/Global_warming (accessed 1 February 2010).
3 IPCC (2007) Summary for Policymakers (PDF). Climate Change 2007: The Physical Science Basis. Contribution of Working Group I to the Fourth Assessment Report of the Intergovernmental Panel on Climate Change http://ipcc-wg1.ucar.edu/wg1/Report/

AR4WG1_Print_SPM.pdf (accessed 1 February 2010).

4 Instrumental temperature record (2010) http://en.wikipedia.org/wiki/File:Instrumental_Temperature_Record.png (accessed 1 February 2010).

5 Pachauri, R.K., and Reisinger, A. (eds) (2007) Climate Change 2007: IPCC Synthesis Report to the Fourth Assessment Report, IPCC, Geneva, Switzerland.

6 IPCC (2007) http://en.wikipedia.org/wiki/IPCC_Fourth_Assessment_Report (accessed 1 February 2010).

7 Atmospheric carbon dioxide (2008) http://en.wikipedia.org/wiki/File:Mauna_Loa_Carbon_Dioxide-en.svg (accessed 1 February 2010).

8 Hegerl, G.C., *et al.* (2007) Understanding and Attributing Climate Change (PDF). Climate Change 2007: The Physical Science Basis. Contribution of Working Group I to the Fourth Assessment Report of the Intergovernmental Panel on Climate Change. IPCC. http://www.ipcc-wg1.unibe.ch/publications/wg1-ar4/ar4-wg1-chapter9.pdf (accessed 7 April 2010). 'Recent estimates indicate a relatively small combined effect of natural forcings on the global mean temperature evolution of the seconds half of the 20th century, with a small net cooling from the combined effects of solar and volcanic forcings.'

9 Ammann, C., *et al.* (2007) Solar influence on climate during the past millennium: results from transient simulations with the NCAR Climate Simulation Model (PDF). *Proc Natl Acad SciUSA*, **104** (10), 3713–3718. doi: 10.1073/pnas.0605064103 PMID 17360418. http://www.pnas.org/cgi/reprint/104/10/3713.pdf. 'Simulations with only natural forcing components included yield an early 20th century peak warming of ≈0.2 °C (≈1950 AD), which is reduced to about half by the end of the century because of increased volcanism.'

10 Royal Society (2005) Joint science academies' statement: Global response to climate change. http://royalsociety.org/Joint-science-academies-statement-Global-response-to-climate-change/ (accessed 7 April 2010).

11 Chemical Industry (2010) http://en.wikipedia.org/wiki/Chemical_process_industry (accessed 1 February 2010).

12 The International Council of Chemical Associations (ICCA) (2009) Innovations for Greenhouse Gas Reductions – A life cycle quantification of carbon abatement solutions enabled by the chemical industry. http://www.icca-chem.org/ICCADocs/ICCA_A4_LR.pdf?epslanguage=en (accessed 1 February 2010).

13 McKinsey & Company (2009) Pathways to a Low-Carbon Economy, Version 2 of the Global Greenhouse Gas Abatement Cost Curve, January 2009.

14 Energy Information Administration (2009) www.eia.doe.gov/ (accessed 1 February 2010).

15 Walthie, T.H., (2005)Hydrocarbons & Energy: Energy Pricing and ETS / JI / CDM – The Chemical Industries' View. The Dow Chemical Co., http://news.dow.com/dow_news/speeches/20051110_walthie.pdf (accessed 1 February 2010).

16 Cefic Industrial Policy (2009) http://www.cefic.org/factsandfigures/index.html (accessed 1 February 2010).

17 American Chemical Council (ACC) (2009) http://www.americanchemistry.com/s_acc/sec_article.asp?CID=144&DID=5956 (accessed 7 April 2010).

18 Almaguer, J. (2005) Managing Energy Efficiency Improvement, The Dow Chemical Co., 16.11.2005 http://www.americanchemistry.com/s_acc/bin.asp?SID=1&DID=1757&CID=485&VID=117&DOC=File.PPT (accessed 1 February 2010).

19 The Dow Chemical Company (2007) Corporate Report, http://www.dow.com/PublishedLiterature/dh_010b/0901b8038010bafd.pdf?filepath=financial/pdfs/noreg/161-00695.pdf&fromPage=GetDoc (accessed 7 April 2010).

20 UNFCCC (2009) Feeling the Heat. http://unfccc.int/essential_background/feeling_the_heat/items/2918.php, http://unfccc.int/resource/docs/convkp/conveng.pdf (accessed 1 February 2010).

21 Netherlands Environmental Assessment Agency (PBL) (2008) http://www.pbl.nl/en/news/pressreleases/2008/20080613ChinacontributingtwothirdstoincreaseinCO2emissions.html (accessed 1 February 2010).
22 Chancellor Angela Merkel launches a new climate initiative, (2007) http://www.bundeskanzlerin.de/Content/EN/Artikel/2007/08/2007-08-30-bundeskanzlerin-in-japan__en.html (accessed 1 February 2010).
23 UNFCCC (1998) http://unfccc.int/resource/docs/convkp/kpeng.pdf (accessed 1 February 2010).
24 UNFCCC (1997) http://unfccc.int/kyoto_protocol/items/2830.php (accessed 1 February 2010).
25 Directive 2004/101/EC of the European parliament and of the Council of 27 October 2004 amending Directive 2003/87/EC establishing a scheme for greenhouse gas emission allowance trading within the Community, in respect of the Kyoto Protocol's project mechanisms (2004) http://eur-lex.europa.eu/LexUriServ/LexUriServ.do?uri=OJ:L:2004:338:0018:0023:EN:PDF (accessed 7 April 2010).
26 World Bank (2009) State and trends of the carbon market. http://siteresources.worldbank.org/EXTCARBONFINANCE/Resources/State_and_Trends_of_the_Carbon_Market_2009-FINALb.pdf (accessed 1 February 2010).
27 UNFCCC (2008) The Nairobi Work Programme, The Second Phase. http://unfccc.int/files/adaptation/application/pdf/nwpbrochurenov2008.pdf (accessed 1 February 2010).
28 UNFCCC (2008) Investment and financial flows to address climate change: an update (Technical paper). http://unfccc.int/resource/docs/2008/tp/07.pdf (accessed 1 February 2010).
29 World Resources Institute (2009) The American Clean Energy and Security Act: Key Elements and Next Steps http://www.wri.org/stories/2009/05/american-clean-energy-and-security-act-key-elements-and-next-steps (accessed 1 February 2010).
30 Hone, D., (2009) Incoming Japanese Prime Minister announces key climate policy choices http://www.earthsky.org/blogpost/energy/new-era-in-climate-politics-may-be-dawning-in-japan (accessed 1 February 2010).
31 The EU climate and energy package (2007) http://ec.europa.eu/environment/climat/climate_action.htm (accessed 1 February 2010).
32 National Action Plan on Climate Change (2008) http://pmindia.nic.in/Pg01-52.pdf (accessed 1 February 2010).
33 World Business Council for Sustainable Development (WBCSD) (2010) www.wbcsd.org (accessed 1 February 2010).
34 World Resources Institute, Green Power Market Development Group (2010) http://www.thegreenpowergroup.org/pdf/case_studies_DuPont.pdf (accessed 1 February 2010).
35 Bayer AG (2009) http://www.climate.bayer.com/en/bayer-policy-on-climate-change.aspx (accessed 1 February 2010).

Part One
Administrative and Cultural Aspects

Managing CO_2 Emissions in the Chemical Industry. Edited by Leimkühler

ISBN: 978-3-527-32659-4

1
Analysis Methods for CO_2 Balances

Martin Wolf, Birgit Himmelreich, and Jörn Korte

1.1
CO_2 Balances and Carbon Footprints

1.1.1
Measuring Impact on Global Warming

Fighting global warming is regarded as one of the key challenges of mankind for the coming decades. Emissions of gases into the atmosphere from natural and anthropogenic sources are major contributors to increasing average temperatures. Numerous activities from all kinds of stakeholders have been initiated to reduce the anthropogenic impact on global warming. The commonly accepted target is to avoid an increase of the average temperature of the Earth's atmosphere beyond 2 °C, which is regarded as the threshold to avoid severe negative consequences for life on Earth. In order to describe and quantify the impact of emissions on global warming, a unified parameter and a unit of measurement are required. In terms of quantity the most common gas leading to climate change is carbon dioxide. Thus, by definition the global warming potential (GWP, sometimes also greenhouse warming potential) of carbon dioxide is used as a reference and emissions are typically measured in mass of CO_2 emitted into the atmosphere (in g, kg, or tons). Gases having a negative impact on global warming are called greenhouse gases (GHGs). Their quantitative impact is referred to carbon dioxide and measured in mass of CO_2-equivalents, typically abbreviated as CO_2e. An important parameter for the GWP is the time horizon under consideration for temperature increase. This is because the residence or lifetime of different chemical components in the atmosphere varies significantly. Table 1.1 shows a subset of important chemical components and their GWP based on the most commonly used 100 years time horizon (taken from [1]).

We do not want to make a digression concerning the science of climate change, which can be found elsewhere (e.g., [2]). However, it should be kept in mind that the numbers have been derived from scientific studies based on climate models. That means that the GWP factors are not physical constants but rather

Managing CO_2 Emissions in the Chemical Industry. Edited by Leimkühler

ISBN: 978-3-527-32659-4

Table 1.1 Global warming potentials (GWP) for gases [1].

Chemical component	Chemical formula	GWP
Carbon dioxide	CO_2	1
Methane	CH_4	21
Nitrous oxide	N_2O	310
Halogentaed hydrocarbons	$C_nH_mCl_xF_y$	From 140 to 11700
Flourinated compounds	C_nF_m	From 6500 to 9200
Sulfur hexafluoride	SF_6	23900

a stipulation in global agreements which is based on today's understanding from scientific climate research.

1.1.2 A Simple CO_2 Balance

Based on the numbers in Table 1.1 we can make very simple calculations of the impact of certain emissions on climate change. Assuming we drive a distance of 100km with a passenger car that has emissions resulting from fuel combustion of 150g $CO_2\,km^{-1}$, yields a GWP of 15kg CO_2e compared with simply staying where you are. This is a very simple form of a carbon balance, which is used to compare different activities with one another. These numbers are only relevant information in the simple cases where we compare certain activities with respect to their direct emissions of GHGs with one another or with no activity (which means staying where you are in this example). A few examples where such calculations can lead to meaningful results:

- Comparison of different power generation technologies, for example, a conventional fossil-fired power plant and a power plant that uses carbon capture and storage (CCS, see Chapter 11) technologies with respect to direct emissions.
- Reducing the impact on global warming of a chemical plant by implementing a N_2O reduction technology (e.g., nitric acid, adipic acid).
- Emissions of methane from cattle: calculation of the amount of CO_2 that needs to be extracted from the atmosphere and fixed to compensate for the emissions of methane from a single beast in one year (approx. 100kg methane = 2100kg CO_2e).

Let's return to the passenger car example. Staying where you are is often not an alternative to driving somewhere. We want to compare alternatives to solve the task of transportation. For example, we want to understand what difference it would make, if we were to use an electrically powered vehicle. This vehicle has no direct emissions at all but it is obvious that the simple comparison does not tell the entire story. We would want to know where the power for charging the battery came from and assign the respective emissions to the trip. And if it is a solar powered vehicle, we maybe want to include the GHGs that have been emitted for manufacturing the photovoltaic panels as we know that this is an energy- and hence emission-intensive process. We now also consider a train as a third alternative for moving forward. If it is electrically powered there are again no direct emissions. The heavy-weight train will certainly have the highest indirect emissions caused by power consumption but shouldn't we compare it on a per passenger basis rather than a per vehicle basis? How many people do we assume then traveling in one train and in one car? Different approaches and assumptions will lead to significantly different results. However, this does not mean that only one result is correct and all others are wrong. Assuming we are doing the math, physics and chemistry right, the results will all be correct. Which one is most reasonable depends on what exactly we want to know, that is, the scope of the problem and how we want to allocate the emissions. Even this apparently trivial problem illustrates that a calculated number standalone is not an adequate answer to the question 'What is the CO_2 balance of this?'

The basic calculations for CO_2 balances rely mostly on mass and energy balances. We will discuss this together with the aspect of data acquisition later in this chapter (Section 1.2.2). The challenges with carbon balances are predominantly in the areas of defining the scope and the base case, choosing system boundaries, making appropriate assumptions, reasonable allocations, and approximations. In the remainder of this section we will highlight a few examples.

1.1.3
Carbon Footprints – A Few Examples

The fight against global warming will finally always have to end up in reducing GHG emissions. This can be partly achieved by measures at the source of the emissions (direct emissions), for example, by increasing efficiencies of fossil-fired power plants to reduce the amount of CO_2 emitted per unit of electrical power produced. Another lever is to influence the demand side that causes indirect emissions, for example, by applying more energy efficient devices, by avoiding or dispensing energy consuming activities. The goal of a CO_2 balance is to illustrate and quantify the overall impact of a certain activity or action.

As the example in the previous section shows, we do not only have to calculate the impact of a GHG emission (although it always goes back to this basic calculation) but carefully analyze the activity with respect to related and allocatable emissions. We have so far used the term CO_2 balances. Popular expressions with a similar meaning are carbon footprint, CO_2 footprint, or carbon balance. All these mean balances of GWPs with a unit of measurement expressing mass of CO_2e. There are multiple attempts and initiatives in the community to make formal definitions for a carbon footprint. Standardization is highly desirable given the diverse meanings and possible breadth of interpretations in this field. Hence we frequently find comparisons of apples with oranges.

Carbon footprints are used in the public domain for various activities, mainly driven by the increasing public demand to understand the impact on global warming. Intensive efforts are under way to determine carbon footprints for all kinds of activities and entities like companies, production and power plants, entire products, food, farming, logistics and transportation, consulting services, insurances, and so on. Some are more meaningful and reasonable than others. In the following we give a few examples which are of interest with a particular focus on the chemical industry.

1.1.4
Company Carbon Balances

Many industrial and business companies today report their impact on global warming. It has become an important aspect of their sustainability reports. Beside aspects of public attention and corporate image, this is becoming more and more business relevant. To be considered 'green' is a competitive advantage particularly in certain end consumer businesses. However, this trend is also pushing upstream into the product supply chain and it is becoming more relevant in business-to-business activities. There is a need for public companies to publish their global warming impact, as stock analysts use these reports to evaluate the position of the company in a future 'low carbon society'. Listing in stock indices such as FTSE-4GOOD [3] or Dow Jones Sustainability Indexes [4] require a comprehensive report and measurable targets on GHG emissions and company's climate related activities.

The standard for company GHG reporting is the Greenhouse Gas Protocol [5]. In principle it follows similar rules to financial accounting. Likewise the reports are verified by certified auditors, similar companies as for financial audits. An important section of the GHG standards is the differentiation of three emission scopes. Direct emissions of GHGs are defined as scope 1 emissions. All direct emissions of GHGs from company-owned assets like production plants, power plants, firing and incineration plants, company cars, and so on belong in this category. Scope 2 contains indirect emissions resulting from energy purchase. Mostly this is electrical power. However, for the chemical industry, purchased heat is often an important contributor due to the nature of its production processes. All other emissions are defined as scope 3. Many companies today report their scope 1 and scope 2 emissions. Reporting of scope 3 emissions is not yet established but under discussion (Figure 1.1).

The calculation basics for the emissions in all scopes are very similar to those presented in later sections of this chapter. The ownership of the emission is determined by the asset owner. Example: A chemical company owns a power plant that supplies energy in form of electrical power and steam to company-owned production plants. The emissions of the power plant will be scope 1 for that company. However, if that power plant is outsourced to an energy supplier that now sells steam and power to the respective production plants, the emissions will

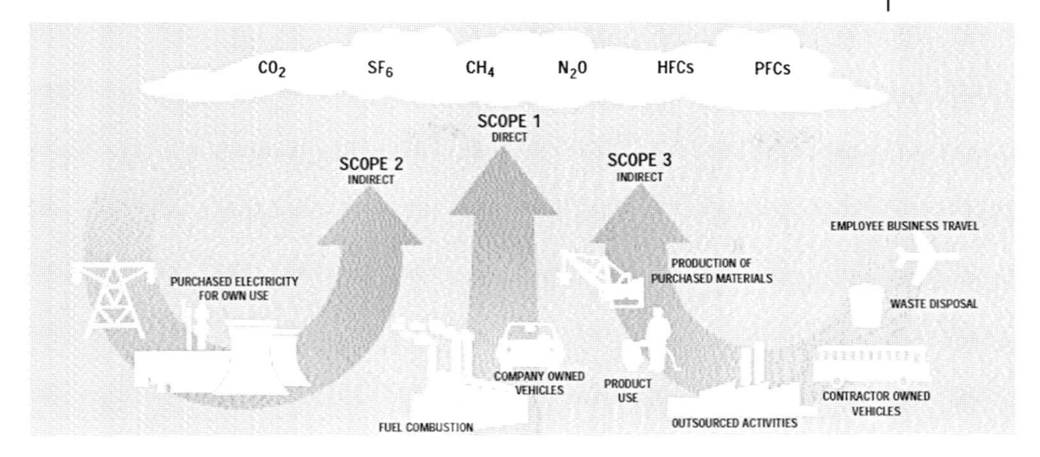

Figure 1.1 Emission scopes according to the GHG Protocol [5].

be scope 2 for the chemical company and scope 1 for the supplier. Similarly with scope 1 and 3 emissions from own or foreign production plants. All legal entities that are more than 50% owned by one company have to be fully accounted to that company.

One of the major goals of the GHG protocol is to track emissions of a company over time in order to quantify the improvements that have been achieved. To avoid offsets in the balance by changes in the legal structure, the GHG protocol has defined the concept of a base year. We take again an example of a chemical company with captive energy production. The base year is assumed to be the year 2000. In 2009 the power plant is outsourced to an energy provider, which is a separate legal entity. From 2009 on the emissions of that power plant will be reported as scope 2 emissions. The emissions of the power plant in the years from 2000 to 2008 that were originally reported as scope 1 emissions will now be adjusted to the new company structure. The historic numbers will be corrected as if the power plant had always been in an outsourced legal entity and regarded as scope 2 emissions. A virtual historic company that reflects the actual structure is created and used to track the emission changes over time. In this simple example we have only shifted emissions from scope 1 to scope 2. It becomes practically complex if we consider today's dynamic M&A activities of larger enterprises and shared ownership. Figure 1.2 shows an example of Bayer AG's published GHG emissions according to the GHG protocol. It is part of the company's published sustainability report [6]. Scope 1 and Scope 2 emissions are shown for the major subgroups over time and compared with a voluntary strategic target. Bayer has undergone significant changes in its structure and portfolio over these years. These emissions have been published for the year 2008. For the years 2005 to 2007 the historic emissions have been recalculated to align with the company structure in 2008 (reporting year).

Greenhouse gas emissions for subgroups and service companies
(total direct and indirect emissions in million metric tons of CO_2 equivalents)

	2005	2006	2007	2008	Target 2020
BMS	4.61	5.09	4.73	4.30	–
BHC	0.59	0.58	0.57	0.56	0.56
BCS	0.89	0.86	0.85	0.87	0.76
Other*	0.02	0.02	0.02	0.02	–
CURRENTA**	2.04	1.69	1.98	1.82	–
Group	**8.15**	**8.24**	**8.15**	**7.57**	8.15
Specific greenhouse gas emissions for BMS (metric tons of CO_2 equivalents per metric ton of product)	1.07	1.08	0.95	0.90	0.80

Figure 1.2 Example of reporting GHG emissions and tracking over time: Bayer AG and subgroups [6].

1.1.5
CO_2 Balances Related to Emission Certificates

The implementation of the Kyoto protocol in the EU requires emission certificates to operate certain defined plants. Examples are large scale incineration plants (power plants), steel plants, cement plants or nitric acid plants. Emissions of these plants are always direct emissions and the balances are straightforward.

However, flexible mechanisms in the Kyoto protocol, like clean development mechanism (CDM) or joint implementation (JI) can also be used to retrieve emission certificates. The framework of this is explained in Chapter 2 of this book. For the flexible mechanisms, more complex carbon balances sometimes need to be performed. The procedure is similar to that discussed here and has to adhere to the guidelines of the UNFCCC (www.unfccc.org).

1.1.6
The CO_2 Abatement Curve

A company that seeks to reduce its GHG emissions will always have manifold measures to do this. A ranking of these measures will not only include the level of CO_2e reduced but also the economic impact. The cost abatement curve is a helpful tool to assess the different measures in both dimensions. It has been introduced for different studies by McKinsey & Co. (e.g., [7]).

Figure 1.3 shows an example for such an abatement curve that has been set-up as part of a study for the German industry association (BDI) to assess the potentials of GHG savings [8]. The *x*-axis shows the amount of GHG saving potential, the *y*-axis the abatement cost or the benefit (negative) of the measure divided by the CO_2e-savings. The measures are sorted according to their abatement cost, that is,

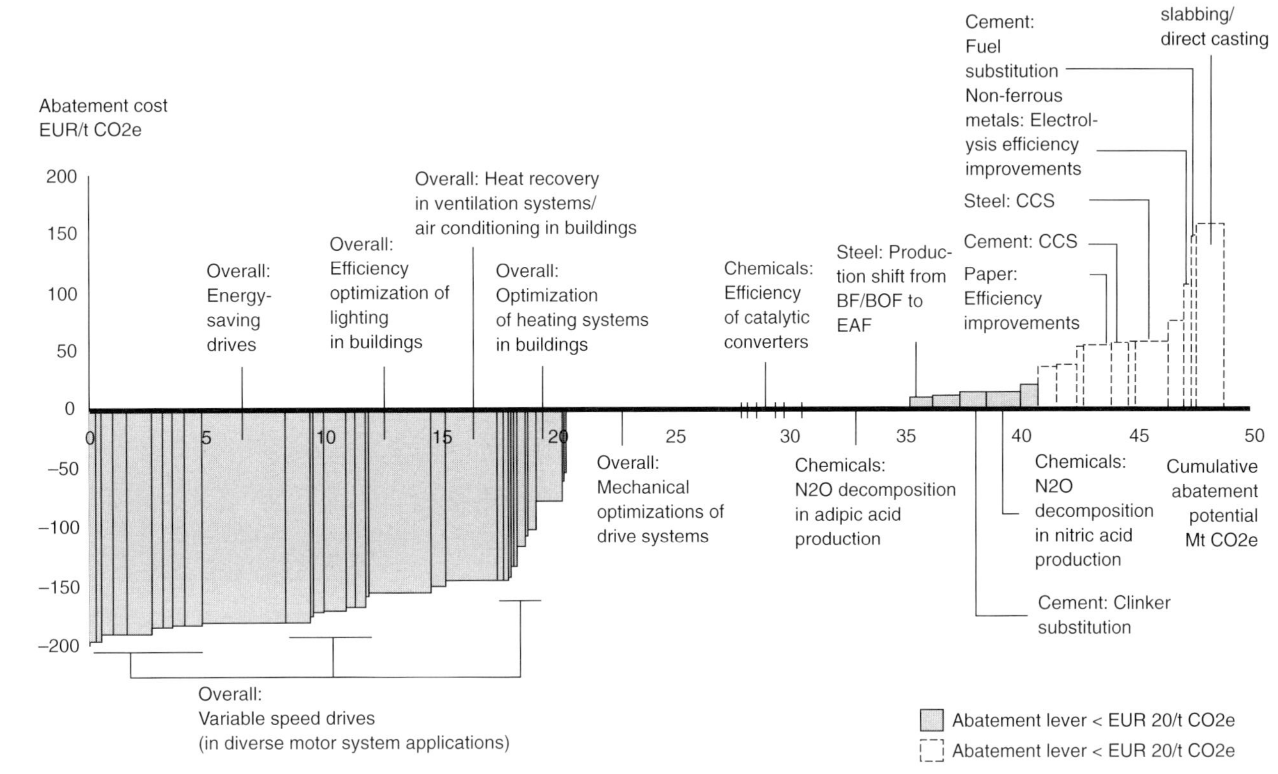

Figure 1.3 Example if a cost abatement curve [8].

the commercially most attractive measure is on the left hand side of the graph. This type of figure ranks the measures. We can easily read from it how much CO_2e we can save, if we implement all measures with positive economics.

Such graphs are also helpful tools for companies to manage their measures, develop a quantitative strategy for dealing with climate change and achieve GHG targets. The CO_2e calculations are done according to the GHG balance of the company (see above). The economic impact is calculated with standard or simplified assessments, for example, annuity methods and typical requirements for amortization periods.

1.2
Product Carbon Footprints (PCF)

Product carbon footprints are published to show the overall climate impact of a product. Usually the calculation methods follow the principles of a life cycle assessment (LCA) (ISO 14040: [9]). An LCA covers many more aspects than just the climate impact and is used to document the overall sustainability of a product. We will only focus on the aspect of climate relevance, which is of major attention today. It should be noted however, that other aspects of an ecological assessment of products must not be neglected, and focusing only on climate impact can lead to overall misleading conclusions.

Some development of standardization initiatives is underway to harmonize and render more precisely the methods and procedures of carbon footprint calculations, for example, a Publicly Available Specification, PAS2050 [10], from the British Standards Institution. We distinguish two major scopes of PCFs; the cradle-to-gate and the cradle-to-grave scope. A cradle-to-gate scope covers all upstream activities, logistics, and production processes. The last step covered is the production unit for the respective product, potentially including a supply and distribution step to the point of sale. The cradle-to-grave approach also includes all steps downstream of the production. That involves supply logistics, downstream processing, packaging and particularly the use of the product and what happens to it at the end of life. Some cradle-to-grave footprints of various products can be found in [11] with explanations.

Many polymers and chemicals are intermediates and have their end use in manifold applications. For example window frames, cable coatings, toys, construction materials, electronic devices are all made of PVC. If we wanted to prepare a cradle-to-grave PCF for PVC we now have also to include downstream production and logistics, the different use phases and the question of end of lifetime, whether it will be recycled, incinerated or deposited. Lifetime and use-phase vary from one-way products with a very short lifetime (e.g., food packaging) up to construction materials that might be in use for more than 100 years. We could use a representative subset of applications and use phases of some major routes. This expansion into multiple product lines is typical for intermediates and can lead to ambiguous results (Figure 1.4).

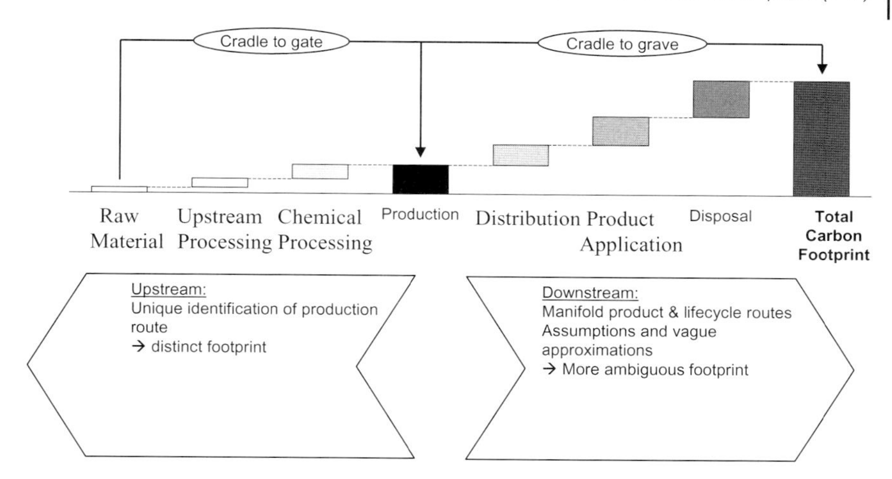

Figure 1.4 Schematic contributions for cradle-to-gate and cradle-to-grave approaches for a typical chemical intermediate.

This can be avoided if the definition of a functional unit is used that determines the application.

Extensive literature on the topic is available elsewhere, for example, [9, 12–14]. We would like to highlight some important aspects for PCFs and focus on example illustrations.

The sections below make frequent use of the term 'product'. Here we focus mostly on a product defined in the conventional way as a tangible matter. However, one finds PCFs applied in a similar way to immaterial products like financial products, logistic services, consulting services, and so on.

1.2.1 PCF Methodology

1.2.1.1 Goal and Scope

The goals associated with the preparation of the carbon footprint should be defined first including the target audience the results are intended for. The scope of the assessment is defined accordingly and describes the system boundaries of the analysis. Example goals for a chemical product are documentation of the overall GHG impact, a comparison of different raw material or sourcing alternatives, a comparison of different production processes or different production sites. The definition of the functional unit is a key element of the scope. As defined in the ISO Standard the functional unit is the quantified performance of a product system for use as a reference unit [9]. For cradle-to-gate analysis the functional unit is mostly a unit mass of the product. For cradle-to-grave analysis it needs to be carefully selected according to the use phase. This is discussed in more detail in Section 1.2.3. For multiple-output processes (MOPs) allocations of emissions are required. If possible the scope should be selected such that these are avoided where possible.

1.2.1.2 Data Retrieval and Data Sources

A PCF is a summation of the individual contributions from its production chain or its total life cycle. We can distinguish between data from publicly available sources (secondary sources), from approximations and analogies, and from detailed calculations of the specific situation (primary sources). The method must be chosen according to the scope and the relevance of the contribution. If we consider a footprint for a chemical intermediate with intensive energy consumption supplied by an on-site power plant, it can make sense to model the specific power plant in detail to calculate the carbon footprint of the respective energies. However, if the energy consumption is of minor relevance it is appropriate to use a value delivered by the energy supplier or a region specific value for electrical power from the grid.

Furthermore we need to pay attention to consistency if data from different sources is used.

1.2.1.3 Calculation Tools

The calculations of PCFs are usually simple and can be done using spreadsheet calculators. There are also numerous commercial tools to support the calculation of PCFs on the market [15]. Although these are less flexible, there are some advantages of using commercial tools:

- Quality assurance, data trace-back, and adherence to standards are easier with commercial tools.
- The tools allow scenario studies and verification of the results with different methodologies.
- Some are LCA tools which allow a full life cycle assessment.
- Some include databases or provide interfaces to commercial databases containing various numbers of inventories from secondary sources (different basic chemicals, energy generation processes, modes of transport, packaging materials, etc.).
- Graphical guidance for model building and result analysis is provided. Documentation is standardized and revision is straightforward.

1.2.2
PCF from Cradle-to-Gate

In the following we will discuss the contributions for a cradle-to-gate footprint. There will be a general part on relevance and options for retrieving the data including a discussion on consequences. Calculations are illustrated using an example from the bulk chemical industry, chlorine production via electrolysis. The data is

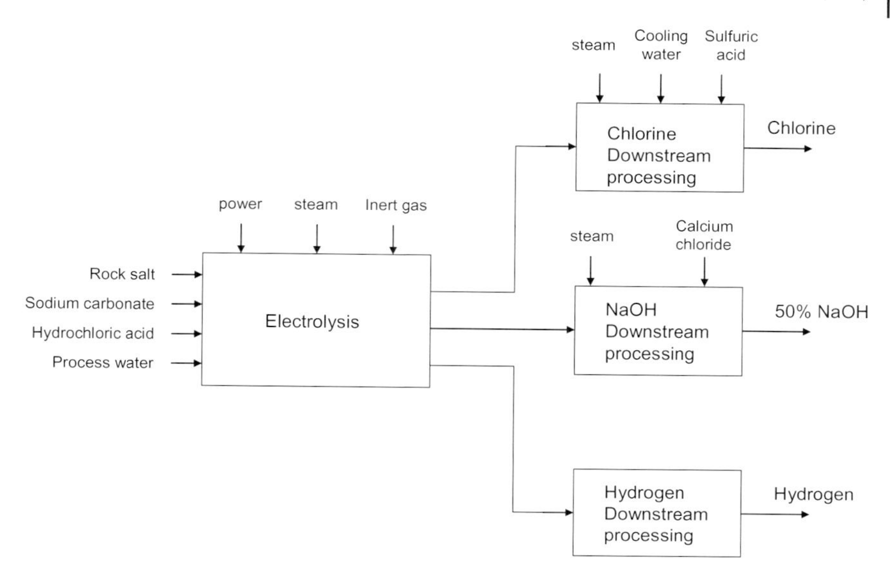

Figure 1.5 Block diagram for the example chlor-alkali electrolysis.

taken from public literature [16] for a diaphragm process with some adjustments. Please note that the example is simplified for the purpose of making it understandable. It is not a comprehensive accurate calculation of a carbon footprint for an electrolysis.

Figure 1.5 shows a block diagram of the production process and Table 1.2 the major consumption numbers for this process. The generation of such a balance is usually the first step and required for the decision on scope and on the contributions that need to be taken into account or neglected. The major raw materials are salt and water. Electrolysis processes have large energy consumptions determined mainly by electrical power and some steam for product processing. The target product is Cl_2, and, per t Cl_2, 1.097 t of NaOH (in 50% water) and 2.817 kg of H_2 (stoichiometric) are produced as coproducts (Table 1.2).

The carbon footprint to be calculated will be used to assess the overall impact of chlorine production on climate change. We apply the cradle-to-gate scope which in this case covers

- raw materials used for the production with their upstream contribution;
- transportation of raw materials to the production unit;
- energy consumption;
- considerations on coproducts.

The example will be used in the following sections to demonstrate the calculation of the individual contributions. For this example we have used the tool GaBi [17] to perform the calculation. GaBi's data sources and methods are used unless otherwise quoted.

Table 1.2 Utility demand for the process steps of the example chlor-alkali electroylsis.

		Electrolysis	Chlorine	NaOH	Hydrogen	Total
Raw materials						
Rock salt	t/t Cl_2	1.67				1.67
Sodium carbonate	t/t Cl_2	0.024				0.024
Hydrochloric acid	t/t Cl_2	0.016				0.016
Calcium chloride	t/t Cl_2			0.013		0.013
Sulfuric acid	t/t Cl_2		0.009			0.009
Utilities						
Steam	t/t Cl_2	0.22	0.02	2.86		3.10
Electricity	MWh/t Cl_2	2.95				2.95
Process water	m3/t Cl_2	4.20				4.20
Cooling water	m3/t Cl_2		69.00			69.00
Inert gas	Nm3/t Cl_2	24.00				24.00

1.2.2.1 **Energy Supply**

Chemical production plants are often located in larger manufacturing sites with on-site energy generation. The heat demand is large and independent of the time of year and hence ideally suited for combined heat and power (CHP) generation. In this situation it makes sense to perform a detailed analysis of the power plant to calculate the carbon footprint of the energies provided. A CHP power plant is a typical example where allocation cannot be avoided. That means we have to decide how to distribute the emissions to heat and to power. There are various options available, for example:

1) Based on enthalpy or energy content. Electricity and heat are regarded as of equal value. This approach is used very often.
2) Based on the exergy. This is a thermodynamic concept to assess different values of form of energy. Electricity has 100% of exergy as it is theoretically fully convertible into any other form of energy, whereas the exergy of heat depends on its temperature. This approach makes most sense from a scientific point of view but is not consistent with some publicly available datasets.
3) The electricity gets its carbon footprint from a publicly available dataset, for example, the country footprint. All remaining inputs and outputs are allocated to steam. A rationale for this could be that electrical power supply is tightly linked with the country grid.
4) Inputs and outputs caused by producing steam in a boiler house are allocated to the steam and the remaining inputs and outputs are allocated to the electricity. This would make sense if the power plant is mainly used for producing steam and electrical power is only generated part-time.

Commercial energy generation datasets are available for many countries for a variety of power plant types (brown and anthracite coal, natural gas, hydroelectric, etc.). The specific energy mix results in a considerably lower carbon footprint for countries with a large fraction of nuclear energy use compared with countries dominated by coal-fired power plants. Thus, when using commercial datasets, it should be verified that the datasets for the respective country fit to the example.

Similar methods have to be applied for other forms of energies, like refrigeration, chilling, cooling water. Compressed air and technical gases like nitrogen or oxygen are usually also treated as energies in chemical plants. Their carbon footprint is normally made up only of the energy contributions required for the generation of these utilities. Like CHP power plants, air separation units require allocations (see also Section 1.2.2.4).

Example: The electrolysis plant is located on production site with a natural gas-fired CHP plant providing electricity and steam. We want to assess the contributions from energy consumption for chlorine production using different methods. Therefore we need to calculate the carbon footprint of power and steam for the power plant. The following parameters of the CHP plant are assumed:

- overall energetic plant efficiency 65%;
- energy mix: 60% power and 40% steam;
- CO_2 emissions are:
  - Direct emissions from combustion: 204 kg $CO_2e\,MWh^{-1}$ natural gas
  - Emissions caused by upstream chain: 21.4 kg $CO_2e\,MWh^{-1}$ natural gas;
- Assumed exergy factors: 1 for power, 0.33 for steam.

Applying the described methodologies leads to the following results (Table 1.3):

As an alternative to this concrete analysis we can use country specific carbon footprints from public sources for power (Database: GaBi):

Germany (2002): 706 kg $CO_2e\,MWh^{-1}$
France (2002): 150 kg $CO_2e\,MWh^{-1}$
Great Britain (2002): 664 kg $CO_2e\,MWh^{-1}$
Norway (2002): 31 kg $CO_2e\,MWh^{-1}$

We can see that the numbers from databases and the numbers from a detailed analysis vary significantly. All numbers are equally correct. Which one should be applied depends on the detailed target of the analysis. For the case of the

Table 1.3 Carbon footprints of power and steam from different methods, kg $CO_2e\,MWh^{-1}$.

	Exergy	Enthalpy	Power from natural gas fix	Steam from natural gas fix
Power	473	347	535[a]	419
Steam	157	347	65	239[b]

a) Database: GaBi Power from natural gas (2002).
b) Database: GaBi Steam from natural gas (94%) (2002).

electrolysis we arbitrarily use the country factor for Germany as the base case and use the enthalpy allocation for a case scenario.

1.2.2.2 Raw Materials

For many chemical products the contribution from raw materials is significant. As described earlier, raw materials (upstream production) have to be tracked up to natural resources. All natural resources themselves have a considered carbon footprint of zero. However, emissions from exploration, mining, and processing need to be taken into account.

For the production of chemicals the most important natural resource is crude oil. It is obvious that products further down the production chain like chemical specialties and active ingredients carry a larger fraction of raw material contribution as these have undergone more intensive processing. Even polymers and many crude oil based chemical intermediates typically have fractions of more than 50% originating from raw materials in their cradle-to-gate carbon footprint.

As in energy generation we have again different options to retrieve carbon footprints for raw materials:

- A detailed model of the individual upstream production steps. Basically each step is analyzed like the actual production step including its logistics (next sections). This can become very tedious and almost impossible as many intermediates which serve as raw materials are purchased from spot markets, where it is not possible to trace back production processes and origin.
- Sometimes a supplier can deliver the carbon footprint of his own products. This is a preferable way but consistency with the selected methodologies must be verified and potentially adjusted.
- Many bulk chemicals are listed in databases with their carbon footprints. A list of databases available can be found in [18]. Data can be outdated as it is often based on older public information. Special care needs to be taken when using this data with respect to consistency with the selected methodology, the dependency on the region, site (see Section 1.2.2.1 for energies) and the respective manufacturing process. These parameters are usually documented in the database. Purity can also be of major relevance. The carbon footprint of highly purified substances can be two or three times higher than standard grades.
- Carbon footprint of precursors can also be approximated. This is recommended if data is not available via reliable other sources and tracing back of production processes is impossible or the effort cannot be justified. There are different methods for generating proxies available. Two alternatives are described below:
  - Selection of an alternative substance that has a similar carbon footprint and where data is already available. Where necessary corrections can be made.
  - Anticipation of a chemical synthesis route to build up the component from substances where data is available. The carbon footprint will then be made

up of the footprint of theses substances and an estimation of the energies required for production.

The use of such proxies is usually very inaccurate and also difficult to reproduce. Application should only be done with critical expert judgment and under careful consideration of the conclusions.

Treating bio-based raw materials can become complex and there are again different options available. The topic has been intensively investigated in various studies of the climate footprint of biofuels. The options differ fundamentally in the way, plant growth is handled:

- CO_2 sequestrated for the growth of the plant can be taken into consideration and credited to the raw material. Meanwhile, technical processes associated with cultivation (e.g., fertilization and release of nitrous oxides) and transportation, as well as the chemical or biochemical processing is regarded as a 'debt' with emissions, similar to exploration and processing of crude oil. The carbon footprint calculated using this approach will mostly result in a negative number for total carbon.
- The CO_2 sequestrated in the growth phase is not taken into consideration and the plant's footprint is set to zero. Cultivation and processing are again included. The carbon footprint will result in a positive number as only fossil carbon is accounted for.

Plant growth is a very sensitive issue as it depends highly on the comparison case for the land use. If waste land is used the first option is applicable. If a rainforest is cut down to cultivate plants, the impact on climate change might be debateable. If the origin of the raw material is not exactly known or verified, we recommend a conservative standpoint and assume a carbon footprint of zero for bio-based raw materials. Please also refer to Chapter 12 for a more intensive digression on bio-based materials.

In our example the handling of raw materials is simple as rock salt is a primary resource. Contributions originate from mining which are small. The data for rock salt, sodium carbonate and hydrochloric acid are taken from the GaBi database. The fractions of calcium chloride and sulfuric acid are very small and we use a substitution proxy for illustration purposes (Na_2SO_4 as another inorganic salt).

Climate impact of the raw materials (CML 96 value):

Rock salt:	0.095 kg CO_2e/kg rock salt
Sodium carbonate	2 kg CO_2e/kg sodium carbonate
Hydrochloric acid:	0.082 kg CO_2e/kg HCl (30%)

1.2.2.3 **Logistics and Supply Chain**

Greenhouse gas emissions generated by transportation and packaging are contribution in the production chain. Transportation itself has a major share of all anthropogenic GHG emissions into the atmosphere. However for most chemicals, particularly for large volume chemicals, the fraction on the total carbon footprint is mostly small. It becomes larger for cradle-to-grave footprints [19]. The impact of transportation has been intensively investigated in the literature and the assessment approach is straightforward. The mode of transportation (road, water, rail or air) and the distance are taken into consideration for each raw material, precursor or product package. For liquids and gases, pipelines have the smallest carbon footprint, followed by ships, rail cars, and trucks. Many commercially available databases have a variety of transportation datasets available based on characteristics such as lorry size or locomotive engine type or the direction for river transportation for ships. Some logistic companies also provide their own information on the carbon footprints of their products.

With regard to packaging, raw materials need to be taken into consideration. The product packaging itself can or cannot be included depending on the scope. Like for raw materials, the carbon footprint of a package takes into consideration all emissions generated along its value chain. If material packaging and transportation are not separated there is a risk of lacking or double counting contributions.

In the case of reusable packaging such as pallets or intermediate bulk containers (IBCs), a rough assessment can be performed of the effort and resources required to account for reuse.

As a rule of thumb, approximations are applicable if the influence on the overall result is less than 5%. Otherwise the effort and resource requirements for transportation and packaging should be analyzed more detailed.

For the example it is assumed that rock salt is delivered by ship, distance 100 km. The impact of transportation by ship is very small. We use a rough estimation: transportation of 1 t rock salt by a fully loaded ship over 100 km means 0.0024 kg CO_2e/kg rock salt. For comparison the impact of transportation by a full loaded lorry would be 0.0065 kg CO_2e/kg rock salt.

1.2.2.4 Manufacturing and Product Allocation

For a cradle-to-gate carbon footprint the final step is usually the production of the respective substance and depending on the chosen boundary limits potentially including filling and packaging.

The production is analyzed based on a balance of all material, utilities and energies flowing into and out of the unit. Usually the balance should be done on average numbers for a certain period of time, for example, one year of production. Non-continuous streams like for batch operations or for cleaning purposes should also be taken into account. The total balance should be representative for the goal and scope of the carbon footprint. Preparing a consistent, sound and representative balance is the most tedious part of the entire exercise for a cradle-to-gate footprint.

We first look at the direct emissions of GHG from the unit. Typical sources for direct emissions in chemical production plants are

- Carbon dioxide mostly originates from incinerations. Some large petrochemical production units have significant emissions because of fuel-fired heating or yield loss in strong oxidation reactions. Examples are steam cracking or ethylene oxide production. Smaller amounts of CO_2 often originate from off gases or side reaction, gaseous purge streams or on-site incineration of organic waste streams. Large scale fermentation processes (e.g., ethanol production) also have larger volumes of CO_2 emissions (note issues with bio-based raw materials).

- Nitrous oxides emissions occur on a larger scale mostly in processes with nitric acid (nitric acid and fertilizer production or nitration processes like adipic acid or other nitro-components). Smaller emissions can also originate from incineration processes.
- Methane can often be found in off gas streams, from incomplete conversion of chemical reactions or from leakage (e.g., natural gas leakage). Rotting of biodegradable matter also causes methane emissions.
- Most plants producing organic chemicals have some emissions of VOCs (volatile organic compounds). Their contribution to the carbon footprint depends on the individual composition. Most relevant are halogenated and fluorinated components. Their GWP is very high.
- Sulfur hexafluoride occurs predominantly in processes for its production. Use of SF_6 with emissions is in magnesium production and mainly as inert gas in electrical high voltage applications.

All input material and utility streams into the production unit contribute to indirect emissions with their individual carbon footprint as described above.

Handling of output streams for the carbon footprint has to be distinguished according to the purpose of the respective streams. We can distinguish between main product stream(s), side product streams, and waste streams.

- Product streams are the desired output of the production unit. In the simple case of a continuously operated one-product plant, the total contribution of the carbon footprint is referred to this output stream and results in the total cradle-to-gate footprint of the respective product (e.g., in kg CO_2e/kg product). Multipurpose plants produce a variety of different substances. If the majority of the consumed raw materials and utilities can be uniquely allocated to the individual product, separate carbon footprints for each product can be generated. This cannot be applied to typical coproduct plants (examples: crude oil refineries, steam crackers, NaCl electrolysis, air separation units, steam reformers, propylene oxide and styrene via the POSM process) where the majority of the consumed resources does not naturally split into fractions for each individual product. Here, allocation is always to a certain extent arbitrary. Typical allocation methods are by mass, mole or value of the products. See the example at the end of this section for more details.
- By-products are undesired outputs of the production unit. In contrast to waste streams these are used. We would differentiate them from coproducts mainly by their relevance. Typical examples are purge streams or residues used as fuel

substitute. Off-spec product can sometimes also be handled like by-products. Handling of those streams is very case-dependent. A conservative approach is to set the carbon footprint of these streams to zero, so it does not reduce the total carbon footprint. Sometimes emissions are allocated to the side product based on a comparative use. An example is a process residue used as a fuel substitute in a power plant. If the incineration of the residue causes less GHG emission than the regular fuel (e.g., coal) the difference can be accounted to the side stream and hence reduces the total carbon footprint. However, this usually exceeds the cradle-to-gate scope as credit to the product is generated from an application downstream of the gate. Furthermore there is a risk of double counting if this effect is already included in the model for the supplied energy.

- Waste streams are of no use for the plant or the process and need to be disposed of. Handling of streams emitted into the environment is already explained above. If the stream does not carry any GHG relevant component, its carbon footprint would be set to zero. If the stream is cleaned or destroyed in a dedicated unit (e.g., off-gas incineration, waste water treatment), it needs to be decided if that unit is inside the system boundary limits. In that case the proportional fraction of emissions needs to be included.

The chlorine electrolysis example does not involve any direct emissions. Electrolysis is a typical coproduct or multi-output process, well suited to describe the options and issues from allocation. Allocation in NaCl electrolysis can be avoided if the impact is referred to an electrochemical unit (ECU) which is the combination of the products Cl_2 and NaOH. Boustead [20] describes five methods for allocating raw materials and another 13 methods of allocating energies. This shows that there are manifold possibilities leading to very different results. It is not a question of correctness but of validity of the respective method. Frequently it is ambiguous which method is most appropriate and different opinions exist. This can only be decided case by case. For our example we will perform a mass and mole based allocation and show the impact on the results (Figure 1.6 and Table 1.4).

In an additional case we treat hydrogen as a by-product to substitute fossil fuel. The impact of the raw materials and energies are solely allocated to chlorine and NaOH and hydrogen is taken into account as a credit. This calculation is based on the assumption, that hydrogen will completely replace natural gas for thermal use (Figure 1.7).

First of all we can see from the graphs that power consumption is by far the major contributor to the carbon footprint. This is true even if we consider the energy carbon footprint from a local CHP plant (Section 1.2.2.1). In the base case raw materials have a share of 8.5% (based on total ECU carbon footprint) and transportation is negligible. The result can look completely different for other chemicals. Variance is too high to give typical or representative numbers for broader classes of products. From the scenarios we can also see the significant impact of different allocation procedures. The carbon footprint of hydrogen varies by a factor of 20 from mass to molar allocation. The credit of hydrogen if treated as a fuel substitute also has a relevant influence. Summarizing we can state that

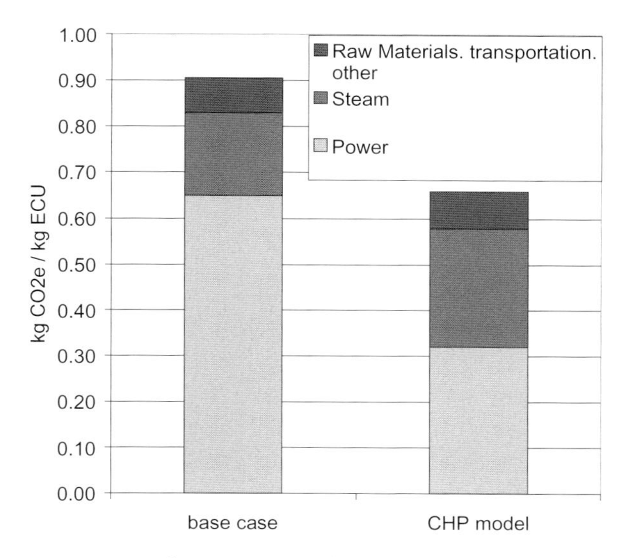

Figure 1.6 Cradle-to-gate carbon footprint for an ECU (electrolysis example).

Table 1.4 Mass and mole allocation of the electrolysis step.

	Mass allocation	Mole allocation
Chlorine	46.41%	25%
NaOH	52.29%	50%
Hydrogen	1.31%	25%

the method applied for the total carbon footprint for an ECU would allow a sound comparison of different electrolysis processes at different sites including the complete cradle-to-gate value chain. Special regard must be given to the approach for modeling energy supply. However, for the single product chlorine, it is absolutely essential to analyze in closer detail the methods applied for the calculations before drawing conclusions from the resulting footprints. Electrolysis is a special example that we selected intentionally to discuss some of the issues. Chlorine is also a basic building component for the chemical industry and hence it has influence on many end products.

1.2.3 Cradle-to-Grave Carbon Footprints

Cradle-to-grave carbon footprints include the contributions downstream of the production. The upstream part is handled as described in the previous section. Downstream contributions can be very diverse and typically come from logistics

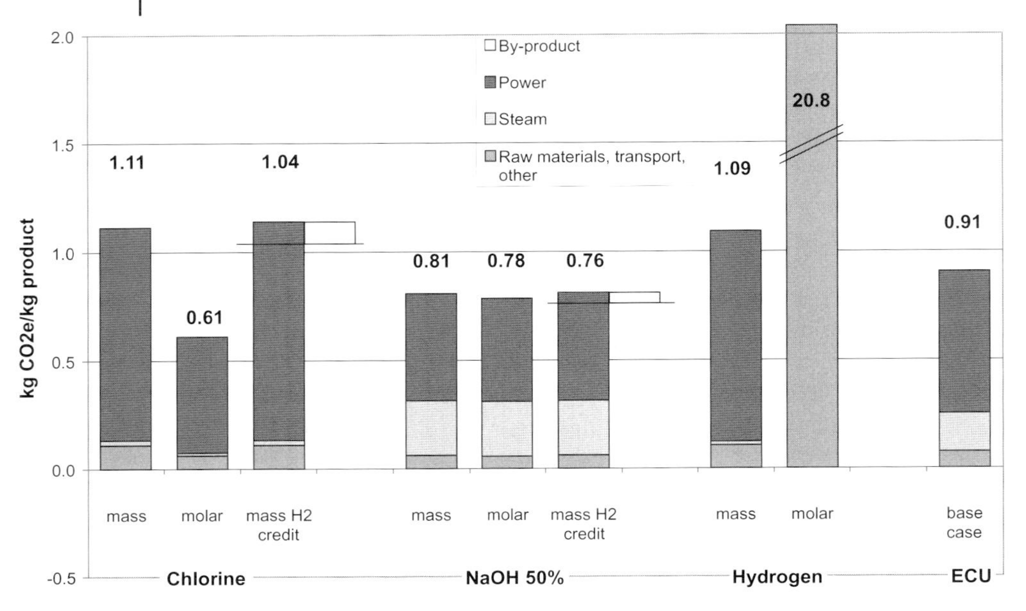

Figure 1.7 Cradle-to-gate footprints (electrolysis example) with different allocation methods.

and packaging, additional production or finishing, purchasing (mostly a logistic effort), product application and disposal. Most of them can be treated similarly to the methodology described in the previous section. The process of calculating the contributions from application will not be discussed here in detail as it varies significantly case-by-case. Conceptually it is similar to the other contributions.

For the majority of cradle-to-gate footprints the functional unit is naturally the product itself expressed in unit mass, volume, or as packaged or delivered (barrel, bottle, blister, ...). The definition of the functional unit for a cradle-to-grave footprint is often more tricky and determines the downstream scope of the assessment. Let's consider an example for an insulation of a building. We want to compare a building without insulation to one with insulation made of different materials (e.g., rock wool, polystyrene foam, polyurethane foam). A cradle-to-gate footprint would compare the production chain and an obvious functional unit is kg or m^3 of insulation material. Alternatively we can use a reference on the material property. For this application it is more meaningful to use the heat transfer resistance of the material as a functional unit. However, this still does not take into account the effort to install the insulation, the energy savings caused by it and the effort for removal and disposal. A reasonable functional unit is the life cycle of the wall to be insulated. This includes considerations on insulation thickness, lifetime of the insulation or the building, building conditions, mode of use (indoor temperatures), type of heating device, location of the building (climate, transport distances), and so on. It is also plausible to extend the scope and use the life cycle of an entire building as the functional unit. However, this scope would be so broad that we loose the close link to the product insulation material. A PCF for wall insulation compared to no insulation can be found quantitatively in [11] (Neopor from BASF). The results are given in Table 1.5 The heat loss through the wall is included in the use phase and hence dominates the result.

Another example is shown in Figure 1.8 for a washing detergent (Persil Megaperls from Henkel). The selected functional unit is a washing cycle. We can see from the graph that again the application dominates the carbon footprint and that the influence of washing temperature is greater than that of raw materials and manufacturing.

The use phase of products is often very vague and ambiguous. It raises the question to which amount the emissions are attributable to the existence or the characteristics of the product. The use phase contributions will dominate for most products that are applied in conjunction with energy consumption. Cradle-to-grave PCFs are frequently prepared to canvass consumers or inform about the climate impact of certain products. Some of these should be regarded critically as a single

Table 1.5 Contributions to the PCF for a building wall in kg CO_2e/life cycle.

Contribution	Wall with insulation	Wall without insulation
Raw materials, production, distribution	17 609	–
Product use	469 166	754 505
Disposal	3 804	

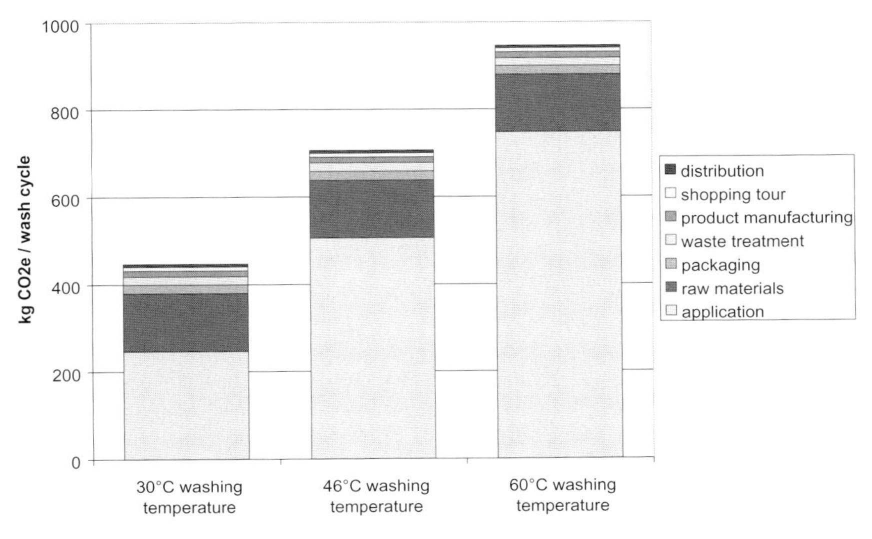

Figure 1.8 Cradle-to-grave footprint of a detergent at three different washing temperatures [21].

CO_2e-number (like on a can of beans in a supermarket) usually does not have a relevant meaning without a comparison against an alternative and additional information on scope, assumption on use phase and so on.

1.3 Remarks and Summary

In this chapter we have discussed different types of carbon balances, given examples of typical applications and calculation procedures. Calculation of direct emissions is simple and straightforward. The basics for the calculation of indirect emissions will always be mass and energy balances. The challenge for these balances lies particularly in defining an appropriate scope for the application and gathering the data. Meaningful conclusions from carbon footprints can only be derived if numbers are compared that have been calculated based on a verified scope, with similar methodologies and standards. Thus, one can find substantially differing numbers for the carbon footprint of a unique chemical product. Correctness and accuracy are a matter of doing the balances right and even more of appropriate assumptions and choice of calculation procedures. Arbitrary results are typical where the scope is broad, allocation methods are undefined, and manifold and ambiguous product pathways are included. Carbon footprints are meaningful when they deliver helpful and tangible indicators for decisions and for the search of alternatives. This can be purchasing options like the type of product or material (plastic vs. metal), regional sourcing, or a selection for different producers. For a production company these are typically technical alternatives (synthesis route, efficient equipment), or again sourcing options for example, for energy and raw materials including the decision on production site or region.

We will often not be able to retrieve clear and unique results for carbon footprints without complex and time-consuming calculations and discussions on scopes and methods. However, in most of the cases it is obvious what is beneficial and what is harmful to the climate. We should not take the occasional ambiguity of carbon footprints as an excuse for not acting against climate change.

References

1 Global Warming Potentials, Climate Change, (1995) http://unfccc.int/ghg_data/items/3825.php (accessed 8 April 2010).

2 IPCC Working Group I, (2009) http://ipcc-wg1.ucar.edu/wg1/wg1-report.html (accessed 8 April 2010).

3 FTSEGood Index,(2007) http://www.ftse.com/Indices/FTSE4Good_Index_Series/Downloads/FTSE4Good_Climate_Change_Criteria.pdf (accessed 8 April 2010).

4 Dow Jones Sustainability Indexes, www.sustainability-index.com (accessed 8 April 2010).

5 The Greenhouse Gas Protocol, www.ghgprotocol.org (accessed 29 September 2009) (http://www.ghgprotocol.org/standards/corporate-standard).

6 Bayer AG sustainability report, http://www.bayer.com/en/Sustainable-Development-Report.aspx. (accessed 8 April 2010).
7 Innovations for Greenhouse Gas Reductions – A life cycle quantification of carbon abatement solutions enabled by the chemical industry, (2009), ICCA International Council of Chemical Associations, July 2009.
8 McKinsey & Co., Cost and Potential of Greenhouse Gas Abatement in Germany, A report behalf of BDI initiative – Business for Climate; October 2007, http://www.mckinsey.com/clientservice/ccsi/costcurves.asp. (accessed 8 April 2010).
9 ISO (2006) 14040, *Environmental Management – Life Cycle Assessment – Principles and Framework*, International Organisation for Standardisation, Geneva.
10 BSI,PAS2050, (2008) *Assessing the life cycle greenhouse gas emissions of goods and services*, British Standards Institution, London, UK.
11 Product Carbon Footprinting – Ein geeigneter Weg zu klimaverträglichen Produkten und deren Konsum? (2009) PCF Pilotprojekt Deutschland, c/o THEMA1 GmbH; Torstr. 154; 10115 Berlin; Germany.
12 PlasticsEurope, Eco-profilesand Environmental Declarations (2009) PlasticsEurope, Brussels, Belgium, http://www.plasticseurope.org/Content/Default.asp?PageName=openfile&DocRef=20071012-002 (accessed 29 September 2009).
13 The Carbon Trust (2010) http://www.carbontrust.co.uk/default.ct (accessed 7 April 2010).
14 Product Carbon Footprint (2008) http://www.pcf-projekt.de/main/product-carbon-footprint/?lang=en (accessed 29 September 2009).
15 European Commission: Joint Research Centre, LCA Tools, Services and Data, List of tools, (2009) http://lca.jrc.ec.europa.eu/lcainfohub//toolList.vm (accessed 29 September 2009).
16 PEP yearbook, (2002) SRI Consulting, Menlo Park, CA, USA.
17 GaBi Version 4.3; PE International, http://www.gabi-software.com/gabi/gabi-service-support/gabiversion/ (Accessed 8 April 2010).
18 European Commission: Joint Research Centre, LCA Tools, Services and Data: List of databases, (2009) http://lca.jrc.ec.europa.eu/lcainfohub//databaseList.vm (accessed 29 September 2009).
19 Schulte, M. (2009) Über Die Nachhaltige Logistik in der Chemisch-Pharmazeutischen Industrie, 14th Symposium on Sustainable Logistics, Magdeburg, Germany, 26-27 February 2009.
20 Boustead, I. (2005) Eco-profiles of the European Plastics Industry; Methodology, Last revision March 2005, PlasticsEurope, Brussels, Belgium.
21 www.pcf-projekt.de (October 9, 2009); http://www.pcf-projekt.de/files/1241103260/lessons-learned_2009.pdf

2
Managing the Regulatory Environment

Nathan Steeghs

2.1
Introduction

Global concern over climate change is placing pressure on industrial sectors to reduce greenhouse gas (GHG) emissions as more countries establish targets to reduce emissions through internationally binding agreements such as the Kyoto Protocol. Policy measures aimed at mitigating GHG emissions rely heavily on technology based solutions to produce renewable energy and conserve fossil fuels, as well as target harmful emissions such as nitrous oxide (N_2O) and methane (CH_4) which pose significant concern to climate change because of their high global warming potential (GWP).[1)] While the chemical industry continues to play an integral role towards achieving such objectives through the development of light weight composites, renewable fuels, PV cells, high performance insulation, and specialized emissions catalysts, the use of energy intensive products such as ethylene, ammonia, and chlorine makes the industry inherently exposed to policy measures that impose carbon pricing[2)] from emissions caps on combustion installations [1].

This chapter examines climate change policy instruments such as emissions trading, taxation, and regulation in the context of the chemical process industry and focuses on risk and opportunity management for industry operating in a globally regulated environment on GHG emissions. Recognizing the highly integrated and energy intensive nature of the chemical industry, this chapter focuses on carbon pricing throughout each stage of the supply chain from the production of energy and feedstock to high value downstream chemicals.

While the European chemical industry has nearly achieved a 30% reduction of absolute emissions between 1990 and 2005 [2], the industry still accounts for

1) Global Warming Potential (GWP) is a measure of how much a given mass of greenhouse gas is estimated to contribute to climate change relative to carbon dioxide which has a GWP of 1 on a 100 year time horizon.

2) Carbon pricing refers to the cost of greenhouse gas emissions under policy measures such as emissions trading.

Managing CO_2 Emissions in the Chemical Industry. Edited by Leimkühler

ISBN: 978-3-527-32659-4

approximately 12% of the of total energy demand in the European Union (EU) [3]. Drawing on examples from the European chemical sector, this chapter explores the challenges faced by the industry's limited ability to convert to more efficient processes within the timelines of an emission trading scheme. The analysis places emphasis on the most vulnerable products to carbon price absorption, while considering factors such as exposure to international markets, transportation costs, and dependency on carbon intensive upstream products.

The chapter analyzes carbon pricing for a select group of common basic chemicals and downstream products such as ammonia, ethylene, pesticides, varnishes and pharmaceuticals. The analysis takes into consideration the costs associated with purchasing emission allowances, as well as cost pass-through[3] from upstream feedstock and electricity producers that also fall under emissions caps. It examines carbon price permeability on the supply chain and the downstream impacts on production cost relative to gross value added. Energy costs already represent up to 60% of chemical production costs and can be as high as 80% for ammonia production [4].

Section 2.4 examines international carbon offset mechanisms within the chemical context and how they can be used to lower the cost of emissions compliance. Building on the basis of the introductory chapter, this section elaborates on international carbon offset markets through the *Flexible Mechanisms* of the Kyoto Protocol, and identifies what options are currently available to the chemical industry. Outlining the basic guidelines for the project development cycle, the chapter discusses how the chemical industry can use the flexible mechanisms through the clean development mechanism (CDM) and joint implementation (JI).

Recognizing the integral role of the offset market for achieving reductions to emissions levels outside of emissions capped economies, the section on international offsets examines where the largest opportunities exist. A study by the World Bank revealed that 10% of the CDM potential in China rests with the chemical sector [5].Within this context, the chapter places emphasis on where transnational companies can lower their cost of compliance through market mechanisms, drawing on specific examples such as the French chemical company, Rhodia, who developed CDM projects to abate N_2O emissions from their adipic acid facilities in Brazil and South Korea.

While long-term agreements to mitigate GHG emissions are currently under negotiation in most of the largest economies of the world, the final section examines the future of climate policies and the positions of leading chemical companies and industry associations such as the European Chemical Industry Association (Cefic),[4] towards managing the risks and opportunities of operating within a global emissions market. In this context, discussion around how *performance based allocations* and *sub-sector division* will play an integral role avoiding emissions

3) Pass-through refers to carbon cost pass-through associated with emissions compliance as it relates to the chemical supply chain.

4) Cefic is the Brussels-based organization representing the European chemical industry.

leakage[5] to 'hot spots',[6] market distortion, and border tax adjustments that risk retaliation from trade partners.

2.2 Overview of Climate Policy

2.2.1 Economics of Climate Change

The impacts of global warming as described in the introductory chapter are expected to have far-reaching consequences on global economies. The IPCC Fourth Assessment Report published in 2007 examined the aggregate economic impacts of climate change by studying the net economic costs of damage from climate change across the globe. In 2005, the estimated social cost of carbon was approximately 12 US$ per tonne of carbon dioxide (CO_2) but reached as high as 95 US$ per tonne across more than 100 estimates [6].

The Stern Review, a 2006 report by the former Chief Economist and Senior Vice-President of the World Bank, revealed that the cost of climate change will have grave consequences on economies unless mitigation measures are pursued. The report suggests that investments of one percent of global GDP are required to mitigate the effects of climate change, while failure to do so could result in a recession worth up to 20% of global GDP [7].

Recognizing the overwhelming cost benefit of mitigating emissions rather than adapting to climate change impacts, this section examines the role of climate policy as an instrument to initiate carbon pricing measures in order to achieve a shift towards low carbon intensive products. This section examines the types of policy measure that exist for reducing emissions, and draws on examples in the EU and USA.

2.2.2 Policy Measures to Mitigate Greenhouse Gas Emissions

2.2.2.1 Cap-and-Trade

The concept of a cap-and-trade scheme was first introduced by the United States Environmental Protection Agency (US EPA) through the Clean Air Act Amendments of 1990 as a cost effective mechanism to regulate emissions of nitrogen oxides (NO_x) and sulfur dioxide (SO_2) in the power sector; the primary causes of acid rain[7] [8]. The overwhelming success of the pollution control scheme led to

5) Leakage refers to the undue shift in production to countries with little or no regulation on GHG emissions.

6) Hot spots refer to areas that fall outside of emissions compliance countries where industry can shift production to avoid regulations on GHG emissions.

7) Acid rain refers to the deposition of precipitation containing acidifying particles and gases. The acid is derived from sulfur oxides (SO_x) and nitrogen oxides (NO_x) which enter the atmosphere from the combustion of coal and other fuels from industrial processes.

the adoption of emissions trading as one of the key flexible mechanisms of the Kyoto Protocol, conceived in 1997 as a means for countries to achieve their greenhouse gas reduction targets.

Emissions trading schemes operate by establishing an absolute emissions limit, usually done by a government or international authority, and issue emission allowances to industries covered under the scheme. This limit serves as a cap on emissions permitted under the framework during the compliance period. Following the establishment of a cap, emissions allowances are distributed to participating entities through free allocations, auctions, or a combination of both. In a cap-and-trade system for regulating CO_2 emissions, a quantity of allowances are surrendered at the end of each compliance period that are equivalent to the emissions for which the installation is responsible according to the governing authority. Each allowance generally refers to one tonne of carbon dioxide equivalent (CO_2e). The distribution of allowances can be based on historical emissions, known as grandfathering,[8] or can be output based and linked to a product of a given sector where benchmarking is used set emissions levels [9].

To prevent a negative economic burden to industries which are competitively disadvantaged by an absolute cap on emissions levels, the start-up process of emissions trading schemes generally relies heavily on free allowances and gradually incorporates auctioning of allowances by the governing body. In 2005, the launching of the European Union emissions trading scheme (EU ETS) took place under the guidelines of the Kyoto Protocol. The scheme relies heavily on the free allocation of emissions allowances through a national allocation plan (NAP)[9] during Phase I (2005–2007) and Phase II (2008–2012). During the compliance period, the EU ETS enables member states to trade allowances issued through the NAP to meet compliance obligations corresponding to national targets under the Kyoto Protocol.

Auctioning of emissions allowances, such as proposed under the US Climate Bill, generates revenue for the governing body to redistribute and lessen the economic burden on society resulting from pass-through cost to consumers. This is particularly attractive for governments to lessen taxes in certain areas which could otherwise curb or prevent economic growth. Auctioning of allowances also prevents some of the key criticisms surrounding a cap-and-trade, namely where early entrants can achieve windfall profits[10] through the sale of free allowances.

The cap-and-trade approach uses market efficiency by taking advantage of the least marginal abatement cost (MAC)[11] to reduce the overall economic burden of

8) Grandfathering is method of allocating emissions allowances based on historic emissions levels.

9) A national allocation plan (NAP) defines basis on which free greenhouse gas emission allowances are allocated by national governments to individual installations covered by the emissions trading scheme.

10) Windfall profits refer to large profits that occur due to unforeseen circumstances in a market, such as unexpected demand or government regulation.

11) Marginal abatement cost (MAC) refers to the marginal cost of reducing one tonne of CO_2 equivalent and is generally expressed in terms of €/tCO_2e.

reducing emissions. Most notably, a cap-and-trade scheme places a value on CO_2 which effectively drives innovation for technologies that reduce emissions. Entrants to the scheme which are capable of reducing emissions at a price lower than market value may generate a profit by selling excess allowance for surpassing their compliance obligation.

One of the fundamental advantages of a cap-and-trade scheme is the flexibility to use carbon offsets, which depending on the scheme, may include both domestic and international offsets. Domestic offsets are sourced from initiatives within the compliance country to reduce unregulated emissions, while international offsets are generated outside of the compliance country. Section 2.4 of this chapter further examines the role of international offsets and what mechanisms are currently available to develop and trade these offsets.

Emissions trading is often criticized by environmental non-governmental organizations (NGOs) and governments because it is viewed as a disincentive for 'own action'.[12] While this argument holds merit for some installations which have a high cost of abatement, supporters of a cap-and-trade dismiss the argument suggesting that installations will invest in their own reductions when they are more economical than the price of allowances in the market.

Further criticism attacks the unpredictable nature of market mechanisms, arguing that long-term price signals are needed to justify investment in low carbon technology. The EU ETS faced severe price volatility in the start-up Phase I where the price of European Union allowance (EUAs) fell to zero because of oversupply in the market. Market analysts attributed the price collapse of the Phase I EUA to a learning phase directed at gaining participation from installations and establishing baseline data for the following Phase II.

2.2.2.2 Command-and-Control

Command-and-control measures seek to influence a market by imposing regulations and taxes aimed at equalizing the negative externalities attributed to the production or trade of a specific product or service. This concept is widely adopted in regulating pollutants such as ozone depleting substances (ODS) and holds promise for achieving reductions of fossil fuel consumption through carbon tax or emissions regulation of high GWP gases such as hydrofluorocarbons (HFCs) or SF_6.

Carbon Tax Carbon tax is a form of indirect tax used as a price instrument to regulate greenhouse gas emissions resulting primarily from the combustion of fossil fuels. The release of CO_2 into the atmosphere is considered as a negative externality of a product or service that enters the economy where the producer or distributor is not required to pay the marginal cost of damage to the environment. English economist, Arthur Pigou, first proposed the concept of taxing goods that

12) Own action is a term used to describe a company undertaking its own initiatives to reduce emissions from their own installations or operations.

are the source of negative externalities in 1912 in order to accurately reflect the true cost of production to society [10].

While a carbon tax is an effective instrument for achieving a shift in consumption patterns, there is concern that as a stand-alone measure, it is not sufficient to drive innovation beyond the compliance obligation. For large industrial sectors that fall under a carbon tax, the danger is that industry will pass-through the cost of carbon to consumers in circumstances where end products such as ammonia, experience limited exposure to international markets. It then becomes a tremendous burden on the regulating authority to redistribute revenue to lessen the impacts on the economy. Particularly for multinational companies, taxation does not deliver flexibility to maximize global assets and minimize the cost of carbon compliance [11]. However, taxation is an effective instrument to equalize negative externalities and promote conservation in sectors such as transportation where it is difficult to monitor the end use of fossil fuel consumption.

Regulations This form of command-and-control measure has been particularly effective for managing ODS under the United Nations Montreal Protocol on substances that deplete the ozone layer. This internationally binding treaty is designed to protect the ozone layer by phasing out the production of a number of ODS [12].

A governing authority to reduce or phase out specific emissions or activities linked to emissions, establishes regulations to manage GHGs. Discussions around the introduction of regulations to manage emissions generally target high GWP gases such HFCs, perfluorocarbons (PFCs), SF_6, and N_2O. This is an effective instrument to reduce high GWP gases where a low MAC is available; however, it is difficult to apply the regulation corresponding to a particular gas. For instance, an adipic and nitric acid production facility both produce N_2O as a by-product from the production process but have significantly different abatement solutions resulting in a large discrepancy in the MAC.

2.2.2.3 Hybridization of Taxation and Trading

There are certain sectors of the economy, such as transportation, that are difficult to incorporate into an emissions trading scheme. These types of sectors are largely managed through taxation measures on liquid fuels aimed to promote conservation along with biofuels such as ethanol. Governments can promote other initiatives to reduce emissions, such as the investment in public transportation, and redistribute revenue collected by such schemes.

The combination of taxation in parallel to emissions trading is regarded as hybridization; however, a true hybridization of carbon tax and emissions trading is when an emissions trading scheme introduces a price cap, a price floor, or both. This enables government to sell credits at the price cap (which is effectively a tax) or purchase credits at the floor when the market price falls below the floor [13].

2.3 Carbon Compliance for the Chemical Process Industry

2.3.1 Carbon Pricing and Industry Exposure

The fundamental purpose of climate policy, whether through emissions trading, taxation, or a combination of both, is to introduce an economic value on global warming gases and reduce the negative externalities of industry on the economy. The most efficient tool for accomplishing this goal is through carbon pricing, either by penalizing industry for polluting or by economically incentivizing the reduction of emissions.

As discussed previously, a cap-and-trade scheme distributes emissions allowances in the form of free allocations, auctioning, or a combination of both. From an industry perspective, auctioning of emissions allowances imposes a similar economic burden as taxation: factoring the cost of every tonne of CO_2 into the overall cost of production. It is a useful technique for governments to place pressure on industrial sectors in order to shift to lower carbon intensive processes and technologies. It can however be problematic for products such as PVC that rely heavily on carbon intensive upstream products and sell on the international markets, as discussed further in this section.

Apart from direct emissions from production installations, there are two important aspects to consider in the analysis of an industry's level of exposure to carbon pricing. The first relates to energy and raw materials sourcing and risks associated with upstream carbon pass-through. The second is the industry's ability to pass-through carbon cost to the downstream consumer.

Energy-intensity[13] and trade-intensity[14] are useful indicators for analyzing industry exposure to carbon pricing. Figure 2.1 examines the intensity indicators for a select number of industries in the USA. The figure illustrates that while some sectors such as lime face high energy consumption, the relatively low trade intensity limits exposure to carbon price absorption. Conversely, other sectors which are both energy and trade intensive, such as nitrogenous fertilizers, experience the highest exposure to climate policy measures. Based on this high-level analysis, the basic chemical industry is one of the most exposed sectors, considering the relatively high level of energy and trade intensity compared with other sectors.

In order to further analyze carbon pricing in the chemical context, Figure 2.2 is a simplistic representation of carbon price exposure, taking into consideration the exposure to international trade markets and likelihood of carbon cost pass-through for each subset of the chemical supply chain. The figure illustrates how basic inputs of electricity and feedstock sell in the regional market while basic chemicals such as ethylene and ammonia are more exposed to international markets.

13) Energy-intensity is the sum of energy and fuel costs divided by the value of shipments.

14) Trade-intensity refers to the sum of the imports and exports divided by the sum of shipment value and imports.

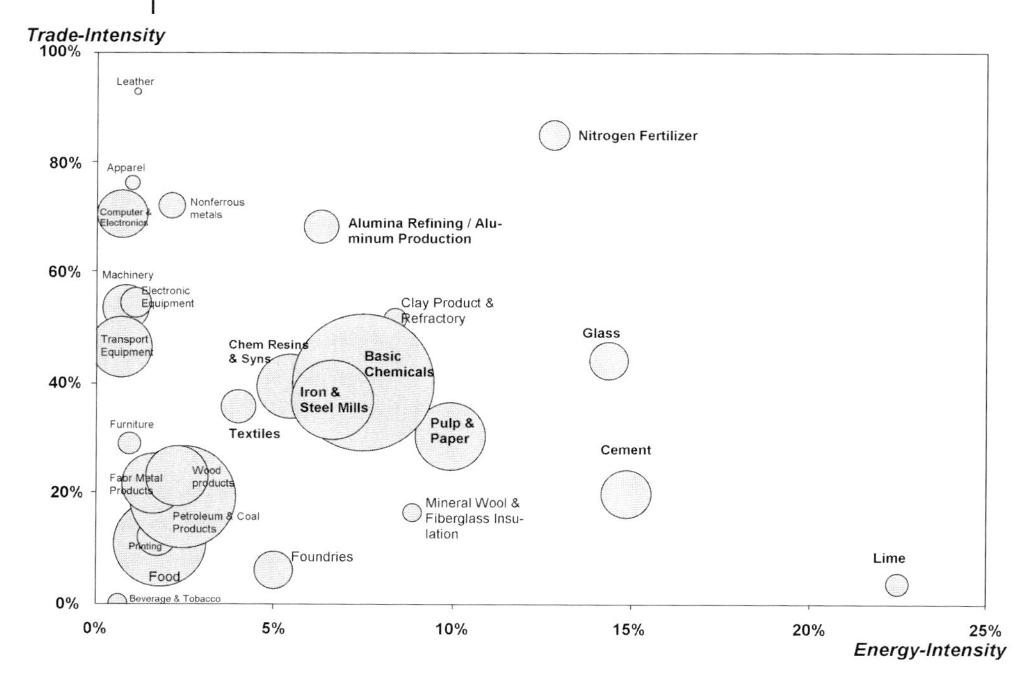

Figure 2.1 Energy and trade intensity for a select group of USA industries [14].

Assuming a direct correlation between exposure to international markets and carbon pass-through, the downstream chemical and sub-sector branches such as PVC, pesticides, varnishes and pharmaceuticals are considerably more exposed to trade and thus to carbon price absorption.

Although trade exposure generally increases significantly with each subsequent stage of the supply chain, the relative impact of carbon pricing to gross value added also decreases. In order to adequately assess whether an industry is financially at risk to carbon pricing measures, it is important to consider how the relative carbon cost affects gross value added or profitability of the end product. A detailed analysis of carbon pricing in the following section examines a select group of basic chemicals and sub-sector chemicals based on an analysis conducted in 2008 by Climate Strategies, a division of the Electricity Policy Research Group at Cambridge University. The analysis reveals that most chemical sectors are affected by climate policy that targets electricity and feedstock producers, and low value downstream substances such as PVC and soda ash are among the most at risk to carbon pricing [15].

2.3.2
Applying Carbon Pricing to the Chemical Production Chain

Policy measures aimed at mitigating emissions generally target point sources of emissions such as the current EU ETS which covers CO_2 emissions from

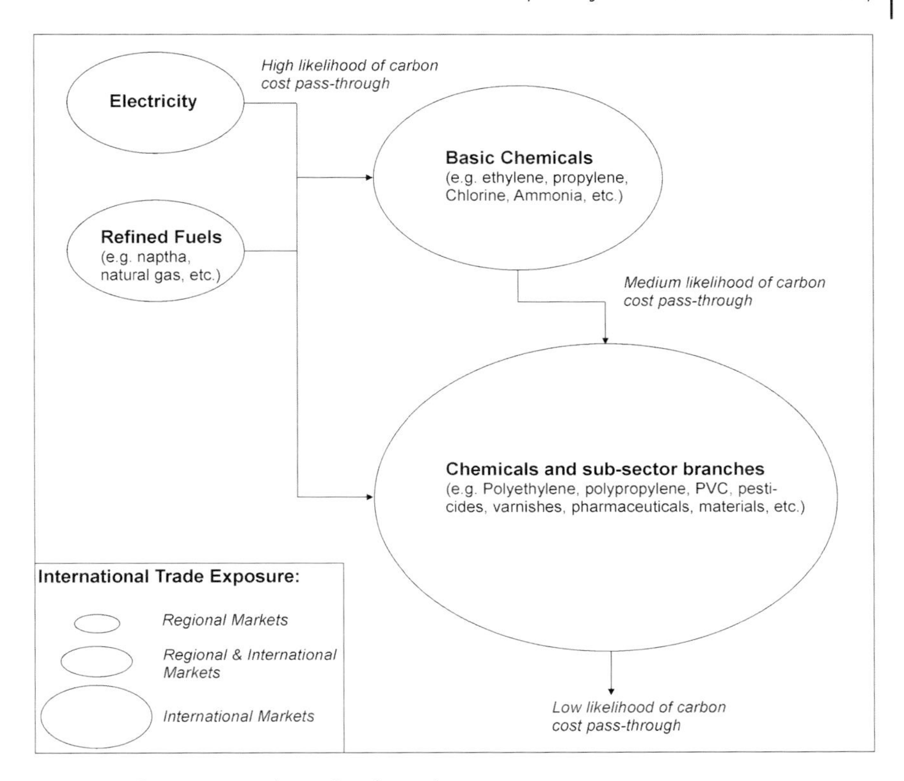

Figure 2.2 Exposure to carbon price absorption.

combustion installations. This section examines the direct cost and pass-through costs associated with 100% auctioning of emissions allowances for the chemical industry at each stage of the production process.

While it is unlikely that full auctioning of emission allowances is imposed on the chemical industry during the early phases of an emissions trading scheme, there are already provisions to target sectors such as power generation for auctioning because of its low level of exposure to international trade. Figure 2.3 demonstrates how the cost of carbon is applied to the gross production value under an emissions capped economy. This aspect of cost increase is referred to as the *value at stake* because of the diminishing gross value added relative to gross production value. This section examines how the value at stake is affected under various scenarios for a range of basic and sub-sector chemicals.

Certain aspects of the chemical industry are particularly vulnerable to cost increases attributed to carbon since the industry relies heavily on fossil fuels as feedstock and for process energy. A study conducted by the UK based Centre for Economic and Business Research (CEBR) on the impact on UK and EU chemical industry suggests that even at a carbon price of 20€ per tonne of CO_2, the percentage of the chemical industry's value at stake relative to value added could be as

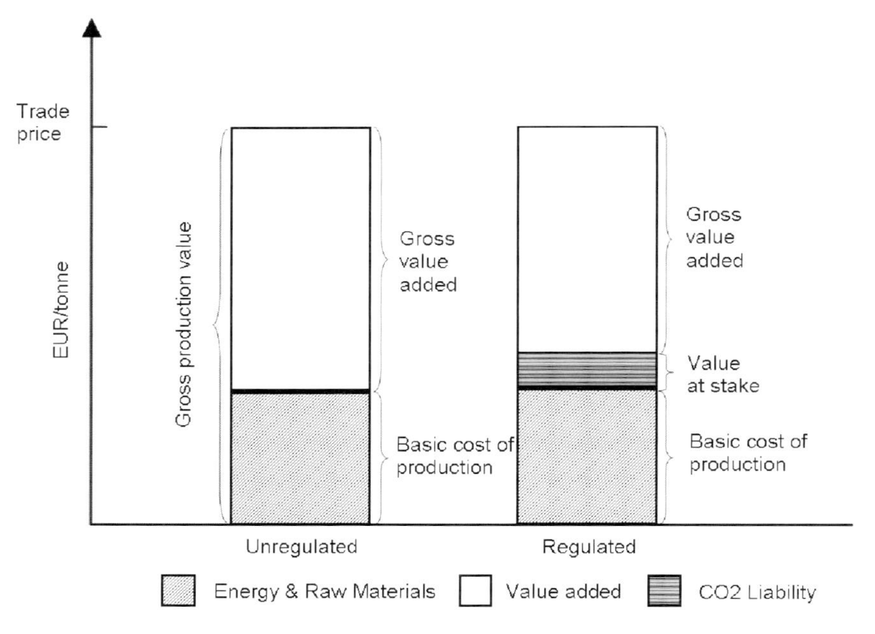

Figure 2.3 Value at stake.

high as 4.7% [16]. While in the USA, the petrochemical manufacturing industry could see short-term cost increases of more than 4% even with a modest carbon price of 10 US$ [17].

For the purpose of this analysis, carbon pricing is applied to direct emissions from the production process as well as indirect emissions from upstream products such as electricity and naphtha with a high likelihood of carbon cost pass-through. The next section also touches on indirect impacts of carbon pricing such as competition for low carbon fuels and cost of compliance for direct emissions associated with combustion installations.

2.3.2.1 Electricity Generation and Supply

While examining the impact of carbon pricing on electricity, the power sector is one of the least exposed sectors to international trade and therefore more capable of transferring carbon prices to end consumers, as was observed under Phase I and II of the EU ETS [18]. While this is an effective means toward achieving a reduction of point source emissions from power generation, electro-intensive processes such as the chlor-alkali are particularly vulnerable to indirect increases in electricity costs from pass-through carbon prices. Under a scenario where power producers are forced to purchase 100% of emissions allowances at 20€ under a compliance market, the cost of electricity could rise by approximately 15€ per MWh.[15]

15) Assuming a grid carbon emissions factor of 0.75 $tCO_2e\,MWh^{-1}$ for regions that rely heavily on thermal based generation.

To reduce the impact on consumers and maintain competitiveness in the regional market, power stations can shift to lower carbon content fuels, namely natural gas, when the 'clean spark spread'[16] exceeds the 'clean dark spread'.[17] While this is an effective measure to reduce the pass-through cost to consumers, the increased demand for natural gas in the power sector, referred to as the 'dash-for-gas', may result in higher natural gas prices, consequently affecting the downstream chemical industry that relies heavily on gas as a primary feedstock.

The price of raw materials as well as escalating electricity prices from pass-through carbon costs threatens the marginal cost of production, particularly for the inorganic industry that produces chlorine, nitrogen, hydrogen and oxygen. For chlorine producers in Europe which are subject to rising electricity prices under a 20€ per tonne of CO_2 scenario, this could result in a 16% increase in production cost [15].

2.3.2.2 Feedstock Extraction, Transportation, and Preparation

In addition to the increased energy costs driven by the power sector, the preparation of feedstock used for chemical production is also energy intensive which results in a high level of direct emissions associated with the refinery cracking process. Approximately 4% of the world's oil and equivalent fossil fuels are used as feedstock for plastics and chemicals [19].

Under an emissions capped economy, the inclusion of carbon pricing on this segment of the supply chain is likely to increase chemical production costs, especially under a scenario of taxation or auctioning of emissions allowances. While oil and coal are globally traded commodities, refined feedstock such as naphtha are significantly less exposed to international markets and therefore more likely to transfer carbon costs to the petrochemical industry. For example, under a carbon pricing scenario of 20 € per tonne of CO_2, the petrochemical industry experiences a price increase of approximately 8€ per tonne of naphtha.[18] Naphtha obtained from crude oil refineries represents approximately 75% of the feedstock used for ethylene production in Western Europe [20].

It is estimated that feedstock will represent the largest cost increase for certain products, and a study conducted by the Plastics and Chemicals Industries Association of Australia estimates that carbon pricing on feedstock will have four to five times more impact on the chemical industry than carbon pricing on electricity.

It is also important to consider the feedstock route and whether they are derived from gas, oil, ethane, coal, or biomass. When examining the energy consumption

16) The spark spread refers to the theoretical gross margin of a gas-fired power plant from selling a unit of electricity, having bought the fuel required to produce this unit of electricity. The clean spark spread refers to the spread indicator which includes the price of carbon allowances.

17) The dark spread refers to the theoretical gross margin of coal-fired power plants from selling a unit of electricity, having bought the fuel required to produce this unit of electricity. The clean dark spread refers to the spread indicator which includes the price of carbon allowances.

18) Assuming an emissions factor of 0.405 t CO_2 to produce 1 tonne of naphtha.

of each feedstock route compared to conventional route which uses oil or ethane in the production of high value chemicals (HVC), methane-based routes are 30% higher while coal and biomass-based routes are 60–150% higher. However, the total CO_2 emissions of conventional and methane based routes have a similar intensity of 4–5 tonnes CO_2 per tonne of HVC. Coal-based routes produce the highest quantity of CO_2 ranging from 8–11 tonnes of CO_2 per tonne of HVC. Preliminary research discovered that biomass-based routes range from 2 to 4 tonnes of avoided CO_2 per tonne of HVC depending on the source of the biomass and whether combined heat and power (CHP) maximizes efficiency [21].

2.3.2.3 Basic Chemical Preparation

The production of basic organic and inorganic chemicals requires the use of feedstock and energy (both electricity and fuel). This section examines how the marginal cost of basic chemical production is affected by carbon pricing on direct emissions associated with the production process, as well as on the upstream carbon pass-through from feedstock and electricity producers. Steam cracking of feedstock for chemical preparation is also very energy intensive and installations which operate their own CHP facilities are potentially subject to carbon pricing which incurs a direct cost for the installation.

A report by Climate Strategies on estimating carbon costs for the chemical sector outlines a basic methodology to determine the impact of carbon pricing for a range of chemicals under a scenario where 100% of emissions allowances are auctioned to the chemical and power sector at a price of 20€. The report estimates the value at stake for a select group of chemicals by calculating the production cost increase relative to gross value added; where value added is determined by subtracting the cost of raw materials and energy from the gross production value for the German chemical industry using 2003 as the base year.

Figure 2.4 illustrates the three elements of cost increase imposed on the basic chemical industry. While all basic chemicals experience cost increases from direct emissions, substances such as alumina, soda ash, methanol and ammonia experience the highest effect. Feedstock pass-through is experienced most notably by olefins such as ethylene, propylene, butadiene, and butane while electricity pass-through is more prominent for alumina, methanol, acetylene, and aromatics.

Figure 2.5 illustrates the same analysis for electro-intensive industries which are highly susceptible from cost pass-through from the power sector. While all electro industries are vulnerable to such cost increases, chlor-alkali for chlorine production experiences the highest rise in cost. Calcium carbide also experiences a significant increase in production cost relative to added value when the power sector is faced with auctioning of allowances.

The analysis by Climate Strategies suggests that while basic chemicals experience the highest production cost increase relative to gross added value, substances such as chlorine and ammonia generally sell within a regional market because of the high transportation costs associated with potentially hazardous materials. This logic suggests that products such as alumina, soda ash, and basic nitrogen

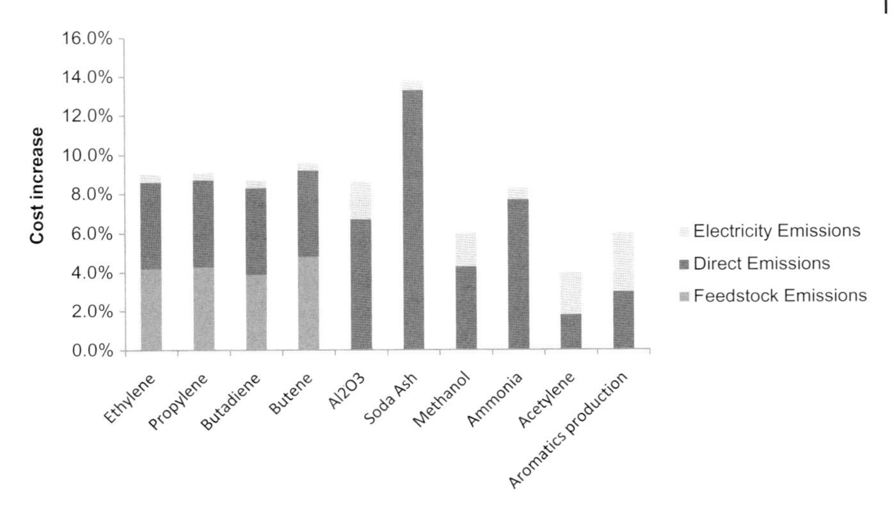

Figure 2.4 Cost increase relative to gross added value for basic chemicals [15].

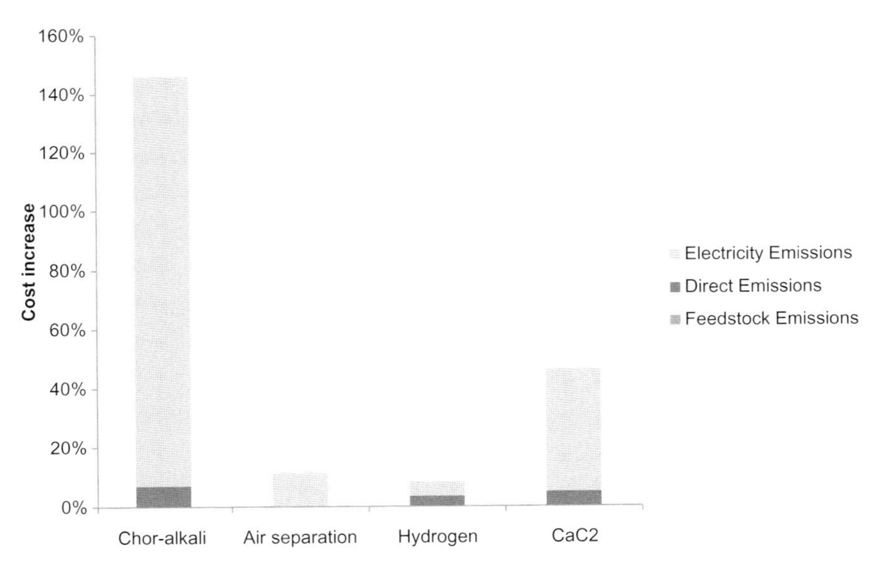

Figure 2.5 Cost increase relative to gross added value for electro-intensive basic chemicals [15].

fertilizers derived from ammonia are the most at risk to leakage where exposure to international markets prevents carbon pass-through from producers.

Inter sector trading of basic chemicals makes across border risk analysis difficult where carbon intensive intermediate processes are subject to leakage from undue shift in production to countries with little or no regulations on carbon emissions.

Since much of the downstream chemical industry relies heavily on chlorine and olefins as inputs for the production of products such as PVC, polyethylene, and ethylene, the next section examines how pass-through cost affects the relative value at stake for a select number of downstream chemical products.

2.3.2.4 Subsector Chemical Preparation

In an earlier discussion around industry exposure and carbon pricing, it was concluded that although downstream chemical products such as pesticides and synthetic materials have the most exposure to international trade, they are not necessarily the most affected by carbon pricing measures, given the high added value relative to production cost increase. The value at risk for high value chemicals is particularly low even with a carbon cost transfer at each stage of the production process.

Here the analysis extends for downstream chemicals substances polyethylene, polypropylene, and PVC. Using the same methodology as derived by Climate Strategies, the cost increases are examined relative to added value, as illustrated in Figure 2.6. Under a carbon cost scenario of 20€ per tonne of CO_2 imposed on the power sector, PVC experiences the highest cost increase relative to added value. This estimate assumes that chlorine producers pass 100% of the carbon cost through to PVC producers; a realistic assumption based on low trade exposure in the chlorine industry.

Where two-thirds of chlorine trade occurs in the downstream PVC industry, this represents a significant concern for regulating authorities of a cap-and-trade scheme where product demand in emissions capped countries will not decrease. The Chlor-Alkali Association of Europe is particularly concerned about the PVC

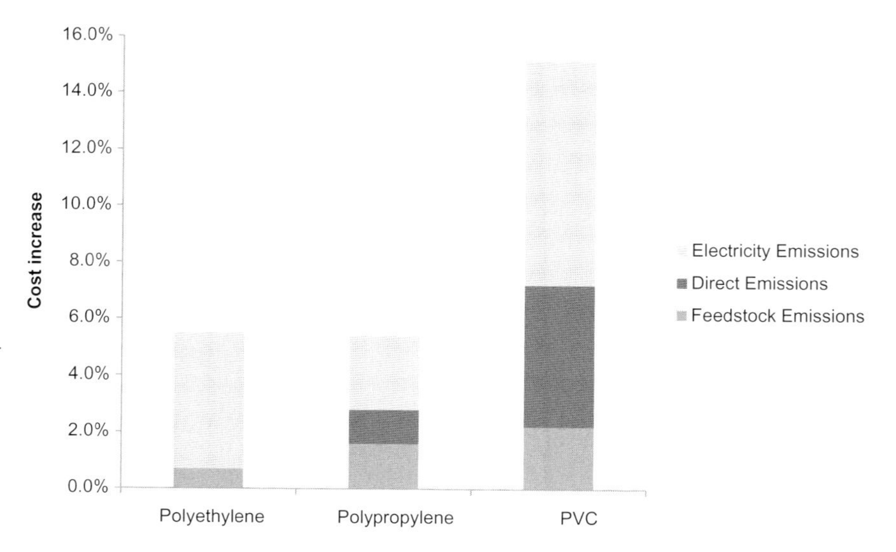

Figure 2.6 Carbon cost increase relative to gross added value [15].

industry as there is little ability within the European market to transfer carbon cost associated with chlorine production. Since the demand in Europe will remain whether or not production is local, there is concern that policy measures could lead to leakage where PVC production shifts to countries with no regulations on CO_2 emissions [22].

While carbon cost pass-through from upstream industries remains particularly high relative to added value for low value downstream products such as polyethylene, polypropylene, and PVC, higher value products such as pharmaceuticals and pesticides experience very low cost increases in the range of 0.2% relative to value added under an identical scenario of carbon pricing at 20€ per tonne [15].

2.3.3 Opportunities within a Compliance Market

While emissions caps may lead to increased production costs for industrial installations that fall under compliance, emissions trading offers companies the ability to deploy new technologies when the MAC at their own facilities is below the market value of allowances. Where technology measures reduce the emissions level below the target for the facility, surplus allowances represent value to installations that fall short of their commitments.

Emissions trading can also offer provisions for companies to voluntarily enter facilities into the scheme referred to as 'opt-in' to undertake measures that reduce emissions and sell surplus allowances from the facility. The opt-in rule in the EU ETS allowed France and the Netherlands to include N_2O emissions from fertilizers as of 2008 [23].

In order to adequately assess the economic drivers for implementing low carbon technologies, it is important to consider the price point for each abatement solution in the chemical industry. An analysis by McKinsey examined the MACs for a number of abatement solutions for the chemical industry ranging from CHP to carbon capture and storage (CCS) for ammonia production. The analysis concludes that while business and society view the cost of abatement differently, solutions such as fuel switching and catalyst optimization are immediately effective without carbon price signals. However, other solutions such as CCS examined in Chapter 11 require carbon prices as high as 100€ or more per tonne of CO_2e to be feasible [2].

2.4 Carbon Offsetting in the Chemical Industry

2.4.1 Concept of Offsetting

International carbon offsets further take advantage of the least cost of abatement or MAC to lessen the economic burden for industry and governments to achieve emissions targets. Similar to an emissions allowance, a carbon offset is measured in tonnes of CO_2e and represents one certified metric tonne of CO_2 reduction. While the voluntary market participates in this arena, it is largely led by the compliance market which reached 118 billion US$ in 2008 [24].

Under the Kyoto Protocol, carbon offsets represent an opportunity for companies that have across border operations in countries with emissions caps (Annex-1 countries) and those without (Non Annex-1 countries). This is very common in the chemical industry and investigated further in this section in the chemical context. There are also large opportunities for the chemical industry to generate offsets for sale to other companies that fall under compliance.

2.4.2
Flexible Mechanisms of the Kyoto Protocol

The flexible mechanisms of the Kyoto Protocol, introduced in the introductory chapter, are the only existing methods to produce carbon offsets acceptable for compliance under nationally recognized cap-and-trade schemes such as the EU ETS. Recognizing the benefits of carbon reduction in the atmosphere, the mechanisms were established to lessen the economic cost of emissions compliance for signatory countries to the Protocol and promote the transfer of clean technologies and investment. There are two mechanisms for generating offsets under the Protocol; the CDM between an Annex-1 country and a non Annex 1 country; and JI between two Annex 1 countries [25].

These mechanisms provide additional flexibility to companies operating within a global context to maximize assets and achieve companywide carbon reductions at the least cost. This mechanism acts as a principle financial incentive for com-

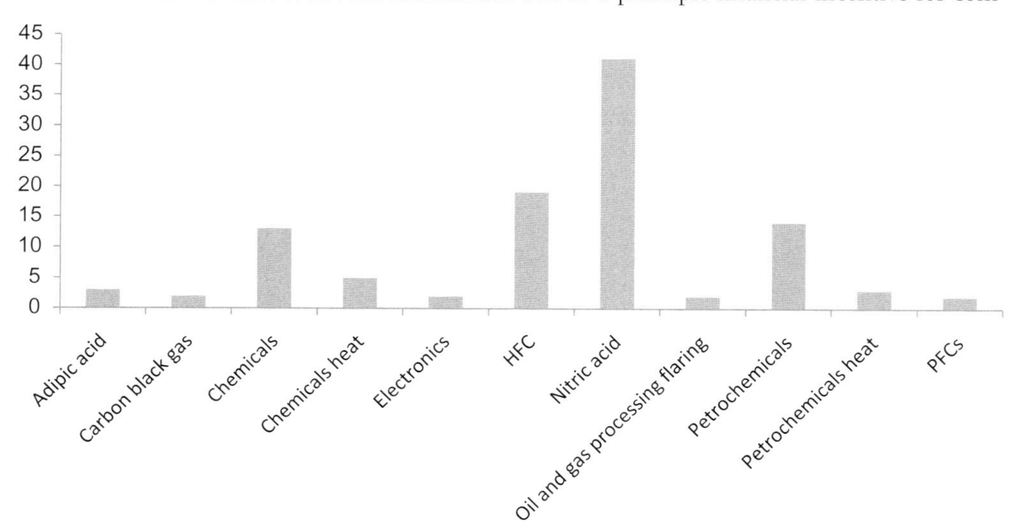

Figure 2.7 Number of approved CDM projects in the chemical and petrochemical industries [27].

panies to generate revenue through the sale of emission reductions (so called carbon credits).

2.4.2.1 Developing a CDM Project

The CDM defined in Article 12 of the Kyoto Protocol is the most established carbon offset scheme and enables the implementation of project-based emission reductions in developing countries (non Annex-1 countries). This section examines the CDM registration cycle and what steps must be undertaken by project developers when applying for salable certified emission reductions (CERs). One CER represents a tonne of reduced CO_2 compliant within a cap-and-trade system such as the EU ETS.

The CDM represents a good opportunity for an industrial installation to claim emission reductions for activities undertaken to conserve energy, switch to lower carbon fuels or abate harmful greenhouse gas emissions. There are a number of project types in the chemical industry that are eligible under the CDM and the most successful initiatives thus far have targeted high GWP gases, such as HFC23 and N_2O.

For example, companies such as Rhodia have undertaken CDM projects to reduce N_2O emissions from adipic acid production plants in South Korea and Brazil. Rhodia's two projects combined are expected to produce 12 000 000 CERs per year, at an estimated market value of more than half a billion Euros up to the end of the Kyoto commitment period [26]. Figures 2.7 and 2.8 show the

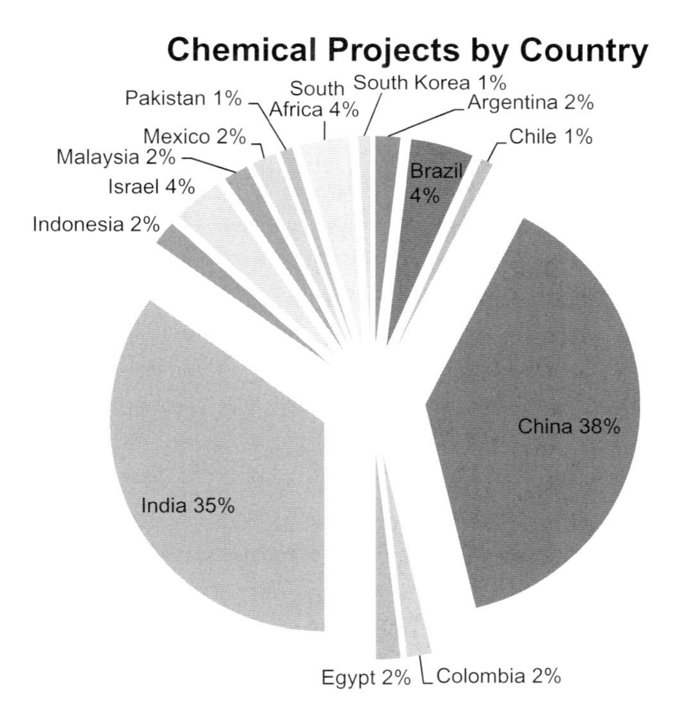

Figure 2.8 Approved CDM project distribution by country for the chemical sector [27].

distribution of CDM projects by type and geography for the chemical and petrochemical industries. It is interesting to note that while there are a number of different types of projects for the chemical industry, projects to reduce HFCs represent nearly 70% of the registered emission reduction volume [27].

Project Types and Applicability Criteria The CDM provides a framework to develop a wide range of project-based opportunities that reduce gases covered under the Kyoto Protocol.[19] Each new project type must comply with a baseline and monitoring methodology approved by the United Nations Framework Convention on Climate Change (UNFCCC) for calculating emission reductions achieved by the project activity. A baseline methodology defines the applicability conditions and establishes a procedure to determine the baseline scenario and estimate emission reductions that result from the project activity. The monitoring methodology outlines the requirements for monitoring equipment, monitoring parameters, and quality control and quality assurance measures. There are three types of methodologies; two large scale types and one small scale type. AM refers to large scale 'Approved Methodology' where ACM refers to large scale 'Approved Consolidated Methodology'. Small scale methodologies are referred to as AMS for 'Approved Methodology for Small Scale Projects' [28].

This section examines the eligible project types within the chemical process industry. Table 2.1 outlines the existing approved UNFCCC methodologies and their applicability criteria. While more emission reduction opportunities exist at a chemical complex, new project types that do not conform to existing methodologies must undertake an application process with the UNFCCC for their inclusion.

Project Design Phase Chemical facilities that identify an emission reduction opportunity that complies with the requirements of existing baseline methodologies are able to begin the application process by completing a project design document (PDD) in accordance with the guidelines of the UNFCCC [30].

A PDD is completed in a standardized template issued by the UNFCCC and must clearly demonstrate that the proposed project activity will generate additional GHG reductions in accordance with the local host country's requirements on economic and sustainable development objectives. At this stage, the project developer decides whether to apply for a fixed 10 year crediting period or a renewable crediting period of 3 × 7 years. The PDD is a comprehensive document which covers the following areas [31]:

1) Description of the project activity and technology to be installed at the facility.
2) Identification of all plausible baseline alternatives in accordance with the approved baseline and monitoring methodology.

19) Gases covered under the Kyoto Protocol include carbon dioxide (CO_2), methane (CH_4), nitrous oxide (N_2O), sulfur hexafluoride (SF_6), hydrofluorocarbons (HFC) and perfluorocarbons (PFC).

Table 2.1 Approved baseline and monitoring methodologies applicable to the chemical process industry [29].

Ref	Methodology	Applicability
AM0001	Incineration of HFC23 waste streams (version 5.2)	• Brownfield projects[a)] only • Limited to status quo activity level until the lifetime of existing equipment • Only for existing facilities that have been in operation (emitting HFCs and not CFCs) for at least 3 years between 2000 and 2004 and have remained in operation after 2004 until the project start date
AM0021	Baseline methodology for decomposition of N_2O from existing adipic acid production plants (Version 2.2)	• Brownfield projects only • Limited to status quo activities until the lifetime of existing equipment • Only for existing capacity installed before 31 December 2004
AM0027	Substitution of CO_2 from fossil or mineral origin by CO_2 from renewable sources in the production of inorganic compounds (Version 2.1)	• Limited to status quo activities until the lifetime of existing equipment • All carbon in the inorganic compound is from CO_2 added during production • No change to plant output • No change to process and related to emissions due to energy consumption • No biomass crowding out
AM0028	Catalytic N_2O destruction in the tail gas of nitric acid or caprolactam production plants (Version 4.2)	• Brownfield projects only • Limited to status quo activities until the lifetime of existing equipment • Only for existing capacity installed before 31 December 2005 • Destruction of N_2O emissions by catalytic decomposition or catalytic reduction of N_2O in the tail gas of nitric acid or caprolactam production plants • No change in plant output • Plant does not already have a Non-Selective Catalytic Reduction (NSCR) unit
AM0034	Catalytic reduction of N_2O inside the ammonia burner of nitric acid plants (Version 3.2)	• Brownfield projects only • Limited to status quo activities until the lifetime of existing equipment • Only for existing capacity installed before 31 December 2005 • Destruction of N_2O emissions by catalytic decomposition or catalytic reduction of N_2O inside the burner of a nitric acid plant • Not for process with new ammonia oxidizer • Plant does not already have a Non-Selective Catalytic Reduction (NSCR) unit

Table 2.1 *Continued*

Ref	Methodology	Applicability
AM0037	Flare reduction and gas use at oil and gas processing facilities (Version 2.1)	• Brownfield projects only • The tail gas from an oil or natural gas processing facility was flared for at least 3 years prior to the project start • The tail gas is used to produce useful energy or useful products and replace the fuels/feedstock with at least the same CO_2e impact • Also includes activities that vent in the baseline • Limited to cases where associated gas substitutes feedstock.
AM0050	Feed switch in integrated Ammonia-urea manufacturing industry (Version 2.1)	• Brownfield projects only • Limited to status quo activities until the lifetime of existing equipment • Limited to existing production capacity • Historical operation of at least three years prior to the implementation of the project activity • Thermal energy for processing the feed is the combustion of fossil fuels in boilers in baseline and project • The quantity of steam and electricity is the same for naphtha and natural gas
AM0051	Secondary catalytic N_2O destruction in nitric acid plants. (Version 2)	• Brownfield projects only • Limited to status quo activities until the lifetime of existing equipment • Only for existing capacity installed before 31 December 2005 • Destruction of N_2O emissions by catalytic decomposition or catalytic reduction of N_2O inside the burner of a nitric acid plant • Not for the process with new ammonia oxidizer • Plant does not already have a Non-Selective Catalytic Reduction (NSCR) unit
AM0053	Biogenic methane injection to a natural gas distribution grid (Version 1.1)	• Methodology can be used in conjunction with methodologies for capture and destruction/use of biomethane
AM0063	Recovery of CO_2 from tail gas in industrial facilities to substitute the use of fossil fuels for production of CO_2 (Version 1.1)	• Substitution of CO_2 production from fossil fuels with CO_2 recovered from an industrial process • Applicable to either activities that consider the difference between highest level of historic emissions from existing plant and project activity emissions or the case where old plant is no longer in operation and thus historic data from operations of the old plant prior to the CDM project are used to define activity level • If project results in end to operations of the existing CO_2 production plant, the remaining lifetime of the old plant prior to the project will define the crediting period

Table 2.1 *Continued*

Ref	Methodology	Applicability
AM0069	Biogenic methane use as feedstock and fuel for town gas production (Version 01)	• Brownfield projects only • Project activities where biogas is captured at wastewater treatment facility or landfill is used to substitute natural gas or other fossil fuels of higher content as feedstock and fuel for the production of town gas • Must be 3 year record of venting or flaring of biogas. • Town gas factory must have no history of using biogas prior to project and must have 3 years data on quantity and quality of fossil fuels used
AMS.III.J	Avoidance of fossil fuel combustion for carbon dioxide production to be used as raw material for industrial processes (Version 3)	• Limited to 60 Kt annual emission reductions • The generation of CO_2 from fossil or mineral sources in the baseline is only for the purpose of CO_2 production to be used for the production of inorganic compounds. • There is no energy by-product of CO_2 production from fossil source and its consumption in the baseline • All carbon in the CO_2 produced under the project shall come from a renewable biomass source • The residual CO_2 from the processing of biomass was already produced but was not used before the project • CO_2 from fossil or mineral sources that is used for the production of inorganic compounds prior to the project will not be emitted to the atmosphere when the project activity is in place
AMS.III.M	Reduction in consumption of electricity by recovering soda from paper manufacturing process (Version 2)	• Energy savings resulting from project activities that reduce caustic soda that would be purchased from in country production or imported from facilities located in Non-Annex 1 countries
AMS.III.O	Hydrogen production using methane extracted from biogas	• Up to 60 Kt annual emission reductions • Activities that install a biogas purification system to isolate methane from biogas for the production of hydrogen displacing LPG as both feedstock and fuel in a hydrogen production unit • Not applicable to technologies displacing the production of hydrogen from electrolysis. Must prove no diversion of biogas occurs.

a) Brownfield projects refer to projects at existing facilities.

3) Estimation of baseline and project emissions as well as identification and quantification of any sources of leakage in accordance with selected methodology.
4) Evidence that the project is additional based on the latest tool for the demonstration of additionality.[20]
5) Proof of CDM consideration.
6) Monitoring plan for the operational phase of the project activity.
7) Evidence of environmental and social benefits of project activity including stakeholder consultations.

Host Country Approval Each CDM project must gain approval from the host country in order to gain registration with the UNFCCC. Host country approval is conditional on the national criteria and requirements established by the Designated National Authority (DNA) acting as the national approval body for CDM projects. Host country requirements vary considerably but generally relate to the demonstration of environmental, social, and economic benefits for the country [32].

Validation and Registration The validation and registration cycle shown in Figure 2.9 is the CDM approval process before the project is eligible to receive CERs. Validation is an independent third-party evaluation of the CDM project documents against the international requirements established by the UNFCCC. The independent third-party accredited by UNFCCC is referred to as a designated operational entity (DOE) and checks that all information and assumptions in the PDD are accurate and reasonable. Following the validation process, the PDD is posted on the UNFCCC website for public comments [31].

Once a project is validated, the registration process with the UNFCCC begins. Throughout the validation and registration period, information about the project is made publically available. The registration of the project occurs in the eight week period after the date of receipt by the Executive Board (EB) of the UNFCCC. During the eight week period, the EB determines whether the project activity meets the CDM requirements and may pull a project for review if there are any particular concerns with the project [33].

Monitoring and Verification Project developers must monitor various parameters during the project activity according to the monitoring methodology approved by the EB. Monitoring is initially considered during the project design phase where a monitoring plan is included in the PDD. Information pertaining to the monitoring plan is included in a monitoring report prior to the verification of emissions by an accredited DOE [33].

20) Additionality addresses the question of whether the project activity is additional to the business-as-usual scenario and merits the environmental benefits of reducing greenhouse gas emissions.

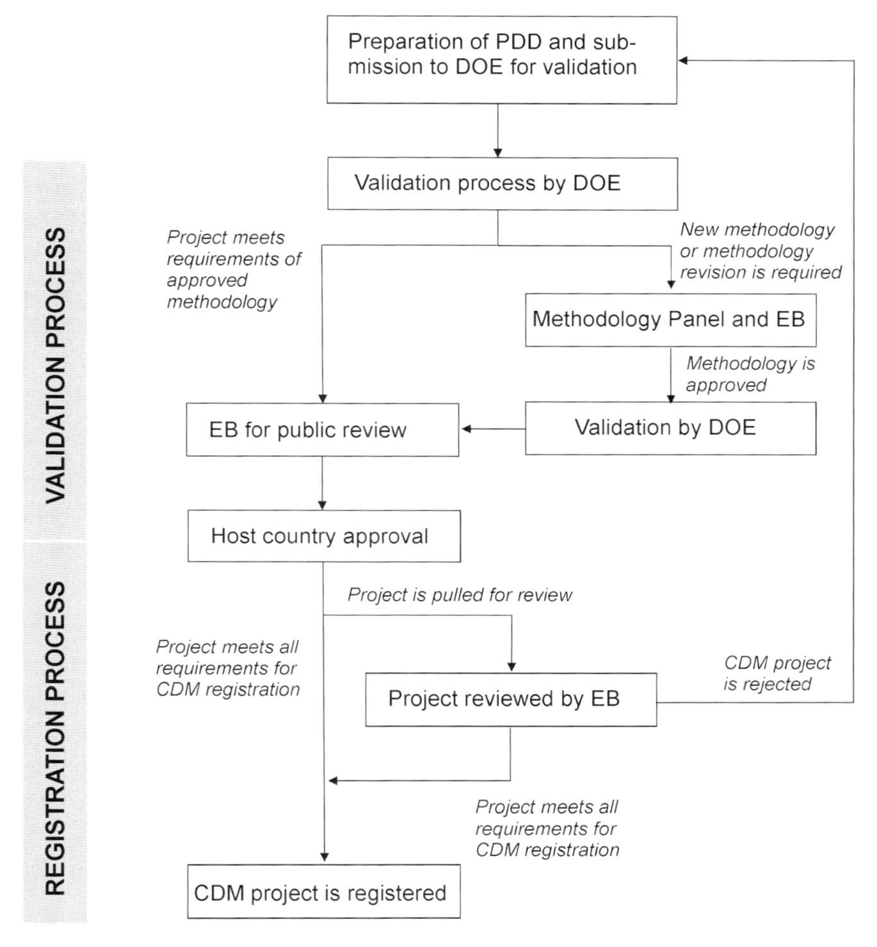

Figure 2.9 Validation and registration cycle.

Verification during the operation of the project activity is required to demonstrate that project is achieving real and quantifiable emission reductions in accordance with the requirements of the methodology. Verification refers to the periodic and independent review of emission reductions by a DOE before CERs are issued by the UNFCCC. The DOE must provide certification in the form of written assurance that the project activity achieved the emission reduction verified during the project crediting period [31].

Transacting Emission Reductions CERs are issued to the account of the project developer held with the UNFCCC. Emission reductions are typically transacted through an Emission Reduction Purchase Agreement (ERPA) which contractually

Table 2.2 Example of clean development mechanism project case study.

Project title	**Tianji Group Line 1 N_2O Abatement**
Project description	The project activity involves the abatement of nitrous oxide (N_2O), an undesired by-product formed in the catalytic oxidation of ammonia for the nitric acid (HNO_3) production process. N_2O abatement is achieved through the installation of a secondary abatement catalyst inside the ammonia burner where the emissions are formed. N_2O is a potent GHG emission with a GWP of 310 tCO_2e/tN_2O.
Host country	China
Sectoral scope	Chemical industries
CDM methodology	AM0034 (version 2) – Catalytic reduction of N_2O inside the ammonia burner of nitric acid plant
Estimated annual emission reductions	502 194 $tCO_2e\,y^{-1}$
UNFCCC fee level	US$ 98 939

binds the project developer and credit buyer at the time credits are issued by the UN. Credit buyers comprise of a range of different entities such as emissions compliance buyers as well as traders and credit aggregators. The purpose of CERs is to eventually retire them in the place of emissions allowances to meet compliance obligations under a cap-and-trade regime. For the case of the EU ETS, this is done through the 'linking directive'[21] which enables the compliance installation to surrender CERs in the place of allowances.

Example of CDM project case study (Table 2.2) [34]:

2.4.2.2 **Developing a JI Project**

Joint implementation (JI) is defined in Article 6 of the Kyoto Protocol and allows Annex-I countries to undertake emission reduction projects in other Annex-I countries. Each emission reduction is referred to as an emission reduction unit (ERU) and represents one tonne of CO_2 equivalent. Similar to the CDM, JI offers the potential to take advantage of least cost of abatement or MAC across countries and sectors.

The JI project cycle follows a similar structure to the CDM cycle but is more streamlined with project eligibility and approval process. Under JI, all CDM methodologies are entitled to calculate emission reductions in the PDD and make modifications to existing methodologies or propose new methodologies to serve

21) The 'Linking Directive' allows an operator to use a certain amount of CERs or ERUs from flexible mechanism projects in order to cover their emissions.

the needs of the project. The independent approval of project documents, referred to as validation under the CDM, is called determination and is conducted by an accredited independent entity (AIE).

If the host Party meets all of the eligibility requirements for credit transfer under JI 'Track 1' procedure, the host Party is entitled to receive the appropriate quantity of ERUs if the project is deemed additional to what would otherwise have occurred.

If the host party only meets a limited set of eligibility requirements, verification is done through the verification of procedure under the Joint Implementation Supervisory Committee (JISC) under what is referred to as a 'Track 2' procedure. Under this procedure, an independent entity accredited by the JISC has to determine whether all of the requirements have been met in advance of issuing ERUs to the host party.

Similar to the CDM, verification is conducted by an accredited third party referred to as an AIE, and written certification is provided and used to demonstrate that emission reductions were achieved in accordance to the rules governing JI procedures. ERUs are issued by the host country rather than by the CDM executive board.

2.4.3
International Offsetting in a Post-2012 Context

While the CDM and JI have offered an initial pathway for companies to invest in low carbon technologies, there are key limitations of the mechanisms which prevent the wide-spread deployment of technology based solutions that reduce energy consumption and GHG emissions. These drawbacks are primarily attributed to the 'project-based' nature of the existing offset mechanisms, making it difficult to demonstrate additionality and select the appropriate baseline scenario, particularly for highly integrated industries such as the chemical process industry.

Although there are several proposals for a post-2012 agreement, there is, at the time of this writing, no clear pathway that defines a way forward for international offsets. It is clear that, based on earlier discussion in the introductory chapter, many of the imminent schemes such as Phase 3 of the EU ETS and the United States Climate Change Bill include provisions for international offsets. This section examines existing proposals to scale up investment in low carbon technologies through international offset mechanisms.

2.4.3.1 Scaling up the CDM via Benchmarking

The concept of benchmarking,[22] already widely used by industry, is a powerful tool under discussion for streamlining the CDM to demonstrate additionality. This approach to demonstrating additionality allows for certain project types to always

22) Benchmarking in the manufacturing sector refers to a measured performance indicator for which to compare a group of facilities.

gain additionality approval, providing more certainty to investors by securing carbon revenues at the project design phase. Procedurally, this could be undertaken in a variety of ways by establishing benchmarks on an international, regional or national level agreed between the host country and UNFCCC.

Benchmark based additionality is already proposed in certain sectors such as the cement industry and would enable substantial advantages to the chemical industry to unlock a range of energy efficiency initiatives at one facility without assessing the business-as-usual scenario of each stand-alone activity. Proposals that include the expansion of the CDM through benchmarks are largely accepted by NGOs and governments because of their transparent design but are being met by some degree of resistance from developing countries who feel that it lays out the groundwork of an emission reduction target that could burden their economy. Other issues surround data availability where developing countries do not always feel comfortable with releasing industry performance data.

2.4.3.2 Sectoral Crediting Mechanisms (SCM)

Sectoral crediting, sometimes referred to as 'sectoral no-lose target', works on the premise of installations emitting less GHG emissions than a so-called baseline. The sectoral baseline is a negotiated target which lies below the business-as-usual scenario for a specific region and sector. This allows government to distribute credit to industries that achieve a reduction of emissions levels below the baseline. The following two criteria are important design features of sectoral crediting:

1) The carbon units are issued at the end of the period; and
2) There is no obligation on the government to purchase carbon units from elsewhere if it did not achieve its baseline

2.5 Positioning Industry for a Global Framework on Climate Change

The chemical industry operating within a global framework on climate change is now faced with a broad base of risks and opportunities that threaten key segments of the supply chain while driving efficiency and innovation in others. The previous sections examined various policy instruments used by government to mitigate GHG emissions through emissions trading and taxation; this section outlines various positions taken by leading companies and chemical associations such as Cefic to limit vulnerability and preserve integrity of emissions trading.

Key lessons taken from the existing Kyoto Protocol and EU ETS form the basis for companies to better formulate their corporate position and influence climate policy to lessen the economic burden of carbon compliance while generating new revenue streams from technology-based solutions to climate change. This section focuses primarily on how policy is designed to minimize market distortion and leakage while imposing a similar level of responsibility across industry and consumers operating within an emissions capped economy.

2.5.1 Defining Sectors within a Regulated Environment

In order for the chemical industry to adequately manage risks and contribute to the design of an effective emissions trading scheme, it is important to understand various climate policy instruments and how to categorize the business in terms of energy and GHG emissions as well as end products and trade markets. Cefic strongly advocates for dividing chemical processes into eight major categories in order to appropriately allocate emissions allowances to avoid putting unilateral cost burden on the chemical industry. According to Cefic, the inclusion of the following sub-sectors would represent approximately 80% of chemical industry emissions [3]

- ammonia for the production of fertilizers;
- ethylene (cracker products);
- methanol (synthesis gas);
- soda ash;
- carbon black;
- nitric and adipic acid;
- energy installations in chemical production process (CHP, boilers, captive power, etc.);
- chlor-alkali (compensation for indirect emissions from electricity).

While emissions trading schemes such as proposed in Europe, Australia, and the USA are different in their scope and approach to allocating allowances, each scheme must define the mechanisms to address market distortion and leakage.

The inclusion of sub-sectors as outlined above enables industry to qualify for certain exemptions from an emissions trading schemes such as the case in Europe where the Commission and the European Council are determining which sectors and sub-sectors will benefit from leakage protection based on the following criteria [35]:

- Additional production costs resulting from the ETS will exceed 5% of the gross value added.
- The total value of exports and imports divided by the total value of turnover and imports (a measurement called the non-EU trade intensity) exceeds 10%.

Emissions trading schemes can also incorporate border tax adjustments as an instrument to prevent leakage of emissions to hot spots. While this is a possible measure protect domestic industries, it poses concern with trade partners outside of the regulated country. An effective emissions trading scheme that considers sub-sector exposure can be designed to avoid such border tax adjustments.

2.5.2
Allocating for the Chemical Industry

As discussed in Section 2.2, allowances under a cap-and-trade scheme are either allocated for free, auctioned, or allocated through a combination of both. This section examines each scenario for installations covered under an emissions cap. In order to examine allocations in the chemical context, there are three possible approaches to consider for the chemical sector; *installation based allocations, benchmark-based allocations, and auctioning of allowances.*

Figure 2.10 draws a comparison between each method for distributing allowances. Combustion installations covered under the EU ETS are currently subject to *installation based allocations* according to a national allocation plan for each country where the facilities operate. While this method has proved effective under the EU ETS, discussion of future auctioning of emissions allowances threatens industry with unjust cost distribution. Further criticism is placed on installation based allocations because it does not adequately credit early action and does not impose a similar emissions abatement cost curve under the assumption that the MAC increases with each additional initiative.

Benchmark based allocations are strongly promoted by industry associations such as Cefic for their ability to recognize high performing facilities that operate below the intensity benchmark; thus limiting the economic burden to the least efficient facilities which cause the most damage to the environment. While this is arguably more effective at achieving the overall objectives of an emissions trading scheme, it is particularly unjust for low performing facilities which risk severe lack of competitiveness issues.

Full *auctioning of allowances* is criticized by manufacturing industries but is particularly disadvantageous for some sectors of the chemical industry such as PVC, nitrogenous fertilizer, and soda ash producers, as demonstrated earlier in the chapter. Industry suggests that this approach does not drive market efficiency

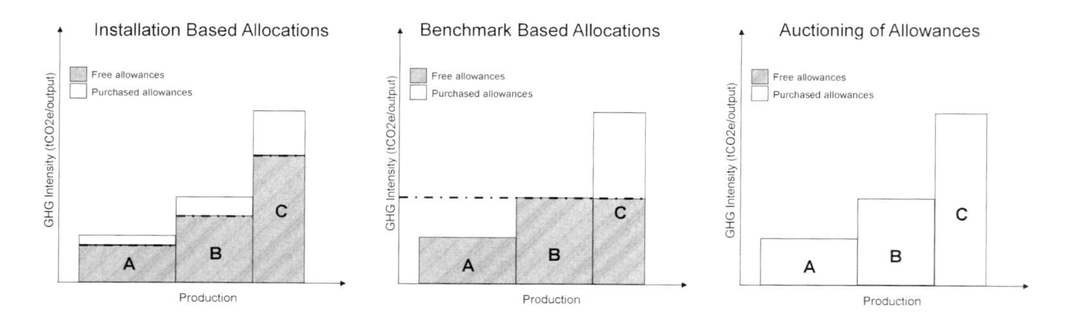

Figure 2.10 Comparison of allocation plans.

and threatens price reduction of emission allowances, preventing industry to reduce beyond the reduction target. This system is also difficult to operate from an administrative position and results in large sums of revenue for the governing body to redistribute.

2.5.3 Key Messages Moving Forward

Evidence is mounting to support the need to better understand and manage carbon liability at a corporate level within the context of climate policy. It is becoming ever more critical for chemical companies to form a corporate position on key issues such as emissions trading, taxation, allocations and border tax adjustments, and to develop general strategies to maximize asset management under a global carbon trading framework.

This chapter examined the key vulnerabilities and opportunities of the chemical industry operating within a regulated environment on GHG emissions. It was revealed that a cap-and-trade can threaten the cost of production for basic and sub-sector products such as PVC, soda ash, calcium carbide, polyethylene, and polypropylene because of their high dependency on carbon intensive upstream inputs such as chlorine and olefins and especially for downstream products that are highly exposed to international trade. While climate policy can incur costs for industry, emissions trading and international offsets offer new revenue streams for the chemical industry which, for example, can deploy waste heat recovery or abate harmful emission from nitric or adipic acid production. Evidence also supports that a price of carbon drives innovation in the chemical industry to develop technology based solutions to climate change such as PV cells and lightweight composites.

The chemical industry must continue to advocate for the inclusion of sector and sub-sector consideration within a cap-and-trade to communicate the risks of products that face market distortion and leakage imposed through climate policy. Options discussed earlier such as benchmark based allocations, opt-in provisions, and international offsets offer a way forward for energy intensive sectors to remain competitive while achieving the overall goal of carbon reduction.

Long term strategies which evaluate the cost of abatement across facilities enable companies to maximize the advantages of emissions trading, particularly where international offsets are available from non-compliant facilities. In regions where biomass is available, further research in the areas of biomass-based routes for high value chemicals which incorporate CHP to maximize energy efficiency will deliver low carbon or carbon negative products to the market. An effective strategy will not only better position companies from climate policy risks but also enable a competitive advantage as more companies fall under compliance obligations. The chemical industry operates at the core of clean technology innovation and will continue to play an integral role towards achieving low carbon solutions in the fight against climate change.

References

1 Saddler, H., Muller, F., and Cuevas, C. (2006) *Competitiveness and Carbon Pricing – Border Adjustments for Greenhouse Policies*, The Australia Institute, Manuka.

2 McKinsey & Company (2009) *Innovations for Greenhouse Gas Reductions*, International Council of Chemical Associations, Brussels, Belgium.

3 Botschek, P. (2008) *Short Cefic Position on ETS Review*, European Chemical Industry Council, Brussels, Belgium.

4 International Energy Agency (2007) *Tracking Industrial Energy Efficiency and CO_2 Emissions*, IEA, Paris, France.

5 Niederberger, A.A., and Saner, R. (2005) Exploring the Relationship Between FDI Flows and CDM Potential, *Transnational Corporations*, **14**, (**1**) 1–40.

6 Pachauri, R.K., and Reisinger, A. (eds) (2007) Climate Change 2007: IPCC Synthesis Report to the Fourth Assessment Report, IPCC, Geneva, Switzerland.

7 Stern, N. (2006) *Part VI: International Collective Action. Chapter 22: Creating A Global Price for Carbon in the Stern Review on the Conomics of Climate Change*, HM Treasury, Cambridge, UK.

8 Gehring, M.W., and Streck, C. (2005) *Emissions Trading: Lessons From SOx and NOx Emissions Allowance and Credit Systems Legal Nature, Title, Transfer, and Taxation of Emission Allowance and Credits*, Environmental Law Institute, Washington DC, USA.

9 The Pew Center on Global Climate Change (2009) *Climate Change 101 – Cap and Trade*, Pew Center on Global Climate Change and the Pew Center on the States, Arlington, VA, USA.

10 Baumol, W.J. (1972) On taxation and the control of externalities. *American Economic Review*, **62**, (**3**), 307–322.

11 Hepburn, C. (2006) Regulation by prices, quantities, or both: a review of instrument choice. *Oxford Review of Economic Policy*, **22**, 226–247.

12 United Nations Environmental Programme (2000) *The Montreal Protocol on Substances That Deplete the Ozone Layer*, UNON, Nairobi.

13 Wilcoxen, P.J., and McKibbin, W.J. (2009) Climate Change after Kyoto: A Blueprint for a Realistic Approach. The Brookings Institution, http://www.brookings.edu/articles/2002/spring_energy_mckibbin.aspx (accessed 12 October 2009).

14. US Census Annual Survey of Manufacturers (2006) US International Trade Commission Tariff and Trade DataWeb.

15 Marscheider-Weidemann, F., and Neuhoff, K. (2008) *Estimation of Carbon Costs in the Chemical Sector*, University of Cambridge, Cambridge, UK.

16 Isted, J., and Long, S. (2008) Carbon Crush. ICIS Chemical Business, (Jul 14: 2008).

17 Ho, M.S., Morgenstern, R., and Shih, J.-S. (2008) *Impact of Carbon Price Policies on U.S. Industry*, Resources for the Future, Washington, DC, USA

18 EPRI (2006) *The Impact of CO_2 Emissions Trading Programs on Wholesale Electricity Prices*, Electric Power Research Institute (EPRI), Palo Alto, CA, USA.

19 Plastics and Chemicals Industries Association (2008) Garnaut Climate Change Review. http://www.garnautreview.org.au/CA25734E0016A131/WebObj/D0847678ETSSubmission-PlasticsandChemicalsIndustriesAssociation/$File/D08%2047678%20ETS%20Submission%20-%20Plastics%20and%20Chemicals%20Industries%20Association.pdf (accessed 12 October 2009).

20 Enviros Consulting Limited (2006) EU ETS Phase II New Entrant Benchmarks, London, UK.

21 Ren, T. (2009) *Petrochemicals from Oil, Natural Gas, Coal and Biomass: Energy Use, Economics and Innovation*, Utrecht University, Utrecht, Netherlands.

22 Euro Chlor (2008) *Revision of the EU ETS – Electro Intensive Sectors – Inclusion and Carbon Leakage Exposure Criteria*, EuroChlor, Brussels, Belgium.

23 CAN-Europe (2009) Emission Trading in the EU. http://www.climnet.org/

EUenergy/ET.html (accessed 12 October 2009).

24 ClimateBiz Staff (2009) ClimateBiz. Carbon Market Worth. http://www.climatebiz.com/news/2009/01/11/carbon-market-worth-118b-2008 (accessed 23 October 2009).

25 UNFCCC (2009) The Mechanisms under the Kyoto Protocol: Emissions Trading, the Clean Development Mechanism and Joint Implementation. http://unfccc.int/kyoto_protocol/mechanisms/items/1673.php (accessed 5 October 2009).

26 Perspectives Climate Change (2007) Perspectives: Providing a full toolbox to meet your CDM/JI challenge. http://www.perspectives.cc/CDM-JI-project-manag.61.0.html (accessed 13 October 2009).

27 Fenhann, J. (2009) Overview of the CDM pipeline (Excel sheet), UNEP Risøe Centre, Roskilde, Denmark.

28 UNFCCC (2009) UNFCCC. Approved Baseline and Monitoring Methodologies. http://cdm.unfccc.int/methodologies/PAmethodologies/approved.html (accessed 12 October 2009).

29 World Bank Carbon Finance Unit (2009) Carbon Finance Unit. Methodologies Database.

30 UNFCCC (2009) UNFCCC. Guidance–Project Design Documents. http://cdm.unfccc.int/Reference/Guidclarif/pdd/index.html (accessed 12 October 2009).

31 EIB (2005) Clean Development Mechanism for Energy Sector. Project Design Phase. http://cdm.eib.org.my/subindex.php?menu=8andsubmenu=80#PDD (accessed 12 October 2009).

32 Fenhann, J., Halsnæs, K., Pacudan, R., and Olhoff, A. (2004) *CDM Information and Guidebook*, UNEP Risøe Centre, Roskilde, Denmark.

33 Foundation for International Environmental Law and Development (FIELD) (2009) The CDM project cycle. http://www.cdmguide.net/cdm15.html (accessed 16 October 2009).

34 UNFCCC (2009) CDM Projects. http://cdm.unfccc.int/Projects/DB/SGS-UKL1195477297.02/view (accessed 21 October 2009).

35 Lomas, O., Townsend, M., and Tredgett, R. (2009) The Chemicals Sector and the EU Emissions Trading Scheme after 2012, Allen & Overy, London, UK.

3
Implementation of Energy Awareness in Plants

Markus Röwenstrunk and Susanne Mütze-Niewöhner

Environmental awareness issues and energy conservation have increased throughout many societies in recent years. The World Commission on Environment and Development (Brundtland Report, 1987), the Kyoto Protocol (1998), an international environmental treaty to the United Nations Framework Convention on Climate Change (UNFCCC), numerous other conferences, treaties, organizations, and activists have made tremendous efforts to raise awareness with the intention of reversing the process of climate change. Their ultimate goal is the '... stabilization of greenhouse gas concentrations in the atmosphere at a level that would prevent dangerous anthropogenic interference with the climate system' (UNFCCC, 1992, Article 2). In relation to this goal, many portray energy efficiency as the most cost-effective measure to lessen the process of climate change. The International Energy Agency (IEA) states that energy efficiency accounts for more than half of the potential to bring CO_2 emissions back to today's levels by 2050. Even if energy consumption is not drawn from sources that produce greenhouse gases, efficient energy use or preservation of this resource is desirable for more reasons than just essential environmental behavior. Besides advantages for the environment and with it for human well-being, reduction in energy use results in less cost. For example, chemical plants could decrease their spending on energy. This leads to a business case that is possibly driven by altruism and responsible sustainable behavior but can be based on monetary and strategic advantages. Even though stakeholders might be interested in saving the planet, executives, employees, costumers, stock exchange etc. are all presumably interested in reducing the costs needed to keep the plants innovative, productive, efficient, a safe employer and valued partner. However, many companies, plants as well as private people have yet to implement measures to preserve the environment, for example through reducing energy usage even though much information, technical devices, and support has been offered.

Natural sciences have developed many technical appliances and products as well as process solutions which use less of the Earth's natural resources and therefore sustain the environment. This book presents and discusses a great number of these technical measures for example, heat integration (Chapter 6), efficient unit operations (Chapter 7), equipment (Chapter 8) or utilities (Chapter 10) and the

Managing CO_2 Emissions in the Chemical Industry. Edited by Leimkühler

ISBN: 978-3-527-32659-4

assembly of these approaches in a systematic procedure for energy reduction (Chapter 4). There have also been attempts to change the way resources are used through modifications in legislation and pricing systems (e.g., taxation). However, no technical and supportive measures will be effective, unless operators and humans use them significantly and change their behavior (see also [1–3]). This means that it will not be enough just to do the right thing every once in a while, but rather a consistent, persistent effort is needed to get away from wasteful and damaging behavior towards energy conservation and more general environmentally sustainability. But what is and how exactly can energy awareness and environmental sustainable behavior be promoted and implemented? The following sections will answer these questions and discuss theories, methods, and measures to develop a state-of-the-art energy awareness program.

3.1
Energy Awareness and Environmental Sustainability

Individual people and entire organizations are not likely to actively preserve energy or in general act in an environmental friendly way, unless they are aware of the issue and its problems. But what does being aware mean? Even though there are many sophisticated definitions of awareness, just in the field of psychology for example, referring to consciousness, the Oxford Dictionary (fourth edition) offers a simple but very useful definition. It states that being aware means having knowledge or realizing something; being well informed or interested in something. In case of energy awareness, it means that someone has knowledge about energy itself, its constitution and properties as well as being informed about ways to reduce energy consumption. Further, the single person needs to realize the importance of energy conservation and should be motivated to act accordingly. Partanen-Hertell *et al.* [4] defines energy/environmental awareness as a combination of motivation, knowledge, and skills.

According to Wong [5], awareness is the seed for tomorrow's changes, suggesting that the first step to promote energy saving is to raise its awareness. Many other researchers agree that energy awareness is highly significant for energy conservation programs [5–7] and further state that it is the most successful measure to motivate employees to conserve energy [8]. However, many organizations and their managers remain skeptical of human centered approaches, prefer technological solutions and therefore do not consider nor implement programs to raise energy awareness. Because managers often lack knowledge about the effectiveness of awareness programs, employees cannot become aware of the issue, which is one reason for energy inefficiency. This situation highlights the need for information about energy and awareness programs even further, leading to the overall goal to integrate energy or even better all environmental issues into professional as well as everyday life to promote sustainability and its advantages [4].

Saunders defines environmental sustainability as a relational term, which describes a '... viable and harmonious relationships between humans and nature over long periods of time' [9]. This means that the term includes any concerns regarding the quality of life for humans and other species as well as the quality of the human–nature relationship itself. For example, energy conservation but also CO_2 emissions, water pollution, the preservation of any animal and other species as well as the well-being of all humans throughout the entire world, sum up to a very complex subject to be managed. The integration of environmental sustainability into an organization's structure and management is therefore an immense burden on traditionally isolated departments and functions. An improvement of co-ordination and integration of corporate structures becomes necessary for a holistic organizational environmental sustainability culture to be successful.

Nevertheless, environmental sustainability or the smaller scaled energy management '... should not be seen as an additional cost for companies, but as an opportunity to improve competitiveness in a win–win logic' [10]. Advantages of the integration of sustainability into business are:

- increased efficiency in the use of resources;
- development of new markets;
- improved corporate image;
- product differentiation and enhanced competitive advantage.

A first step for an integrated environmental sustainability management with its benefits is the raising of awareness because without being aware of such issues, there will not be any relevant behavior change and actions taken.

3.2 How to Raise Awareness and Change Behavior?

Because environmental degradation can be looked at as the outcome of damaging and wasteful human behavior, a movement started to renew conceptualizations of environmental problems with regard to psychological, social, and behavioral factors. Social scientists have begun to draw attention to the necessity of human variables in any potential solution to environmental problems such as the conservation of energy (e.g., [2]). Even though this human or behavioral approach has often been referred to as the counterpart of the technological approach, it is more an addition because its measures are the guarantee that technological advances are being implemented.

In the past, there have been two central theories to explain human behavior with regard to the environment: the rational-economic theory and the attitude theory. These two theories, as well as other approaches such as goal-setting and feedback will be discussed in the following sections to develop a combination of methods and measures, which will be successful in raising awareness and changing energy use.

3.2.1
Rational-Economic Theory

The rational-economic theory is based on the principle that people behave in a certain manner primarily determined through their financial interests. Individuals will engage in a process of cost-benefit analysis to determine which actions to take, which determines their subsequent behavior [11]. Therefore it is obvious, which interventions have to follow this approach not just to foster energy awareness and sustainability but any attitude or change in behavior.

In using this model, the structure of energy pricing, whether it is electricity itself or devices which need energy, have to be altered to benefit the individual financially. For example, there has been publicity that energy saving light bulbs cost more in purchase than the old fashioned incandescent light bulbs but over time save tremendously on energy and therefore on spending as well.

However, and this is the main point of criticism towards the rational-economic theory, many studies have shown (e.g., [12]) that people do not solely decide and behave on their financial interest. Situational factors and/or personal interests such as comfort and luxury (e.g., slow reaction times of the new energy saving light bulbs) but also convenience or resistance to change, habits or behavior pattern, often outweigh logical rationalizations that are motivated by monetary cost [13]. Results show that the rational-economic theory and measures, which follow this approach, often do not succeed in producing the desired behavioral responses of individuals (e.g., [14]). However, in combination with other designs and actions, financial benefits can be appealing.

3.2.2
Attitude Theory

Another theory often used to explain human behavior with regard to the environment and sustainability is the attitude model. Almost two-thirds of all environmental-psychology publications include the concept of environmental attitude in one form or another [15]. The model assumes that energy saving and further all (pro-environmental) behavior will follow automatically from favorable attitudes towards the particular issue [16]. This means concretely that those individuals, who are concerned about the preservation of nature and specifically about energy conservation, will act accordingly. They will use energy saving light bulbs, switch off the light, and turn off the computer or other electrical devices when not needed, because otherwise their behavior would contradict their attitude. Any inconsistency in belief and action is not pleasant and would lead to distress and discomfort and therefore will be avoided through somewhat automatic responses.

As much as the attitude theory has been the basis of many environmental-psychology studies, it has been even more controversially discussed. Even in 1969 Wicker stated in his article that, in the field of social psychology, the prediction of behavior from attitudes had been much debated [17]. This discussion has not decreased since then and has spread to the specific area of environmental sustainable behavior. Leung and Rice [18] and Vogel [19] have found that environmental attitudes do predict corresponding behavior, however little predictive power has been shown in studies for example, from Archer *et al.* [11].

Similar to the rational-economic model, the attitude model fails to consider that peoples' actions are influenced by factors other than their attitudes towards the environment, such as through situational circumstances, social and cultural contexts or government regulations. Furthermore, much research shows (e.g., [20]) that there is rarely a strong, direct, or consistent relationship between pro-environmental attitudes and people's subsequent actions. Even with pro-environmental attitudes, people do not necessarily know which steps are needed to act upon those attitudes, which leads to the suggestion that information on how to behave should always be provided.

3.2.3 Behavioral Theory

The next psychological theory, the behavior model, stands in conflict with the previously described attitude model, because a strict behavioral approach marginalizes the influence of cognitive concepts such as attitudes. The behavioral approach represents the application of behavioral analysis, which originates from the behavior modification work (operant conditioning) of Skinner and his colleagues and is based on learning theory [21]. The focus of this theory is on direct antecedents and consequences of behavior. Briefly, behaviors that are reinforced will increase in frequency, whereas not reinforced or punished behavior will decrease in its occurrence.

Although behavior modification interventions have shown notable positive effects on environmental sustainability during the phase of reinforcement, these short-term behavior changes often diminish after the antecedents or especially when consequences are withdrawn [21]. Unfortunately, this is logical since an intervention relies exclusively on positive consequences, for example, energy preservative behavior and penalties for wasting resources. The removal of these consequences after the completion of the intervention will consequently lead to a return to the unwanted original behavior. Additionally, research from Deci and Ryan [22] implies that intrinsic or altruistic motivation to engage in certain activities could be weakened through reward systems and ultimately lowers the frequency of altruistic behaviors in general. Thus, long-lasting changes to a more energy conserving behavior are therefore very problematic with measurements relying solely on strict behavior models.

3.2.4 Goal-Setting Theory

The goal-setting theory is very efficient and could overcome constraints of strict behavior interventions, in specifying and targeting the desired behavior through goal contracts after the reinforcement through rewards has been withdrawn. However, a mixture of theories and methods will be discussed later (see Section 3.2.6). First, what is the goal-setting theory and how does it work?

Goal-setting is a motivation based cognitive theory, with the fundamental assumption that people's needs can be seen as goals or desired behavior they strive for [23]. The primary representatives with the most complete statement of goal-setting theory are Locke and Latham[24]. According to them, goals direct and sustain individuals' effort towards activities that are goal-relevant and away from those that are irrelevant [25], therefore giving meaning to human behavior [24]. Goals can be understood '... as the object or aim of an action' [26]. A description of goals can be made through the two attributes or dimensions: content and intensity [25]. The attribute content refers to for example, difficulty and specificity, meaning of the goal itself. Whereas intensity is the process of setting and accomplishing goals [25] mediated through factors such as individual cognition, motivation and commitment.

The following findings are essential for goal-setting to be effective in changing behavior [26]:

- The more difficult the goal, the greater the achievement.
- The more specific or explicit the goal, the more precisely performance is regulated.
- Goals that are both specific and difficult lead to the highest performance.
- Commitment to goals is most critical when goals are specific and difficult.
- High commitment to goals is attained when (i) the individual is convinced that the goal is important; and (ii) the individual is convinced that the goal is attainable (or that, at least, progress can be made toward it).

With regard to energy and environmentally sustainable behavior, we may assume that goals will direct awareness onto energy conservation and further foster concrete behavior which is relevant in reducing the use of this resource. However, feedback about the progress of reaching certain goals needs to be provided because otherwise behavior corrections or even the matching of the goals cannot be assessed. Written lists and contracts increase the commitment towards reaching the set goals.

3.2.5
Theories About Feedback

Feedback refers to the process of gaining information about for example, the status of a system, the quality of a product, someone's performance or behavior, which allows an assessment, evaluation and reaction with possible changes. As early as in the 1950s and 1960s, research about feedback showed increases in learning new tasks [27], as well as positive effects of feedback on motivation to perform tasks [28] and performance achievement [27].

With regard to sustainable behavior or energy use, feedback interventions provide information about resource use or abuse with the inherent chance to become aware of the problem as well as the encouragement to act as desired. Literature shows convincing confirmation for the effectiveness of feedback measures to promote environmental sustainable behavior. Studies, which compared daily energy consumption of households with and without feedback, found energy savings up to 20% (e.g., [29–31]).

In most cases, technical devices such as displays are used for feedback because they provide the relevant information (e.g., energy consumption in kilowatt or € per hour) directly and on time. This creates the chance to almost automatically modify behavior and change energy use patterns. However, interpersonal communication with oral praise or critical reactions of superiors, coworkers, customers or friends is equally suited as feedback to change behavior.

Nevertheless, the efficiency of feedback depends much on the motives and goals of the feedback receiver (e.g., [32]). Even qualitatively high feedback does not automatically lead to behavior change unless a need or goal exists, with a personal desire to act accordingly. Feedback that energy is wasted is only helpful in the process of conservation, if there is a committed goal to do so. Goals without any feedback about the progress and achievement are at the same time neither very efficient (e.g., [24]). If employees are unaware of their work performance or effort concerning energy conservation, it becomes complicated or even impossible for them to initiate corrections or change strategies to accomplishing the set goals. Therefore a systematic combination of goals and feedback is desirable [33–35].

3.2.6 Combination of Methods

Just as a combination of goal-setting and feedback unites the strengths and decreases the weaknesses of each individual behavior change method (see above), so does a comprehensive mixture of interventions of the theories described in Sections 3.2.1–3.2.5. Many researchers and numerous studies have shown that a combination of strategies and methods is generally more effective to increase energy awareness and sustainable behavior than applying solely one strategy or a single intervention (e.g., [36–38]). Gardner and Stern [39] found that for example, the provision of education to inform and change attitude as well as the alternating of material incentives for specific behavior through rewards and penalties can change behavior. However, programs which include combinations of interventions lead to much more success than individual interventions. An evaluation study about major investments in home energy conservation highlights the effect marketing and communication have within its combination with financial incentives of 93% subsidy offered by electric utility companies. It was shown that the number of consumers taking the subsidyn increased from 1% to nearly 20% when adequate marketing and communication was used [40].

An explanation of why a variety of approaches are to be chosen over single measures could be individual differences and the fact that '... behavior is determined by multiple variables, sometimes in interaction' [38]. Therefore, it is crucial to identify personal motives, reasons for certain behavior and potential barriers to act as desired for each individual or use a combination of methods to be able to reach as many as possible.

Stern [38] provides the following list of principles, which should be considered while intending to change environmentally destructive behavior.

- Use multiple intervention types to address the factors limiting behavior change (combination of methods).
- Understand the situation from the user's perspective.
- When limiting factors are psychological, apply understanding of human choice processes.
- Address conditions beyond the individual that constrain pro-environmental choice.
- Set realistic expectations about outcomes (goal-setting).
- Continually monitor responses and adjust programs accordingly (feedback).
- Stay within the bounds of work staff's tolerance for intervention.
- Use participatory methods of decision making (active participation).

Besides the combination of strategies and methods, Stern [38] also suggests having focus groups actively participating in the decision making process, which includes agreeing on realistic expectations about outcomes, meaning setting goals together (compare Section 3.2.4).

With these characteristics considered and the resulting consequences implemented, a behavior change program will be able to reach a great variety of individual people, create awareness as well as provide the necessary information and motivation to change behavior and engage in energy conservation. It will become from a practical and theoretical point of view a 'state-of-the-art' energy awareness program.

3.3 Individual and Organizational Change Processes

The implementation of a state-of-the-art energy awareness program with its ultimate goal to reduce energy use through modification in human behavior, concentrates on an increased awareness and behavior change of the individual. At the same time, each employee should also carry through their everyday work as well as social energy conservation into the entire plant or company. With this approach, changes on individual levels can foster the change of an entire organization towards energy preservation or environmental sustainability. The credo is organizational change through individual change!

However, change is a difficult process and people may not wish to change or need good reasons to change (see Section 3.2.1–3.2.6). As said earlier, whoever is required to change should be included in the change process. This means that the particular person should understand why change needs to occur and its importance as well as have an active (participative) part in the planning and execution. Therefore, the following planning of the energy awareness program through a planning committee is a guideline which remains open and relies on information and support from employees of every level.

3.3.1
Planning, Organizing, and Preparing the Program

Before an energy awareness program can be planned, organized and implemented, it is necessary to have a concrete assignment for it. The general/ultimate goal, to reduce the amount of energy consumption through the implementation of an energy awareness program, has to be clear as well as who is responsible and carries out the project. What department is in charge, is it a team or single person effort? Who leads, to whom will the process be reported and how does the decision process work? Are responsibilities within the planning committee of the project shared and if so how? Many answers to the above questions depend on the size of the organization and the scale of the program.

Further, who is to be involved and what support or even obstacles can be expected? Key players who could enhance or inhibit the program's process and results have to be identified and motivated to support the project. What specific constraints and policies are relevant and in place for the particular organization? These and more questions have to be answered in order to thoroughly plan, organize and finally implement a state-of-the-art awareness program.

Additional prearrangements that have to be made are for example, to identify relevant information and its sources. Sub-goals and timeframes have to be defined. It is also important to decide on further personnel as well as material resources needed to support the planning committee. And last but not least, methods and instruments for analyses and evaluations should be decided on.

3.3.1.1 Prearrangements and Pre-analyses

Prearrangements that have to be done to develop the design of an awareness program start with the identification and gathering of relevant information on any energy matter in the organization. For example, is detailed information about the level of energy use and cost already available or if not, who can provide it? When this information is on hand, it has to be assessed what theoretical potentials of energy saving exist and which measures can be beneficially implemented to reach the overall goal of conservation. Consider typical work activities, equipment types, hours of operation, review energy bills and transportation fuel consumption as well as the size of staff and in particular behavioral energy use patterns. Additionally, the company's specific opportunities and constraints have to be assessed. For example, local policy may require night-time lighting for security purposes, and some equipment may need to run continuously. Most often document analysis can provide this kind of information but also meetings and interviews could be relevant techniques. The results will be specific depending on the organization or company, the industry sector, the products manufactured or the services provided.

Generally, all relevant information have to be used to ensure that challenging but obtainable goals and well-suited measures can be developed.

These analyses can be conducted together with internal or external experts but need to be open to discussions with the upper/top management (energy audit, Section 3.3.1.2) as well as with department and plant staff for further input and suggestions (see Section 3.3.2). In addition, it is very important to fully understand the organization's energy management and particular policies and constraints. A complexity reducing summary of the outcome of the analyses can be provided through the use of an energy matrix. The energy matrix provides an effective way to gain global insight into a company's current approach to energy (example energy matrix see Figure 3.1).

To use the matrix, a mark needs to be placed in each individual column that best describes where the energy management of the particular organization is currently located. Then the marks should be connected to show the organization's approach and how well balanced energy management is. The aim is to identify potential for enhancement, which can be met with specific interventions to move up through the levels towards current best practice and, in doing so, develop an even balance across all columns. Figure 3.2 shows an unbalanced and relatively

Level	Energy Management policy	Organizing	Staff Motivation	Tracking, monitoring and reporting systems	Staff awareness/training and promotion
4	• Energy management and action plan have commitment of top management as part of a corporate strategy • Energy management fully integrated into management structure	• Clear delegation of responsibility for Energy matter	• Formal and informal communication channels regularly exploited by energy manager/energy staff at all levels	• Comprehensive system sets targets, monitors consumption, identifies faults, quantifies savings and provides budget tracking.	• Promoting the value of energy efficiency and the performance of Energy management through various interventions
3	• Formal energy manage-ment policy, but no active commitment from top management	• Energy manager accountable to energy committee representing all employees, chaired by a member of the managing board	• Energy committee used as main channel together with direct contact with major employees	• Monitoring and targeting reports for individual premises based on sub metering, but savings not reported effectively to employees	• Program of staff training, awareness and regular publicity programs • Some payback criteria employed as for all other investment
2	• Unadopted energy management policy set by energy manager or senior departmental manager	• Energy manager in post, reporting to ad-hoc committee, but line management and authority unclear	• Contact with major employees through ad-hoc committee chaired by senior departmental manager	• Monitoring and targeting reports based on supply meter data	• Energy unit has ad-hoc involvement in budget setting • Some ad-hoc staff awareness and training
1	• An unwritten set of guidelines • Energy management the part-time responsibility of someone with only limited authority and influence	• Informal contacts between energy manager and a few employees	• Cost reporting based on invoice data	• Energy manager compiles reports for internal use	• Informal contacts used to promote energy efficiency
0	• No explicit policy • No energy manager or any formal delegation of responsibility for energy consumption	• No contact with employees	• No information system	• No accounting for energy consumption	• No promotion of energy efficiency

Figure 3.1 Example energy matrix.

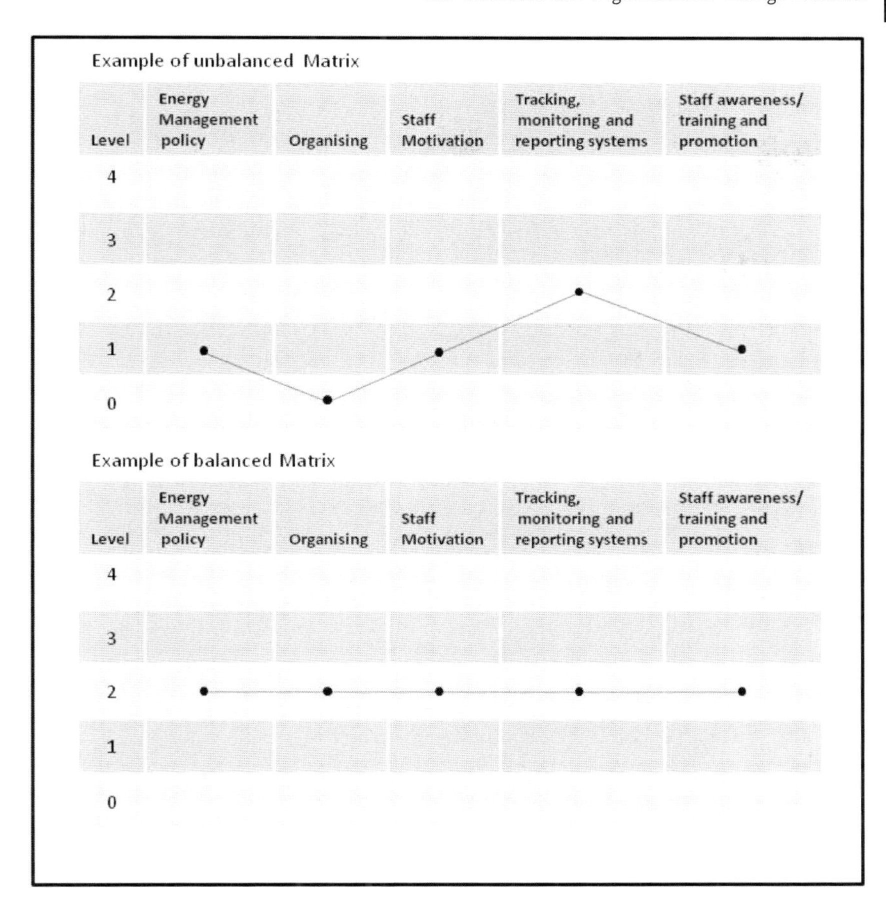

Figure 3.2 Example of a balanced and unbalanced matrix.

low level of energy management of a particular plant, with staff awareness and training lagging behind.

However, it is not necessary to reach a balance before single columns can move up to another level, in this case only awareness and training (behavior change) will be of interest due to the conceptualization of the energy awareness program.

According to the classification of the current energy management level (example energy matrix) and the results of the pre-analyses, a preliminary plan for the energy awareness program has to be developed. It should include specific processes with corresponding events and measures to be taken as well as who needs to be involved (see Figure 3.3).

This preliminary plan, the information from the energy matrix, and the outcomes of the pre-analyses are the foundation for discussion with the upper/top management within the energy audit (Section 3.3.1.2) as well as providing information for the concrete goals, the schedule and explicit interventions (Section

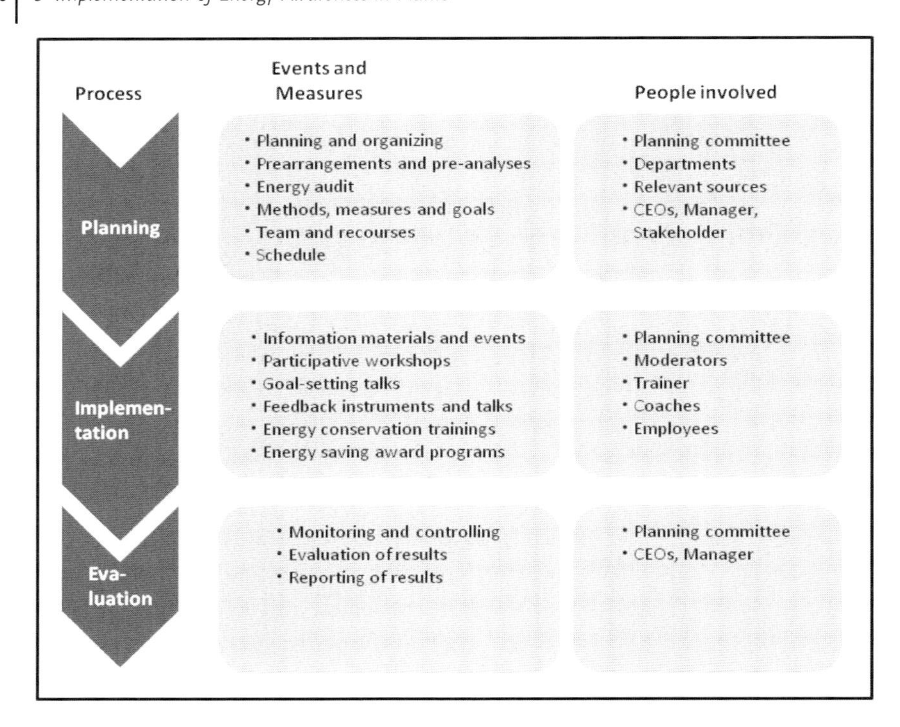

Figure 3.3 Preliminary plan for the energy awareness program.

3.3.1.3–3.3.2). For example, it has to be determined on what scale and how broad the energy awareness program should be implemented. Depending on the stage of energy awareness in a particular company (see energy matrix) and its size and structure, there would be certain advantages and disadvantages to spread and implement the awareness program in all departments and at all levels simultaneously versus consecutively. A pilot program concentrated on only one department could also be very efficient to analyze and evaluate specifications of the organization and/or unusual requirements for the program. But regardless of the company's individual decision on how many departments to include at once, information and support from the upper/top management has to be gained, methods and measures decided on, resources planed, and a schedule developed.

3.3.1.2 **Energy Audit**

After an early assessment of the level of energy use, potential savings, and measures already in place has been conducted and its results collated to give a classification of the particular organization in the energy matrix (compare Figure 3.1), information about the project, its topic and the necessity to reduce energy spending should be circulated. CEOs, managers, key players, the HR department, officials of the workers' council or other employee representative organizations and maybe even stakeholders should be invited to an energy audit. The energy audit is a workshop, which includes information about the project with an aim to reach understanding and support but also its discussion, the gaining of further insight and the agreement on enhancements as well as specifications.

An information presentation on the energy awareness program needs to contain reasons and (dis-)advantages of the project, an analysis of the current energy use, preliminary potential savings as well as measures that could increase energy awareness and ultimately reduce energy usage. The business case for energy efficiency is clear: it reduces energy costs, diminishes vulnerability to increasing energy prices and reduces CO_2 emissions. Overall, the benefits of energy efficiency measures (technical and behavioral) generally outweigh the costs of an energy management system.

The classification of a particular organization in the energy matrix provides further information and is subject to discussion as well as the basis for goal specifications. The participants have to be encouraged to ask critical questions, provide new information and contribute to the analysis and development of the program. Their ideas and involvement are wanted, not only because it is valuable to the quality of the program itself, but also guarantees the understanding and support of the upper/top management. With supportive, or even better enthusiastic, individuals on board, the process and efficiency of the project will improve tremendously. The programs messages and goals will be communicated with more authority as well as through an increased number of formal and informal communication channels.

The relevant and gainful results from the energy audit will be worked into the existing plan and program so that the upgraded version provides a corporate vision and agenda and can be used to decide on methods and measures. However, a participative approach means that the plan and somehow the agenda have to remain flexible and will be adjusted in the course of the implementation. A participative method will lead to greater acceptance of the intervention, with the result of higher outcomes in behavior change (see also Section 3.2.4. and 3.2.6).

3.3.1.3 Methods, Measures and Goals

In general, the methods and measures have to be chosen in compliance with the theoretical approach to awareness/behavior change programs (see Section 3.2) and the preliminary plan with its goal (see Section 3.3.1.1) to guarantee their effectiveness. Methods already used at the beginning of the planning phase (see Section 3.3.1.1) are documented on an interview-based energy analysis conducted with the goal of obtaining information about the current status of energy use, saving potential and possible measures. Document analysis and interviews are valuable techniques for obtaining a broad objective overview, which is then brought into a classification structure; the energy matrix. The energy matrix shows the energy level of a particular organization and can be used as process control in the form of a baseline or benchmark. This matrix is presented within the management energy audit, another method already carried out, for information purposes as well as to be discussed and enhanced. Relevant members of the upper and top management have to be involved in the project and its subject with the goal of gaining further input as well as achieving as much support as possible. Upper management's support and commitment is crucial for the success of the program.

The assessment of the energy use and possible energy saving opportunities will be broadened within the phase of implementation. All employee levels will be included into the pursuit to reduce energy consumption because everyone can and should actively contribute to create an awareness culture and sustainable organization. Information events, participative workshops, and discussions will be carried out with the goal first, to inform and create awareness about and commitment to the subject and second, to gain as much relevant information and ideas as possible. Further, goal-setting combined with feedback will be applied through systematic goal-setting talks as well as technical and interpersonal feedback mechanisms. Additionally, behavioral training on machines and for the accomplishment of processes will be available if necessary. Award programs will lead to recognition and honors as well as create some level of competition and finalize the implementation methods.

This broad mixture of methods and measures combines the strengths and decreases the weaknesses of each individual intervention to become theoretically as well as from a practical point of view a 'state-of-the-art' energy awareness program. It is therefore possible to reach a great variety of individual people, creating awareness as well as providing the necessary information and motivation necessary to change behavior and engage in energy conservation. Nevertheless, the implementation plan and the methods remain somewhat flexible so that additional measures can be applied if need be.

3.3.1.4 **Team and Resources (Budget)**

Now that the support of the upper/top management is guaranteed and the implementation plan with its methods and measures specified, it is necessary to identify and organize required resources.

First, the energy awareness planning committee has to be enlarged because tasks such as marketing, writing and editing, graphic designing, producing informational materials, workshop and training sessions as well as the program evaluation need additional and expert personnel. The new project members should have participated in the energy audit or have to be informed and motivated at this point. Furthermore, it might be beneficial to decide to use the services of outside consultants, contractors or scientific societies such as universities with research interests, but a core group of enthusiastic on-site people is essential for an effective program. Generally, the entire team has to serve as contacts for any information regarding the energy awareness program and should also act as behavior models practicing energy efficiency. Energy sustainable behavior has to be lived authentically and vividly presented through various communication channels.

There should be at least one specific team member in charge of the management of the facility's existing communication technologies, including newspapers or newsletters, radio, closed-circuit TV, websites, as well as any specialized methods for communicating, such as staff meetings. This person is responsible for representing the interests of the project as well as initiate and carry out mass media information. Auditoriums and rooms for meetings, information events, workshops and training purposes have to be managed as well so that their availability is assured.

Another important resource is production capability. Due to the methods and measures used in this awareness program, it is crucial to have the facilities and capabilities (either in-house or through a contractor) for producing all types of printed materials, displays, and videotapes. The organization's public affairs or training offices might be able to assist with many resources.

The personnel and material resources (see below) for the awareness program will have to be financed so that calculations can be done and a budget plan written.

Typical budget items

HR expenses:

- obtaining management approvals, identifying necessary resources, and designing the program;
- conducting surveys, information events, workshops, trainings and award programs;
- research, writing, editing, designing, and printing information materials;
- creating art, including a logo, graphics, etc.;
- shooting, narrating, and editing videos;
- working with media representatives;
- preparing informational materials;
- distributing materials;
- evaluating the program;
- preparing reports.

Materials:

- paper and services for producing printed materials;
- computer programs, CDs;
- video recording equipment;
- display materials for special events.

Direct costs:

- printing
- giveaways such as pins, pens, refrigerator magnets or coffee mugs;
- incentives for staff members and program participants;
- postage for mailing surveys and informational materials;
- food and beverages for meetings and events.

According to the budget items, a budget plan describing each expense and its purpose should be discussed and approved by the finance department. Due to the flexibility of the program, the budget might have to be increased during the execution of the interventions. But the more support from the upper/top management has been reached through the energy audit, the fewer problems will be expected. However, if the designed program exceeds the financial resources and only a limited budget is approved, the scale of the program (e.g., pilot study) should be reduced without jeopardizing the quality.

3.3.1.5 Plan and Timeframe (Schedule)

Since the energy awareness plan has been enhanced through the incorporation of further input from the management (energy audit) as well as through decisions on methods, measures, available resources and budget, it is now possible to specify a schedule. This schedule is based on the preliminary plan, which already has a timely structure (see Figure 3.3) and considers the size of the organization, the scale of the program, and the duration of each individual process, event and measure (see Figure 3.4). Its purpose is to inform anyone interested about the content and timely course of the implementation program, to keep the work on track, and prevent any skips or missed deadlines.

The schedule should also include events such as Energy Awareness Month and Earth Day as well as take advantage of certain times of the year when particular actions for change are necessary–such as just before heating or cooling season, when reducing peak energy use or saving natural gas may be critical. Adequate time buffer for feedback loops, approvals, deadlines, and other scheduling issues have to be integrated for realistic implementation.

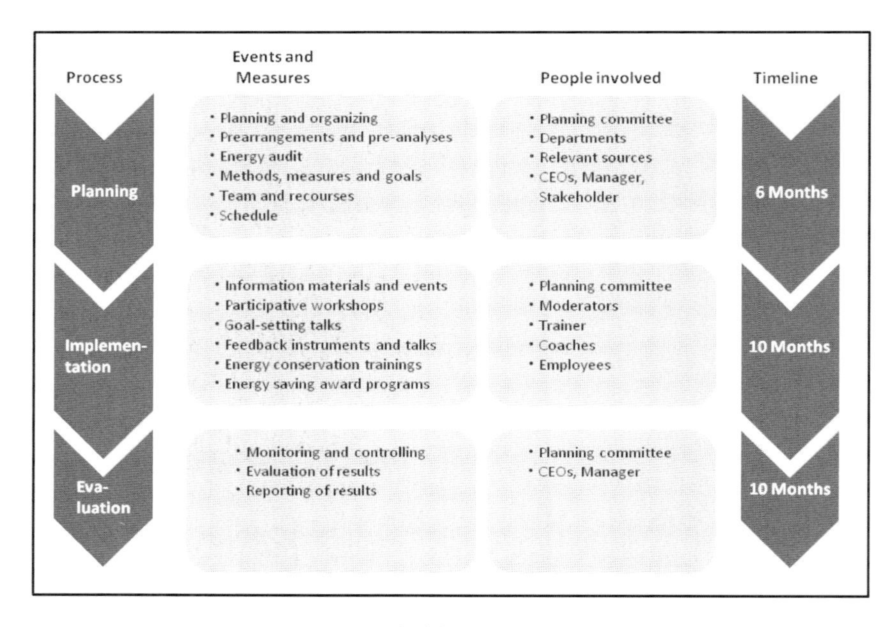

Figure 3.4 Energy awareness program schedule.

3.3.2 Implementation

The implementation is the core process of the energy awareness program, substantially following the developed plan and schedule with its methods and measures (see Section 3.3.1). Thus far, the plan and content have been developed from the top down but it was also decided, and theoretically as well as practically reasonable, to actively incorporate all relevant personnel. It is very important to extensively inform all employees about the energy issue, the project and their part in it, as well as to solicit information about energy use and possible saving potentials. Especially the staff at the base for example, in production, who work on machines that use up vast amounts of energy, need to be consolidated/incorporated.

The following sections describe the sending of informing emails, the execution of information events and also participative workshops. These interventions complete the pre-analyses on energy matter and finalize as well as lead to the execution of measures such as goal-setting talks, feedback instruments, training, and award programs.

3.3.2.1 Information Materials and Events

Information about the energy awareness project, its content, and the necessity to conserve energy needs to be passed on to every employee of the particular organization or the pilot department. First, posters should be placed at gathering point such as cafeterias, canteens, and blackboards. A video on energy or the environment could boost the informative effect through vividness and at the same time function as a role model, showing people acting sustainably. Newsletters, leaflets, and flyers should also be given out. The design and content has to be eye-catching with motivational messages and slogans concerning for example, energy waste, climate change, environmental obligations and/or stating high costs, saving potential, possible chances for the company and the employees. This will raise awareness; awaken interest in the project, and the urge for more information. It is advantageous to send out informing emails afterwards because employees will be primed about the importance of the program and are less likely to ignore or worse still, delete it without reading.

An informing email has to describe the program, the reasons and gains, the plan and schedule, and also the need for active participation in a motivating way. Additionally, it has to include dates about information events and workshops, that the recipient is required to or can attend voluntarily. Obligation has the advantage that high attendance is guaranteed but there could be resentment in return. Optional participation reduces the importance of the issue, risks low attendance but can assume that the participants are truly interested. If mandatory events and/or workshops cannot be enforced, a fictitious restriction on the number of applicants can increase interest to participate in the voluntary interventions.

Information events can include general information about the climate change, energy conservation, possible chances and advantages but also specifics of the pre-analyses of the plant's energy management (e.g., current energy use, potential savings, measures, advantages, and efforts), the energy matrix, benchmarks, and ideas as well as decisions of the management. The business case for energy efficiency is clear: it reduces energy costs, diminishes vulnerability to increasing energy prices and reduces CO_2 emissions. It is equally effective to invite an external expert or celebrity to present climate change and energy issues. The participation and active support of the upper/top management also shows the significance of the program.

Besides solely providing information, these measures and messages already serve as stimuli or cues for raising awareness and promoting desired behaviors (see Section 3.2.3). The attendance, whether obligatory or voluntary, and a (positive) attitude about the issue can be enforced through follow-up constructive discussions, positive responses and social recognition through coworkers, colleagues, and supervisors/superiors.

Depending on the size of the events, workshop settings in the same constellation could follow automatically or large groups have to be split into many separate workshops (<20 person). In the case of large-scale energy awareness programs, workshops should take place with focus groups, leading to more specialized and in-depth results. However, information events and workshops do not necessarily have to be scheduled on the same day but a short time span in between is beneficial since the participants are already sensitized in contrast to the possibilities of disregarding certain contents after longer periods.

3.3.2.2 Participative Workshops and Specific Techniques

Workshops for energy conservation are very efficient in raising awareness, increasing knowledge, and building strong intentions to implement energy saving measure (see also [41]). They should be moderated through well-trained energy awareness planning committee members, who promote the project, answer questions, and make sure that every participant is being involved and contributes. Again, it is vital that all employees are enabled to give input about energy conservation so that all possibilities of reducing the use are regarded and their support is gained. Therefore the size of individual workshops should not exceed 20 persons and contain employees from same departments, productions lines and backgrounds. Homogeneous groups have the advantage that for example, machines, processes, and the language are somewhat similar, simplifying communication and discussion. Heterogeneous groups have the advantage that its members might have new and totally different thoughts and ideas about subjects they are not familiar with. At the same time interdisciplinary workshops lead to confusion and misunderstandings by its participants if subjects are too specific and complex. Nevertheless, the aim is to consider the individual thoughts, opinions, and ideas. Techniques that can be used to facilitate discussions with creative and productive outputs are:

- Brainstorming (thoughts, ideas, and solutions to energy use).
- Mind maps (diagram to visualize and structuralize thoughts and ideas).
- Energy thieves (tasks, processes, behaviors which require much energy).

- Old hats (declaration of undesired behaviors and development of alternatives).
- Transfer walk (walk through the facility and identification of saving potentials).
- Letter to or contract with oneself (catalog of measures and commitment to implement).

Brainstorming, a creative technique designed to generate a large number of ideas, is a very valuable way of approaching the problem of energy conservation. The participants of workshops will be encouraged by moderators to spill out as many thoughts and ideas on how to reduce the energy use at their workplace, in their department or the entire organization. The first spill-out round can be even more effective if it is done anonymously by writing on paper individually, because fewer inhibitions occur due to social dynamics. According to behavior theory (see Section 3.2.3), behaviors or comments such as unpopular ideas, which will not be enforced, or even worse punished, by others through negative remarks, decrease in its likelihood of occurring again. But especially these presumably unorthodox or silly solutions to energy use, which would be held back in front of coworkers, should explicitly be encouraged because later during discussion others might be able to use the ideas, adjust them or make sense of it. This would happen in the second round, when all written statements are being reviewed and discussed by the entire group to further generate feasible solutions.

Since it can be difficult to start with brainstorming about the question 'how to reduce energy use?' the opposite can help. The right question could be reversed to ask for the best way to waste energy. Answers could be: e.g. having a certain machine run without producing, moving parts unnecessarily often, and leaving the computer or the light on when out of the office. This opens up established thinking and behavior structures and gives an opportunity to develop alternatives. Other techniques to identify energy waste are 'energy thieves' and 'old hats'. Tasks, processes, and structures as well as behavior and habits which use unnecessary amounts of energy are searched and identified as thieves or respectively labeled as old hats. The thieves (e.g., standby modus) should be brought into prison or old hats (e.g., leaving the light on while out of the office) need to be disposed of, meaning undesired processes and behavior have to be replaced with sought-after alternatives. In summary, any of the above measures includes the collection of processes and behavior that increase energy levels, which in return can foster ideas and interventions for energy efficiency and environmental sustainability. Since a chemical plant is a complex system and energy saving measures are often not trivial the inclusion of know-how of internal or external experts and consultants is often useful.

The technique of mind mapping can visualize and bring a structure into the information of any brainstorming. A 'mind map' is a diagram used to represent words, ideas, tasks, or other items linked to and arranged around a central key word or idea, in this case energy (see Figure 3.5).

The elements of the energy mind map are organized intuitively according to their importance to the concept, and are classified into groups, branches, or areas, with the aim of representing semantic or other connections.

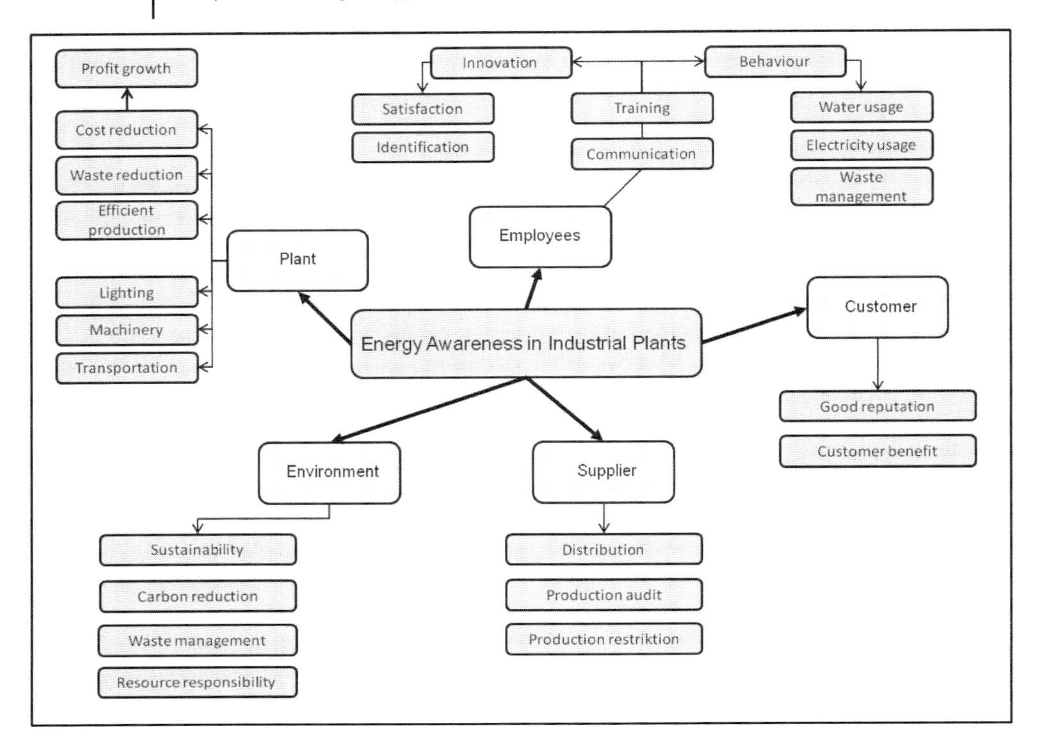

Figure 3.5 Example energy mind map.

The advantage of this radial, graphical, and non-linear organization is that it interrupts the prioritizing of concepts typically associated with hierarchies presented with linear visual cues. Further, this approach towards brainstorming encourages users to name and connect concepts without a tendency to begin within a particular conceptual framework, allowing a higher level of creativity when gathering ideas and information.

A 'transfer walk' can be used as a brainstorming technique itself as well as to verify results from prior brainstorming or mind mapping within the same workshop or as a separate event. By physically walking through a specific department or production line, two coworkers discuss saving potentials and develop suggestions for possible solutions. Their task is to produce concrete ideas and plans, how to change and implement certain processes and behaviors to reduce energy use. They could come up with advancements on equipment, which increase the efficiency, activities that should not be executed anymore, as well as new or restructured behavior patterns which conserve energy. After the pairs have walked through their workspace and made a list of suggestions, the results will be reported and discussed within the workshop to incorporate them into a specific focus group energy change document.

The energy change documents are fixed and provide commitment to transfer the suggestions on energy conservation, comparable with a letter to or contract with oneself. They should contain a catalog of measures and specific instructions about what to do to meet the desired outcomes. After its development, the energy change document will be aligned with the results of the energy audit to become the blueprint guideline for concrete measures such as the following goal-setting talks because workshops alone most often do not change behavior [41].

3.3.2.3 Goal-Setting Talks

The energy change documents, which contain catalogs of measures to reduce energy use developed from the facility and production staff within the above-mentioned workshops, will be reviewed by members of the energy awareness planning committee and (direct) superiors to align them with the corporate vision and agenda agreed on within the energy audit (Section 3.3.1.2) through upper/top management. Eventually, specific information or interests from the pre-analyses and energy audits contradict with the change documents, which would have to be considered and make adjustments necessary. The newly developed energy change guidelines (specific to departments and product lines) will then have to be discussed with each individual employee within goal-setting talks.

The goal-setting talks allow each employee and the direct superior to discuss not only past efforts but especially expected performances related to energy conservation in the future. The core process of this instrument is the agreeing on and setting of concrete goals for the next term. These goals need to be worked out in a participative manner, having the focus on the employee, his/her thoughts, comments, and ideas. This approach and the writing out in form of a contract increases the commitment and the willingness of the particular employee to regard and implement agreed goals (see also [42]).

Further, goals should be phrased according to the SMART-Formula, which is highly recognized in operational practices [43]. Each letter of the formula represents specific advice, which should be followed while developing goals (see Figure 3.6).

In relation to energy awareness and conservation, the ultimate goal: to reduce the use of energy, an overall and very vague goal, has to be broken down into many particular sub-goals. These sub-goals need to correspond with the information and ideas on how to save energy stated in the change documents and enhanced guidelines (see Section 3.3.2.2). They also have to be formulated in compliance to the SMART-Formula and its recommendations.

A sub-goal could be for example, the switching off of production machines instead of keeping them on standby to reduce energy spending. First, it has to be (**S**)pecifically defined what the goal is and what to do to accomplish it. The specific machines need to be named, the sub-goal stated and the instruction given to turn off the machines by pressing the power button, when not in use. One possibility is to set a particular and challenging amount of kilowatt hour (kWh) to be reduced, which could be calculated according to minutes and hours per day when the machines are not running. This goal-setting can make sense, if it does not

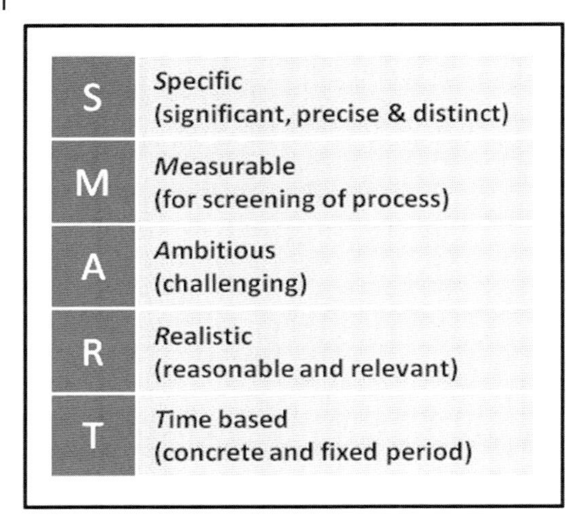

Figure 3.6 SMART-Formula for goal-setting.

contradict with the desire to have the production line work as much as possible and exact numbers are available. In this case, it is more useful to aim at for example, 80% switch off of the time the machines are not producing, which can easily be (**M**)easured as well as communicated via technical feedback. The 80% should be (**A**)mbitious because difficult goals lead to greater achievement [26]. However, a 100% switch off is not (**R**)ealistic because idle periods might not be long enough so that shutting the machine down would not be reasonable. Further, realistic and achievable goals (according to individual performance potentials) are important because tasks that are too difficult can lead to less commitment to attain them [26], de-motivation and even resignation due to high levels of pressure and performance anxiety (e.g., [24]). If people are assigned to tasks that exceed their abilities, it is sometimes better to urge them to 'just do their best' [44] than setting specific performance goals. Nevertheless, a concrete and fixed (**T**)imetable has to be provided (80% every day, week or month) to be able to control the progress and celebrate the achievement of the goal.

This procedure has to be executed through the employee and the direct superior within the goal-setting talk for any sub-goal and fixed in the goal agreement document (see Figure 3.7).

The goal agreement document could be signed by both parties to give it the appearance of a real contract, which will increase the commitment and the willingness of the specific employee to be concerned about and reach the agreed goals. It is also the foundation of the evaluation within the next goal talk after a year. Each goal has to be compared with the current performance and target numbers; however, specific circumstances should be accounted for as well (order situation, workload, stress due to absence of coworkers etc.). Yet, to be able to evaluate, it is necessary to have previously established control processes and feedback mechanisms for comparison (see Section 3.2.5 and [26]).

Employee Name, Given Name: Position:	Goal Agreement Document 2008	Supervisor Name, Given Name: Position:

Goal No.	Goal Description:		
Goal Achievement Criteria	Deadline	Requirements, Participants	Status
The goal will be exeeded when...			
The goal will be achieved when...			
The goal will be partly achieved when...			
The goal will not be achieved when...			
Goal Check Up: **finished:** Implementation successfull **critical:** Implementation critical **open:** Goal not yet achieved, implementation not critical			

Figure 3.7 Example of a Goal Agreement Document [43].

3.3.2.4 Feedback Instruments and Talks

Feedback raises awareness, motivates high performance [28] and positively effects performance achievement [27]. Environmental sustainability studies in households show energy savings up to 20%, when feedback was provided (e.g., [29–31]). It also enhances the effectiveness of goal-setting because progress can be controlled and the final achievement realized and celebrated. Also in households, Becker [34] found that difficult goals with feedback three times a week about how much energy was consumed, was the only measure to reduce energy use significantly (15.1%). Easy goals and difficult goals without feedback did not lead to relevant energy conservation. Therefore technical as well as interpersonal feedback instruments and mechanisms need to be integrated into awareness and change processes as well as goal setting.

Reports and even better displays can be used to provide relevant information for comparisons of current and specified energy levels or certain energy related behaviors. The above-mentioned studies took place in households, however technical devices most often displays, which for example, indicate current or average consumption can also be used in professional organizations such as companies and plants. Displays have the great advantage that they can show status, progress and changes immediately. There are also other machines for human feedback systems, which are able to provide direct information for example, in different colors (white = little energy, blue = much energy) glowing electricity cords [45]. These devices create the chance to immediately modify behavior and change energy use patterns. According to behavior theory, immediate feedback is very desirable and most effective for behavior change due to direct reinforcement of sought-after actions.

Instantaneous as well as delayed feedback can also be given through interpersonal communication with oral praise or critical reactions of superiors, coworkers, customers or friends. It is also perfectly suited as feedback to change behavior within an energy awareness program of a company and could be established similar to the following example.

Between 2002 and 2004 the Deutsche Bahn AG carried out the project 'EnergieSparen' with the aim of reducing the amount of energy use and CO_2 emission of passenger trains [46]. After instructions and driving training to preserve energy were given to the train drivers, the core instrument in the integrated change management process were data-based feedback talks by direct supervisors. Once a year, supervisors gave feedback according to data sheets, which indicated individual driving behavior and possible potential energy saving options. The results show, driving styles changed and 2.7% less energy was used after the first and 3.2% less after the second year of feedback talks [46].

Such feedback talks on energy conservation need to be installed as a continuous intervention following the first goal-setting talks. At the beginning of the program, it might be beneficial to have feedback talks more often than once a year for better guidance. After the energy awareness interventions and processes become familiar, goal-setting and feedback talks can merge into one measure.

The combination of setting goals and providing feedback is most effective, since not only goals with feedback lead to better results than goals alone but feedback by itself is neither as successful as together with concrete goals (see also [47]). However, just as there are the SMART rules for goals, there are rules that should be followed while providing feedback, since their disregard can lead to counterproductive behavior.

Feedback provider:

- feedback only if desired;
- directly address the person;
- own impression (I think ...);
- do not judge but describe (I realized ...);
- remain objective (no nagging and insulting);
- name concrete examples (no generalizations);
- name positives and negatives.

Feedback receiver:

- do not justify or defend;
- let feedback provider finish talking;
- listen first and respond to what is important.

The direct superior, who provides and discusses feedback with his/her plant staff as well as sets the goals together, has to have knowledge about the above rules and the SMART-Formula. A brochure with information and guidelines can be passed out but depending on the qualification and previous soft skill experience, hands-on exercise in form of training is necessary to become proficient.

3.3.2.5 Energy Conservation Training

Goals and feedback are important but instructions and training are sometimes unavoidable especially when new machines have to be operated, functions of change processes be mastered or new behaviors and skills learned. A method that is one of the most effective training methods is behavior modeling [42]. Behavior modeling, based on Bandura's learning theory (1977), focuses on providing examples of recommended behavior. It has been extensively shown that behavior examples will be followed if certain facts are considered. The behavior itself has to be understandable and executable for the specific person, it also should be relevant, meaningful and at best rewarding (in terms of positive results). This means that a behavior modeling training also contains information processes, goal setting, rehearsal (the actual training) as well as feedback, which are methods already used in this energy awareness program. The complexity of behavior modeling requires careful planning and execution following the steps below in this example process of behavior modeling training:

- Introduction to the subject.
- Development of learning objectives/goals.
- Film presentation of the model behavior (positive and negative example).
- Group discussion about the effectiveness of the model behavior.
- Role play to practice the behavior.
- Group feedback.

At first, it is important to introduce and make the relevant subject understandable as well as comprehensible, just like through energy information events (see Section 3.3.2.1) for example, with a celebrity pointing out the impact of the issue. However, the number of listeners or participants in the behavior modeling training should not exceed a regular workshop group. Further, goals and learning objectives for the training session have to be set or at best established with the participants together before actual behaviors to be modeled can be shown. The presentation should include the desired behavior patterns but also give examples on how not to act. The contrast between the favorite performance and maybe even more than one undesired approach, increases comprehension and specifies what is wanted. However, this does not make a thorough discussion on the effectiveness of the modeled behavior dispensable. Following, role play and feedback are used to train the specific behavior as much as necessary. This guided training in workshops or (computer) simulations, in the case of complex or high risk tasks (e.g., chemical production processes, plane flights, etc), could be replaced or enhanced with real life trials such as work on machines or other energy relevant processes.

For example Winett *et al.* [48] used a film presented via cable TV and information booklets containing cartoons to model energy conservation behavior. The visual material depicted various energy saving measures to be copied by middle-class homeowners. In comparison to the control group, the TV modeling group significantly reduced energy use by 10% and enhanced their knowledge about the issue.

A rather unusual but very effective demonstration project on the importance of modeling for energy conservation was conducted by Aronson and O'Leary [49], observing shower-taking behavior in a university field house. The installation of a large sign in the middle of the men's shower room instructing them to turn off water while soaping, resulted in 6 to 19% of people following the request. When the researcher used a student to model the desired behavior, the adoption of the desired action, to turn off water to soap up, increased to 49%. Within the experimental condition of two people simultaneously modeling the behavior, compliance rose to an incredible 67%.

The effectiveness of behavior modeling for energy conservation has been shown through many studies (e.g., [48, 49, 50]). At best, the desired practices are presented by people in similar situations as the participants of the program so that they can identify themselves with the model and the behavior to reach the highest outcome [51]. However the specifics of the programs (e.g., the medium used) depend much on the circumstances and the subject to be trained. Just as the example of Winett *et al.* [48] shows, films, and also prompts or simulations, are suited to initiate behavior modeling.

3.3.2.6 **Energy Saving Award Programs**

The copying effects of behavior modeling are also seen within regular everyday social interaction. Darley [52] found in a study of energy conservation equipment that the installation of clock thermostats spread from the initial users to their friends, colleagues, and co-workers. The adoptions of instruments or measures, referred through social networks, are based on trust or at least knowledge about biases and values, which can be considered. Actions of friends and associates serve as experiments, which can be monitored, evaluated and eventually acted upon.

This social copying effect can also be used within energy awareness programs. Opinions and attitudes about the environment and energy conservation as well as specific actions ensuring sustainability will be communicated and passed on to colleagues and co-workers automatically but can be assisted and enhanced through target-oriented processes. Outstanding thoughts, ideas, and measures of employees need to be recognized, first to show gratitude and second to provide good examples to everyone.

A classical method to recognize a desired performance is a reward system based on behavior theory (see Section 3.2.3). Energy conserving actions are being rewarded with direct or indirect monetary incentives or tokens (incentives that can be exchanged for valuable assets) to reinforce and increase them. However, as discussed in Section 3.3.2 as well, only short-term behavior changes can be reached with interventions of this kind and in most cases no permanent sustainability. Therefore an award program should be used instead of a reward system.

The award program needs to recognize the employee of the week, month or year, who saved most energy or had the best idea how to produce or behave sustainably. With this procedure, gratitude is provided and desirable examples are given to colleagues and co-workers, motivating them to engage in the same or similar behavior (copying effect) or even compete to get the honors themselves. The award should focus on public appreciation and tribute, if at all combined with a small prize. Stern [53] as well as Katzev and Johnsons [54] report that small or weak incentives are often more successful than large or strong ones in inducing increased energy efficiency because the first tend to suggest that the behavior is internally, rather than externally, motivated, which results in a greater commitment to act. A smaller prize enables the copying of desired behavior more by choice than by the pull of a greater incentive.

3.3.3 Evaluation and Report

To ensure that the entire energy awareness program and the above interventions are state-of-the-art and provide quality results, it is necessary to execute evaluations and prepare reports to document the analyses and outcomes.

3.3.3.1 Monitoring and Controlling (Process Evaluation)

The energy awareness program needs to be monitored and controlled in order to guarantee its relevance and quality. Feedback on the program's effectiveness through focus groups and surveys of employees from all levels should be obtained. This process evaluation can be systematic for example, at mid-course as well as at the end of the program or even better continuously within different phases and interventions.

As said earlier, guiding people to change their habits can be very difficult because of many unanticipated aspects and barriers which could block the process (see also Section 3.2.1–3.3). Thus, even if the energy awareness program is well planned, organized, and executed, some surprises are bound to arise along the way. Due to these unexpected but more or less unavoidable problems and challenges, a mid-course or continuous evaluation is very valuable. If the evaluation were to be only at the end of the program, the discovery that certain aspects of the approach are not effective comes too late to intervene. A mid-course or continuous evaluation within specific phases allows the awareness program to be fine-tuned on time to better achieve the desired outcomes. In addition, resources can be shifted to areas that are working well, while cutting back or eliminating activities that are less effective for a particular plant or organization.

A mid-course evaluation does not need be very time consuming or expensive. Strategic surveys or phone calls, personal interviews, or a couple of informational discussions with the energy awareness planning committee can reveal much about what's working and what's not. If possible, a diverse group should be contacted–both demographically different (men, women, young, older) and from different kinds of work levels or locations/plants. With this thorough process evaluation the quality and state-of-the-art of the energy awareness program can be guaranteed.

3.3.3.2 Evaluation of Results

In addition to the process evaluation, which enables corrections while the program is running, an evaluation of the end results is crucial as well. It has to be assessed (i) how the level of energy awareness has changed after the program, (ii) what and how many energy conservation suggestions have been made, and (iii) the eventual behavior changes employees undertook as well as the amount of energy saved. Together, these factors indicate the effectiveness of the program.

The level of energy awareness can be analyzed with before-and-after interviews with certain key personnel or with broad spread surveys. This feedback should directly enable an indication of energy awareness levels but also show problems and indirect reasons for eventual failures in the program after its conduction. The amount of energy suggestions and energy use can also be compared through before-and-after analysis of documentations and data-sheets. Together with the information about energy awareness levels, is it possible to understand the effectiveness of the program well enough to be able to use or adapt its activities for longer term efforts. Certain interventions might have worked excellently while others were rather ineffective and should be eliminated or replaced. It could also be discovered that one department, production line or plant saved considerably more than others due to considerably higher percent of workshops or signed goal agreement documents. A reason could be better efforts from the direct supervisors. Rather uninterested departmental management could have led to employees being very enthused at the beginning of the program, but as time went on, enthusiasm waned and behavior reverted. Or perhaps certain behavior, such as shutting down computers or specific machines, proved uncomfortable or inconvenient over time.

Many findings through the evaluation of the end results, both positive and negative, can shed light on the effectiveness and timing of certain program interventions in contributing to the overall goal to save energy. With adequate resources and time, the best evaluation uses two methods: quantitative (involving data-sheets about energy amounts saved, surveys to show what employees think and feel, or number of energy workshops requested) and qualitative (detailed analysis for example, through interviews to be able to interpret the meaning in what people have said or done). Once the combined data are analyzed and compared, the key findings of the effectiveness of the program will rise to the top and enable enhancement to ensure state-of-the-art for a sustainable future.

3.3.3.3 Reporting of Results and Lessons Learned

The results of the above section have to be documented and reported to various interest groups. In general all employees have to be informed about the outcome of the energy awareness program. However, the upper/top management should be briefed at first and separately by the planning committee to discuss sensitive information and to decide on the specific content and information policy. After the management has cleared the results of the program and the lessons learned, relevant information can be passed on to specific employee groups. The effort of all participants has to be honored by thanking and returning feedback, mentioning

9 Saunders, C.D. (2003) The emerging field of conservation psychology. *Human Ecology Review*, **10**, 2.
10 Porter, M., and van der Linde, C. (1995) Green and competitive: ending the stalemate. *Harvard Business Review*, September–October, 120–133.
11 Archer, D., Pettigrew, T., Costanzo, M., Iritani, B., Walker, I., and White, L. (1987) Energy conservation and public policy: the mediation of individual behavior, in *Energy Efficiency: Perspectives on Individual Behavior* (eds W. Kempton and M. Neiman), American Council for an Energy Efficient Economy, Washington, pp. 69–92.
12 Hirst, E., Berry, L., and Soderstrom, J. (1981) Review of utility home energy audit programs. *Energy*, **6**, 621–630.
13 Yates, S.M., and Aronson, E. (1983) A social psychological perspective on energy conservation in residential buildings. *American Psychologist*, **38** (4), 435–444.
14 Kurz, T. (2002) The psychology of environmentally sustainable behavior: fitting together pieces of the puzzle. *Analyses of Social Issues and Public Policy*, **2** (1), 257–278.
15 Kaiser, F., Wolfing, S., and Fuhrer, U. (1999) Environmental attitude and ecological behavior. *Journal of Environmental Psychology*, **19**, 1–19.
16 Costanzo, M., Archer, D., Aronson, E., and Pettigrew, T. (1986) Energy conservation behavior: the difficult path from information to action. *American Psychologist*, **41** (5), 521–528.
17 Wicker, A. (1969) Attitudes versus actions: the relationship of verbal and overt behavioral responses to attitude objects. *Journal of Social Issues*, **25**, 41–78.
18 Leung, C., and Rice, J. (2002) Comparison of Chinese-Australian and Anglo-Australian environmental attitudes and behavior. *Social Behavior and Personality*, **30**, 251–262.
19 Vogel, S. (1996) Farmers' environmental attitudes and behavior: a case study for Austria. *Environment and Behavior*, **25** (5), 591–613.
20 Vining, J., and Ebreo, A. (1992) Predicting recycling behavior from global and specific environmental attitudes and changes in recycling opportunities. *Journal of Applied Social Psychology*, **22**, 1580–1607.
21 Dwyer, W., Leeming, F., Cobern, M., Porter, B., and Jackson, J. (1993) Critical review of behavioral interventions to preserve the environment: research since 1980. *Environment and Behavior*, **25**, 275–321.
22 Deci, E.L., and Ryan, R.M. (1985) *Intrinsic Motivation and Self-Determination in Human Behavior*, Plenum Press, New York.
23 Locke, E.A. (1968) Toward a theory of task motivation and incentives. *Organizational Behavior and Human Performance*, **3**, 157–189.
24 Locke, E.A., and Latham, G.P. (1990) *A Theory of Goal-Settingand Task Performance*, Prentice Hall, Englewood Cliffs, NJ.
25 Locke, E.A., and Latham, G.P. (2002) Building a practically useful theory of goal-settingand task motivation. A 35-year odyssey. *The American Psychologist*, **57** (9), 705–717.
26 Locke, E.A. (1996) Motivation through conscious goal setting. *Applied and Preventive Psychology*, **5**, 117–124.
27 Ammons, R.B. (1956) Effects of knowledge of performance: a survey and tentative theoretical formulation. *Journal of General Psychology*, **54**, 279–299.
28 Locke, E.A., Cartledge, N., and Koeppel, J. (1968) Motivational effects of knowledge of results: a goal-setting phenomenon? *Psychological Bulletin*, **70**, 474–485.
29 Benders, R.M.J., Kok, R., Moll, H.C., Wiersma, G., and Noorman, K.J. (2006) New approaches for household energy conservation- In search of personal household energy budgets adn energy reduction options. *Energy Policy*, **36**, 3612–3622.
30 EPRI (2009) Residential electricity use feedback: a research synthesis and economic framework. Palo Alto, CA. Report # 1016844.
31 McClelland, L., and Cook, S.W. (1980) Promoting energy conservation in master-metered apartments through group financial incentives. *Journal of Applied Psychology*, **10**, 20–31.

32 Nolen, S.B. (1996) Why study? How reasons for learning influence strategy selection. *Educational Psychology Review*, **8**, 335–355.

33 Bandura, A., and Cervone, D. (1983) Self-evaluative and self-efficacy mechanisms governing the motivational effects of goal systems. *Journal of Personality and Social Psychology*, **45**, 1017–1028.

34 Becker, L. (1978) Joint effect of feedback and goal-settingon performance: a field study of residential energy conservation. *Journal of Applied Psychology*, **63**, 428–433.

35 Erez, M. (1977) Feedback: a necessary condition for the goal–performance relationship. *Journal of Applied Psychology*, **62**, 624–627.

36 Abrahamse, W., Steg, L., Vlek, C., and Rothengatter, T. (2005) A review of intervention studies aimed at household energy conservation. *Journal of Environmental Psychology*, **25**, 273–291.

37 Sheehy, L.A. (2005) Goal-settingfor sustainability: a new method of environmental education. Unpublished dissertation. School of Environmental Science, Murdock University, Western Australia.

38 Stern, P.C. (2000) Toward a coherent theory of environmentally significant behavior. *Journal of Social Issues*, **56**, 407–424.

39 Gardner, G.T., and Stern, P.C. (1996) *Environmental Problems and Human Behavior*, Allyn & Bacon, Needham Heights, MA.

40 Stern, P.C. (1986) Blind spots in policy analysis: what economics doesn't say about energy use. *Journal of Policy Analysis and Management*, **5**, 200–227.

41 Geller, E.S. (1981) Evaluating energy conservation programs: is verbal report enough? *Journal of Consumer Research*, **8**, 331–335.

42 Nerdinger, F.W., Blickle, G., and Schaper, N. (2008) *Arbeits- Und Organisationspsychologie*, Springer, Berlin.

43 Schlick, C., Luczak, H., and Bruder, R. (2010) *Arbeitswissenschaft*. 3. Auflage. Springer Verlag, Berlin.

44 Earley, P.C., Connolly, T., and Ekegren, G. (1989) Goals, strategy development and task performance: some limits on the efficacy of goal setting. *Journal of Applied Psychology*, **74**, 24–33.

45 Gustafsson, A., and Gyllenswärd, M. (2005) The power-aware cord: energy awareness through ambient information display. Proceedings of CHI 2005, April 2–7, 2005, Portland, Oregon, USA.

46 Deutsche Bahn AG (2007) Gespräche zum Energiesparen im Personenverkehr der Deutschen Bahn AG – Feedbackgespräche: überstrapaziert, missverstanden und bedeutungslos? (Effect of feedback Conversations), Veröffentlichung in Personalführung Ausgabe April 2007.

47 McCalley, L.T., and Midden, C.J.H. (2002) Energy conservation through product-integrated feedback: the roles of goal-setting and social orientation. *Journal of Economic Psychology*, **23**, 589–603.

48 Winett, R.A., Leckliter, I.N., Chinn, D.E., Stahl, B., and Love, S.Q. (1985) Effects of television modeling on restidential energy conservation. *Journal of Applied Behavior Analysis*, **18**, 33–44.

49 Aronson, E., and O'Leary, M. (1982) The relative effectiveness of models and prompts on energy conservation: a field experiment in a shower room. *Journal of Environmental Sytsems*, **11**, 219–224.

50 Ester, P.A., and Winett, R.A. (1982) Toward more effective antecedent strategies for environmental programs. *Journal of Environmental Systems*, **11**, 201–221; (b)

51 Aronson, E. (1990) Applying social psychology to desegregation and energy conservation. *Personality and Social Psychology Bulletin.*, **16**, 118.

52 Darley, J.M. (1978) Energy conservation techniques as innovations, and their diffusion. *Energy and Buildings*, **1**, 339–343.

53 Stern, P.C. (1985) *Energy Efficiency in Buildings: Behavioural Issues*, National Academy Press, Washington, DC, USA.

54 Katzev, R., and Johnson, T. (1987) *Promoting Energy Conservation: An Analysis of Behavioral Research*, Westview Press, Boulder, CO, USA.

Part Two
Energy Efficient Design and Production

Managing CO_2 Emissions in the Chemical Industry. Edited by Leimkühler

ISBN: 978-3-527-32659-4

4
Systematic Procedure for Energy and CO_2 Reduction Projects

Hans-Joachim Leimkühler

4.1
Overview

Successful energy and CO_2 reduction projects require a systematic approach. In this chapter we will give an overview over the procedure we recommend for energy saving and CO_2 reduction projects. This procedure was developed and proven in many different projects. Some key aspects for a successful project execution are:

- Clear scope and well defined targets.
- Continuous collaboration between project team and the partners in the plants.
- Systematic, comprehensive and traceable methodology.
- Sustainable implementation of technical and cultural measures.

The systematic approach we choose for our projects follows the Six Sigma procedure (see [1] or [2]).

- Define the scope and the task.
- Measure the actual CO_2 footprint and the specific energy consumption of the products as a baseline.
- Analyze the distribution and consumption of different energies (steam, electricity, cooling media, pressurized gases) in the plants.
- Improve the situation by generating ideas for energy saving measures and implementation of the measures in the plants.
- Control the sustainable success of the project by technical measures (management system) and repeated checks in the following years.

In this chapter we will give an introduction to a systematic procedure of energy saving projects. The basis of this introduction is the high number of projects we executed in different branches, mainly in the chemical process industry (see [3] and others). The general procedure comprises the steps shown in Figure 4.1.

After analysis of energy consumption at the plant or site we generate ideas for energy savings in collaboration with our experts and the customer. These ideas will be evaluated in order to prioritize them. Afterwards the beneficial ideas will

Managing CO_2 Emissions in the Chemical Industry. Edited by Leimkühler

ISBN: 978-3-527-32659-4

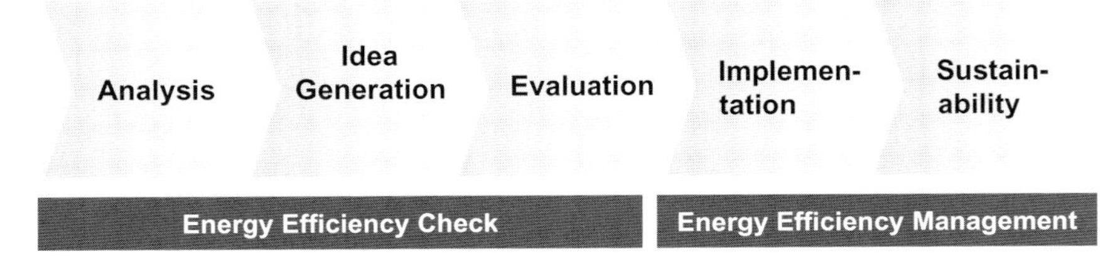

Figure 4.1 General procedure in energy savings projects.

be implemented on site. It is an important, but not an easy, task to assure the sustainability of these energy saving measures in the plants.

All the steps will be presented in this chapter in more detail.

4.2 Definition of Scope and Task

The first step in most projects is a clear definition of scope and tasks. The scope for an energy saving project can be a single plant or a chemical, petrochemical or pharmaceutical site. In many cases enterprises start a global program that consists of different sites worldwide. The advantage of a comprehensive program is that synergies between similar plants in different sites can be exploited. Best practices in energy saving measures are exchanged between the sites.

Definition of the scope for single plants is quite simple: The decision has to be taken as to whether or not logistic facilities such as tank farms will be included in the study. It is recommended that the project scope should be wider from the beginning and all auxiliary facilities belonging to the plant should be included, for the following reason. After the analysis phase, when the energy consumption in different facilities is determined, the project team will concentrate on the main consumers anyway. Therefore the volume of the project will be reduced to the relevant issues. Additionally we notice that sometimes auxiliary facilities show surprisingly high saving potentials (for instance waste gas incineration units). It seems that these facilities are not normally in the focus of the plant and can therefore be neglected. Here we often find the 'low-hanging fruits', where the customer can save money in the short term and with no or moderate investment. When energy saving measures are implemented we anyhow have to check if the proposed utility system is still compatible with a new, lower energy consumption of the site, and whether it can remain efficient at reduced load. There are examples where measures were not feasible because the energy supply system (in this case a turbine) was not designed for the reduced load.

In the case of a worldwide program the scope must be more carefully defined in advance by analyzing the available energy data of the customer. First, the energy consumption of every site is identified. An example is shown in the Figure 4.2.

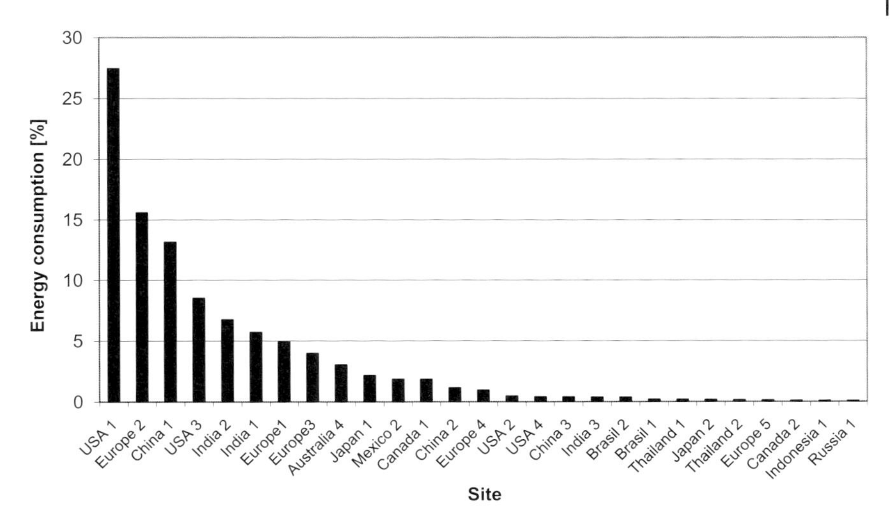

Figure 4.2 Energy consumption of different sites (in % of total consumption).

In this example we can concentrate on eight sites (of 27) and cover 85% of the total energy consumption of the company. The reduction of the number of sites lowers the project costs and helps to concentrate on the biggest energy consumers. In our experience it is recommended that at least 80% of the total energy consumption of a company should be covered in order to derive the maximum advantage from the energy saving project. Smaller sites (in the sense of energy consumption) can be neglected, nevertheless some customers decided to include smaller sites into the program for 'political' reasons. Some customers also use the findings in the energy-consuming sites and transfer these measures to the smaller sites.

Beside the scope it is important to define the goals of the project in advance. What does the customer expect, what can the contractor deliver? On average we identify in our projects 10% energy savings potential without investment and an additional 10% potential with profitable investment (measured in terms of specific primary energy consumption in kWh per t of product). These results differ from case to case depending on the industry branch, the technology, the status of the plant and former energy saving projects (and of course of the project effort). A rough potential analysis of the customer's situation can help to estimate in advance the expected results of the project. That gives the customer an indication whether the intended energy saving project will bring the expected return. For the contractor the potential analysis helps to estimate the effort that will be necessary for an optimal result. That effort depends on different aspects: what data is available on the customer side, what is the automation level of the plants, if there are any process models available and so on. Based on the potential analysis a detailed offer is possible with realistic project targets. Therefore we recommend a potential study before beginning a bigger energy saving program.

4.3
Analysis

4.3.1
Carbon Footprint

In the analysis phase all data is collected and analyzed that is necessary to define the baseline of CO_2 emission and energy consumption and to elaborate measures for reducing CO_2 emission and energy consumption.

The first step in our approach is the determination of the specific CO_2 emission of a product. This step is not mandatory for energy saving projects. Nevertheless many producers need information about the CO_2 footprint of their products because the public and especially their customers ask for that information. We describe here as an example the specific method of Bayer Technology Services to determine the CO_2 footprint. This method we call the 'Climate Footprint®'. Other suppliers may apply different approaches. All approaches are based on the ideas of lifecycle analysis. These ideas as well as the carbon footprint and especially the Climate Footprint have been described in detail in Chapter 1.

The idea of the Climate Footprint is to quantify the amount of CO_2 emitted during the manufacturing of one kilogram of a product and to specify the origin of these emissions. The Climate Footprint follows a 'cradle-to-gate' approach; that means we specify the emissions accrued by the raw materials (production and distribution) and the production in the plants of our customer. This cradle-to-gate approach takes into account all the effects a producing company can influence. The methodology is in accordance with the lifecycle analysis rules. The procedure was certified by TÜV Süd, a German 'Notified Body' for certifications.

A typical Climate Footprint is shown in Figure 4.3.

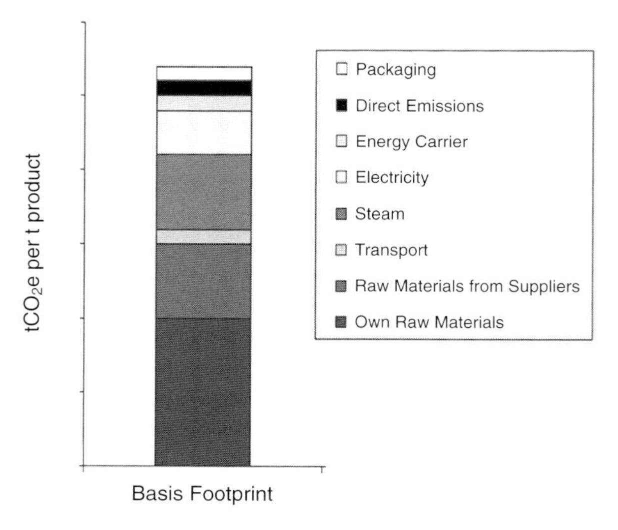

Figure 4.3 Climate Footprint.

The figure shows the specific CO_2 load that the product carries, split into different sources of CO_2 emissions: CO_2 footprint of raw materials, CO_2 from energy consumption during production, direct CO_2 emissions and packaging. Behind the integral numbers detailed values per raw material or per energy (like steam, electricity and so on) are available. The distribution of the product to the customer beyond the gate of the manufacturer is not considered, because he cannot influence this.

The Climate Footprint contains different information:

- The value of the specific CO_2 emission during production (in tons of CO_2 per tons of product) describes the CO_2 relevance of the product. In order to calculate the absolute amount of emitted CO_2 the specific footprint has to be multiplied by the output of the plant (in tons).
- The origin of the CO_2 gives hints for reduction projects. In cases where the footprint is dominated by the raw material, it is recommended that alternative raw materials or an alternative supplier with a low CO_2 production process should be considered in order to reduce the CO_2 load of the product. If the energy consumption is the main CO_2 source then an energy savings project should be initiated.
- The development of carbon footprints over the years shows the progress of the manufacturer in reduction of CO_2 emissions. After implementation of energy saving projects, the value of CO_2 from energy consumption during production will decrease. The change of a supplier can reduce the CO_2 from raw materials.

In the systematic approach the Climate Footprint gives quantitative and comparable information about the total CO_2 emission and the main CO_2 sources. Therefore it can serve as an evaluation tool for strategic decisions and help to answer the following questions:

- Where have the new processes to be developed?
- What investment in new facilities is preferred?
- How efficient are the plants and how can operational excellence be improved?

Additionally it is the basis for the next steps in energy and CO_2 reduction projects.

4.3.2 Energy Distribution per Utility

Normally an energy project includes other utilities, too. Typically the following energy categories will be regarded:

- steam at different pressure levels;
- electricity (sometimes considering different voltages);
- compressed air;
- cooling water;

- chilled water or brine;
- special energies like ammonia, compressed nitrogen and others.

For energy saving and CO_2 reduction projects the consultant needs a comprehensive and, if possible, consistent data set of the real energy consumption in the plants, process steps and equipment. In order to obtain this information he can collect measured and calculated data. Normally every plant, for accounting reasons, has at least one energy metering point at the entrance of the plant for every utility. Therefore the total consumption of the plant is known (assuming that the measurements are correct!).

In order to compare the different energy categories it is necessary to express the consumption in a common dimension. That can be MWh of primary energy or tons of CO_2 emitted or costs (€ or $). The conversion factors between the different dimensions depend on local energy prices, the local primary energy mix and the efficiency of the energy generation. For one of our customers we converted the units for example according to Table 4.1.

Distribution of the total energy consumption for the different categories can then be presented as a chart. In this case the energy consumption is expressed as percentage of the emitted CO_2 (Figure 4.4) and the total energy costs (Figure 4.5). Alternatively, a calculation based on primary energy consumption is of course possible.

It is helpful to look at the reliability of the measured data, because measurement devices can fail. In many cases redundant measurements occur, for example by a main metering point at the entrance of the plant and additional meters for single process steps. In this case the reliability of the measurement can be evaluated. When there is no redundancy in metering points, then the calculation of the main consumers can give a hint on the plausibility of the measurements (see Section 4.3.3).

In the example presented in Figures 4.4 and 4.5 the intention of the depiction of the energy distribution can be easily seen. The consumption of cooling water and air in terms of costs or CO_2 emission is negligible. Here it is obvious that in the following project steps water and air will not be analyzed in detail. An exception would be if the consumption of water or air is caused by only one big consumer. Then we would consider this equipment. In all other cases the consultant

Table 4.1 Conversion of different energy units.

Utility	Unit	Price/Unit	CO_2e/unit	Prim. energy/unit
Steam 6 bara	t	15€/t	200 kg/t	0.765 MWh/t
Steam 31 bara	t	20€/t	200 kg/t	0.778 MWh/t
Electricity	MWh	50€/MWh	500 kg/MWh	1 MWh/MWh
Cooling Tower Water	1000 m^3	26€/1000 m^3	150 kg/1000 m^3	0.348 MWh/1000 m^3
Compressed Air 6 bara	1000 m^3	7€/1000 m^3	50 kg/1000 Nm3	0.095 MWh/1000 Nm3

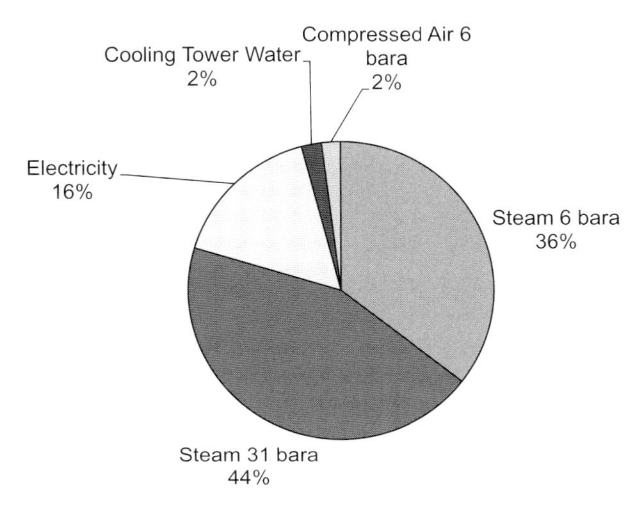

Figure 4.4 Typical energy distribution expressed in % of the total CO_2 emitted.

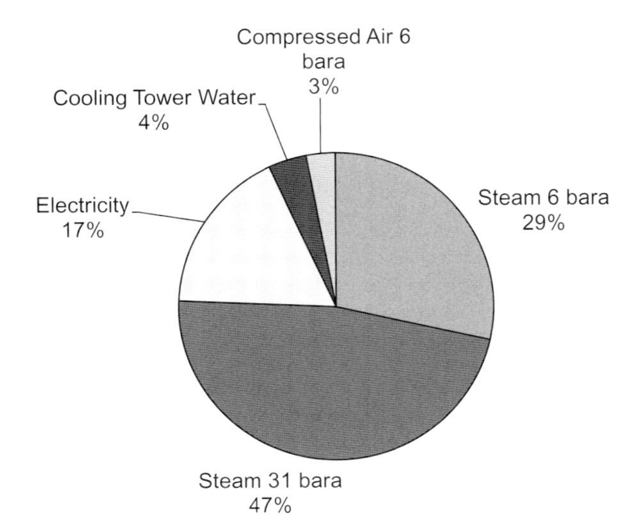

Figure 4.5 Typical energy distribution expressed in % of the total energy costs.

will concentrate on steam and electricity measures in order to reduce the effort of the energy study without loosing relevant saving potential.

4.3.3
Main Energy Consumers

In the last section it was mentioned that the consultant will try to concentrate on the relevant utilities. Equally it is necessary to identify the main energy consumers and find measures to reduce the energy consumption of this equipment. There are different ways to find energy intensive equipment:

1) The main consumers are in many cases equipped with measuring devices. Big evaporators or compressors have dedicated meters for steam or electricity. The availability of data depends strongly on the instrumentation level of the plant. The consultant will analyze the process and instrumentation diagram (P&ID) of the plant in order to identify the energy relevant metering points.

2) If measurements are not available, a thermodynamic simulation of the process (see Section 4.3.5) will quantify the energy consumption of each piece of equipment. The simulation of the process needs some effort, but it gives valuable additional information to be used during the idea generation phase (see Section 4.4). In many cases a thermodynamic model of the plant is available and can be used after actualization.

3) Additionally there is of course the experience and the process know how of the plant manager and of the operators which we use to identify the main energy consuming equipment.

4) The last information source is the design specification of the plant. The expected energy consumption is normally part of this documentation.

The energy consumption of the plant is then depicted in a graph similar to that shown in Figure 4.6. Here the consumption of 31 bar steam is chosen as an example.

In this case there are seven main consumers of 31 bar steam. It is recommended that concentration should be on these consumers in the next projects steps, with respect to steam consumption.

The assignment of energy consumption to the equipment will be repeated for the other utilities. Another example is shown in Figure 4.7. Here the electricity consumption for different consumers is shown in a different visualization.

The depiction of utility consumption shows in an easy and plausible way the energy relevant apparatus and machines to be analyzed in detail.

An important result is the quantification of the 'recovery rate', that means the sum of the energy consumption assigned to all equipment in relation to the total energy consumption. In some cases we find a significant discrepancy here, shown in Figure 4.6 as 'not identified'. If the recovery rate is less than 90% the reason

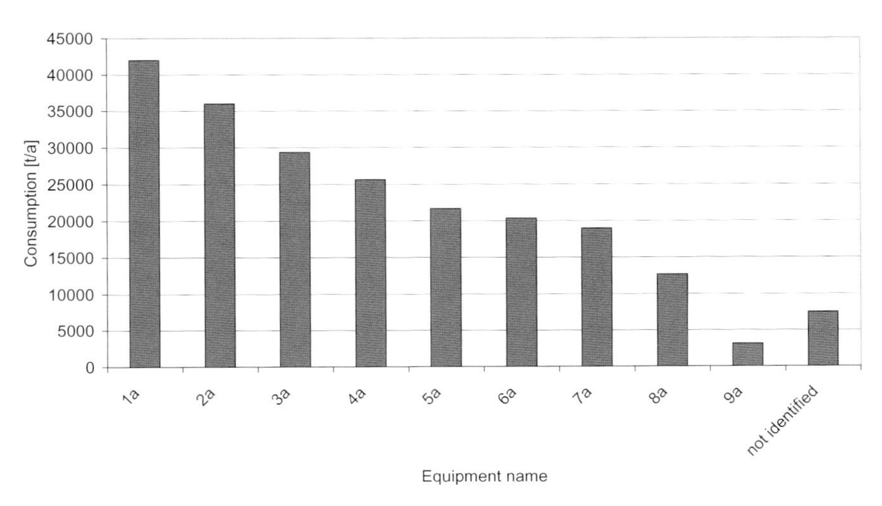

Figure 4.6 Example of 31 bar steam consumption per equipment.

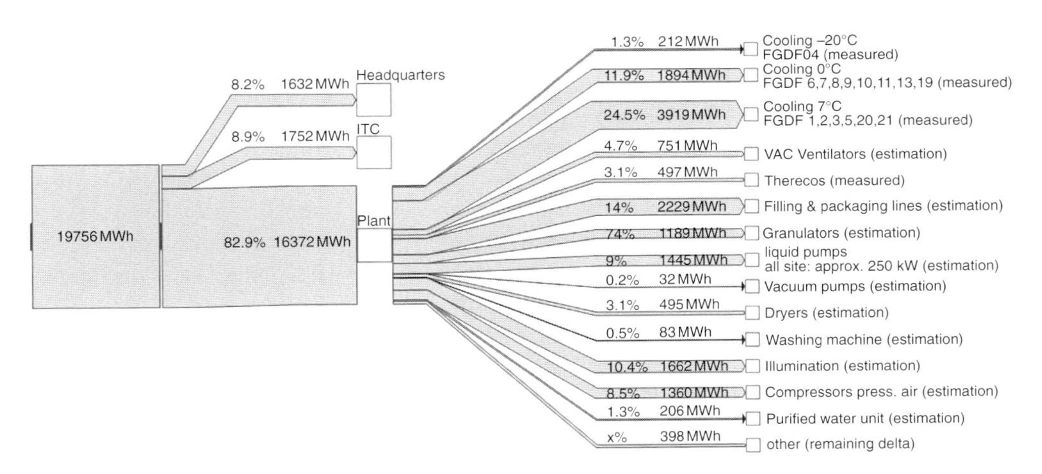

Figure 4.7 Example of electricity consumption per consumer.

for that deviance must be clarified. Possible reasons can be false measurements or unknown or underestimated consumers like trace heating. These unknown or underestimated consumers often are subject to saving ideas.

In total the analysis of the energy consumption per utility and per piece of equipment and its prioritization will typically result in the identification of 50 to 100 main consumers. These consumers now will be analyzed more in detail. The tool for a detailed analysis is the equipment check, which will be described in Section 4.4.3. This stepwise procedure guarantees focused project execution and minimizes the effort and the costs on the energy savings project.

4.3.4
Operational Parameters

Operational parameters are needed for analysis of the process and of the main energy consumers. Typical data for operating parameters are pressure levels and reflux ratios of columns, temperatures in heat exchangers and number of revolutions or pressure ratios for compressors.

The sources for operational parameters are design documents and actual information from the control system. Today many plants have process data information systems. In this case it is easy for the consultant to analyze the history of parameters. If an information system does not exist, then it is necessary to gather the data from local measurements or together with the plant operators.

4.3.5
Process Model

A process model including mass and energy balance of the entire process is an important basis for process optimization projects. A model is useful not only for energy saving projects but also for de-bottlenecking studies, quality improvement projects and operating cost reduction. It provides the operator with information about mass and heat streams, a complete set of operational parameters and internal variables of reactors, columns and heat exchangers. Values that cannot be measured are available from the process model. Nevertheless for an energy saving project we normally do not build up a complete process model, because the effort for a model is relatively high. But if a process model exists, we actualize it and use it for our energy saving purposes. The process model is a good basis for the idea generation phase especially for the pinch analysis (see Chapter 6) and operational improvements in the plant (see Section 4.4.4).

4.3.6
Energy Baseline and Milestone 1

The last step of the energy analysis phase for the plant or the site is the presentation and discussion of the data and the graphs with the customer. In this Milestone 1 of the energy savings project the consultant presents the identified consumption data as described in Section 4.3. A typical Milestone 1 agenda is shown below:

1) Introduction
   - Introduction of the participants
   - Significance of energy efficiency
   - Internal employee suggestion system regarding energy efficiency
2) Results of the analysis phase
   - Cost structure of the site
   - Main energy consumers and recovery rate

3) Discussion of the results of the analysis phase
4) Finalization of analysis phase.

Normally we organize the milestone meeting for every plant. In this meeting the plant manager, plant engineer and foremen are present from the client's side. In some cases we combine the Milestone 1 meeting with the brainstorming session (see Section 4.4.2). The energy distribution, main consumers, recovery rate and first findings of the project team are discussed. The plant staff has the opportunity to comment on energy consumption data and findings and check the plausibility. It is important to inform the participants of the meeting early enough about the results of the analysis phase in order to ensure sufficient preparation. As a result of the meeting customer and consultant agree on the energy baseline, which describes the energy consumption of the plant before the energy savings project. This energy baseline is necessary for later comparisons and to quantify the benefit of the project. It is crucial for the project success that all energy saving ideas after Milestone 1 are part of the project result. Any discussion as to whether an idea was contributed by the plant staff, consultants or other experts participating in the project jeopardizes the collaboration in the team, its productivity and creativity and therefore has to be avoided. All ideas after Milestone 1 must be brought on the table and are owned by the whole team.

4.4 Idea Generation

4.4.1 Fields of Energy Savings

The comprehensive approach that we developed for energy saving projects covers all energy influencing parameters on a production site regarding energy generation, energy distribution and energy demand. The fields are shown in Figure 4.8 and will be explained in more detail in the following sections.

A consultant for energy saving projects therefore needs a broad knowledge in the different fields that have to be covered.

4.4.2 Brainstorming Sessions

The first step in the idea generation phase is a brainstorming session with the partners in the plant. Often we combine this brainstorming session with the Milestone 1 meeting described in Section 4.3.6. This procedure guarantees that the plant staff is involved at an early phase of the project. Operators on the plants know their plant very well and normally have many ideas for energy saving. It is important to involve them and accept these ideas if possible. Our experience shows that ideas generated by the operators have a better chance of being implemented in a sustainable manner than ideas imported from outside.

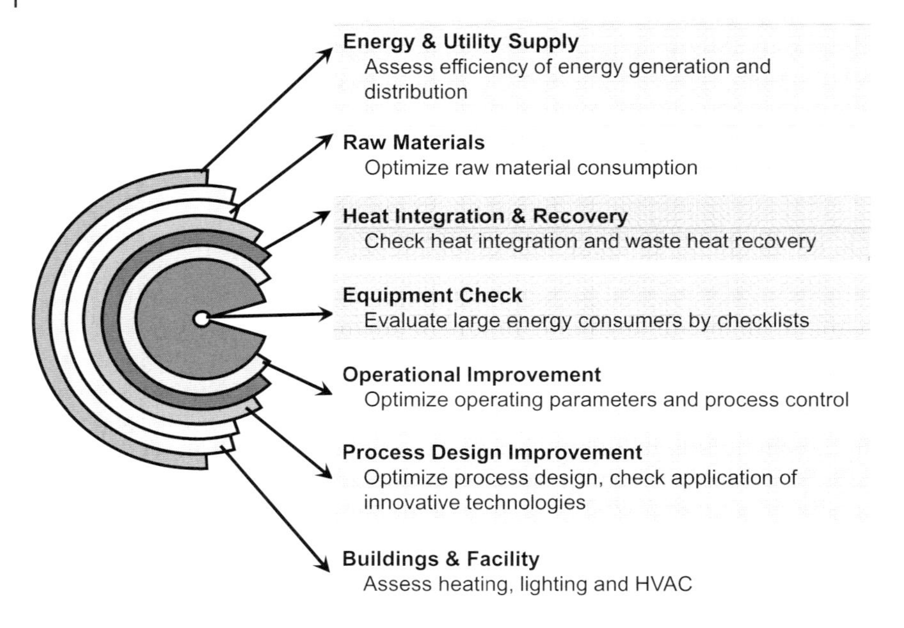

Figure 4.8 Fields of energy savings.

A typical agenda for a brainstorming is shown below:

1) Introduction
   - Introduction of the participants
   - Significance of energy efficiency
2) Brainstorming
   - Rules
   - Procedure
   - Presentation of the process
   - Brainstorming in groups
   - Open brainstorming
3) Discussion and pre-evaluation.

Plant managers, engineers and operators will participate in the meeting. Our consultants act as moderators. After an introduction we go through the process and reflect some findings of the analysis phase. Then we form groups, in which the first saving ideas are developed. These ideas will be presented in the plenary session. The first brainstorming in groups serves as a seed phase. Ideas that were discussed in the plant before will come up in this phase. In the open brainstorming we ask for further ideas which are developed by the participants based on the results of the group brainstorming. In our experience the open brainstorming gets the better results, because the ideas are more creative and not obvious. All ideas then are clustered, discussed and pre-evaluated in the group.

The result of the brainstorming session is a pre-evaluated list of improvement ideas. It is documented in the same way as the results of the following steps in our systematic procedure.

4.4.3 Equipment Check

During the analysis phase and the Milestone 1 meeting we defined the main energy consuming equipment. The energy efficiency of this equipment will be checked now in detail. Together with our experts we developed checklists for standard equipment like pumps, compressors, evaporators, heat exchangers and many more (see Table 4.2).

In these checklists we document the actual operating conditions of the equipment and compare them with best practice and benchmarks. Based on the operating conditions the checklists deliver hints for improvement measures. We observe for example that up to 50% of the pressure increase in pumps is consumed in the control valve or the cooling media flow for heat exchangers is often not controlled and therefore cooling media is wasted. These observations sometimes lead to easily implemented saving ideas. Additionally for special energy consuming equipment we contact our specialists in order to evaluate the actual operating conditions and to come up with saving opportunities. Examples for these special aggregates are spray dryers, refrigeration units or special reactors.

Another source for saving is the detection of thermal bridges or defective insulation. An infrared camera is easy to use and helps to find energy leaks. The repair of the insulation is a 'low-hanging fruit' in our projects.

Table 4.2 Available checklists for standard equipment.

Equipment	Equipment
Boiler	Filter
Centrifuge	Granulation
Chiller	Heat exchanger
Coater	Heat pump
Column	Jet mill
Compressed air system	Kneader
Compressor	Mixing device
Contact dryer	Motor
Convection dryer	Screw
Cooling tower	Steam traps
Dryer	Vacuum pump
Extruder	Ventilator
Fermentor	

All generated saving ideas are documented and evaluated later (see Section 4.5). The equipment check in many cases leads to measures that can be implemented with low or without investment.

4.4.4 Operational Improvements and Process Control

A beneficial source for saving ideas is the improvement of the operating parameters in a plant. In many cases the potential is high and the necessary investment rather low or even zero. Some examples for operating parameters are reflux ratios in columns, recycle and purge ratios, pressure levels and temperatures or solvent concentrations as well as reduction of reactant excess.

The best tool for the improvement of these parameters is the analysis of a process model (see Section 4.3.5). It is not so laborious to optimize the parameters in a model by a sensitivity analysis in comparison with plant or lab experiments. An experienced engineer is able to find improvement ideas (not only for energy savings) performing numerical experiments in the model and assessing the results.

An example: We modeled a complete polymerization process with reaction and processing. The reactor was cooled by boiling off the reaction mixture, and the gas stream from the reactor was condensed and recycled to the reactor. Our model showed a high concentration of by-products in the loop and in the reactor, until then unknown to the plant staff. This by-product came into the plant with one of the monomers. With the help of the model we could find out the optimal localization of a purge stream and its recommended size. The existing purge was modified according to our suggestion. As a consequence the by-product concentration decreased as predicted by the model. This measure reduced the energy consumption and led to a higher capacity of the plant.

Process control is also an important method for improved energy efficiency. The first step in our projects is to check all process control loops. In some cases we see that distillation columns are operated at high, constant reflux flows and the process control is switched off. Here we have to convince the operators to switch on the installed control loop. The development of energy awareness and an energy conscious mindset in the team (see Chapter 3 of this book) is therefore a crucial part of energy saving projects.

In the second step our process control experts identify potential savings which are possible by implementing additional loops.

An example: The reflux ratio determines the energy consumption of a column. We suggested an advanced process control loop in order to come to a more exact control of the product quality. The implementation was successful. The variations of the product concentration decreased considerably. The effect is shown in Figure 4.9.

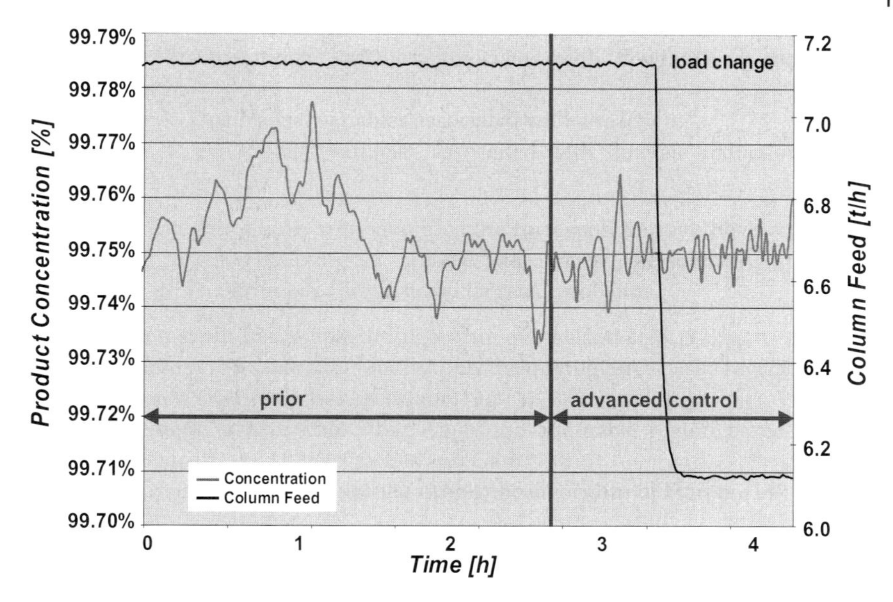

Figure 4.9 Effect of advanced process control loop.

Because of the decreased variation of the concentration curve the operators could reduce the setpoint of the concentration, that means choose a setpoint closer to the product specification. Now the column is operated with a lower reflux ratio and therefore with lower energy consumption. This example shows that the implementation of process control loops, which means an investment, often leads to high energy savings. Within a comprehensive energy project the process control expert is therefore a member of the team.

4.4.5 Process Improvements

In this part of our systematic approach we examine improvements to the process by changing the process itself. Here we consider disruptive innovations therefore the time scale for these measures is medium or long. Our experts who analyze the process come up with novel ideas, which have to be investigated further. Some example for such ideas:

- Change to solvent with better separation characteristics.
- Use alternative raw materials if the CO_2 footprint of the raw materials is high.
- Evaluate the reaction kinetics that may lead to an alternative reactor concept.
- Heat integration by additional or different equipment.
- Advanced equipment types (for example process intensification or micro technology).
- Exchange of columns by membranes.

This is only a small selection of measures that were discussed in our projects.

The savings potential of these ideas will be evaluated. Then we discuss with the customer if a further elaboration of the measure will be pursued.

4.4.6
Heat Integration and Heat Recovery

Heat recovery is described in detail in Chapter 6 of this book. Of course standard methods such as pinch point analysis (see for instance [4]) are part of an energy study. Input data for a pinch point analysis are the heat streams in a plant and their temperature levels. There are some rules deduced from the pinch point theory. The consultant will evaluate if the actual heat exchanger network in the plant fulfills these rules. Then pinch point analysis or use of a software tool that applies the analysis, for example: Aspen Energy Analyzer, formerly Aspen HX-Net (http://www.aspentech.com/products/aspen-hx-net.cfm) or KBC Supertarget (http://www.kbcenergyservices.com/default.energy.asp?id=134) gives proposals for new heat integration measures, that means the exchange of heat between sources and sinks. Normally these measures need additional investments in equipment, piping and instrumentation, therefore their profitability has to be evaluated (see Section 4.5.3). Additionally an experienced energy consultant is able to detect other opportunities for heat integration by analyzing the P&ID diagrams.

There are several examples where measures that are relatively easy to implement have a significant potential:

- The use of bottom product of a column for preheating the feed stream is a classical example. A simple liquid–liquid heat exchanger can save a lot of energy.
- It is worth evaluating the reuse of warm water. Instead of losing the energy the customer can invest in an absorption heat pump and use the energy for the generation of chilled water. The consultant will calculate the profitability of this measure. If the heat streams are smaller, then warm water reuse for building heating sometimes is a solution.
- If steam is generated in the plant, then we try to redesign the process in a way that the steam pressure is high enough to feed the steam into the steam net. If the temperature (that means the pressure) is too low, then the steam might be used for trace heating.
- In pharmaceutical plants a high fraction of the energy consumption is often used for ventilation. Here we check if a heat exchanger, which heats the fresh air with the outlet stream, is profitable.

4.4.7
Raw Materials

In Section 4.3.1 the carbon footprint or, the method developed by Bayer Technology Services, the Climate Footprint was presented. As a part of the result of carbon footprint analysis, the CO_2 load of a product caused by the raw materials is delivered. Now the consultant together with the customer will assess that value and decide if an investigation for alternative raw materials, or an alternative supplier for the same raw materials is a way to decrease the footprint. The search for alternative raw materials can of course imply changes in the process, but that has to be checked. In the field of power generation the substitution of coal by natural gas is a big lever to decrease the CO_2 footprint of the whole site. Another example is the use of renewable feedstock like sugar for the polyol and polyurethane production instead of polyols produced from fossil resources. Further examples for the use of renewable feedstock are given in Chapter 12.

4.4.8
Buildings and Facilities

Heating, ventilation and air conditioning (HVAC) of buildings can have a high impact on the total energy consumption of a plant. Especially for pharmaceutical plants with their high requirements for ventilation, air conditioning and air change rates the HVAC experts play an important role in an energy saving project.

During the analysis phase we normally send out data sheets to the plant manager where all ventilation systems will be recorded by the plant. This list will be completed and evaluated during the site visit of the HVAC expert. The energy consumption of the buildings is analyzed and then compared with known benchmarks, expressed for instance as a building footprint in kg CO_2e per sqm net floor area. Here local weather conditions taken from databases like Meteonorm (http://www.meteonorm.com/pages/en/meteonorm.php?lang=EN) or from ASHRAE (www.ashrae.org) must be considered for the benchmarking process.

In the idea generation phase different modeling tools for buildings are used, for instance to calculate the energy demand for heating, cooling, ventilation, humidification and dehumidification, domestic hot water and lighting (depending on climate data). Other tools can evaluate the impact of solar collectors and photovoltaic plants.

Another important issue is the air change in pharmaceutical plants. Here the discussion between the energy consultant and quality assurance (QA) people on the customer side can find an optimal air change rate for the plants as a compromise between air purity and energy costs.

The workflow for ideas generation is summarized in Figure 4.10.

First the building is modeled and the calculated energy consumption adjusted with the measured consumptions. Then saving potentials are detected by software tools and the expertise of the consultant. The measures will be elaborated to a certain extent. The feedback of these measures into the model gives information about feasibility and impact of the designed measures.

An example: We analyzed an office and laboratory site for a customer. The energy costs before the project were more than 2 Mio. €. We found profitable measures

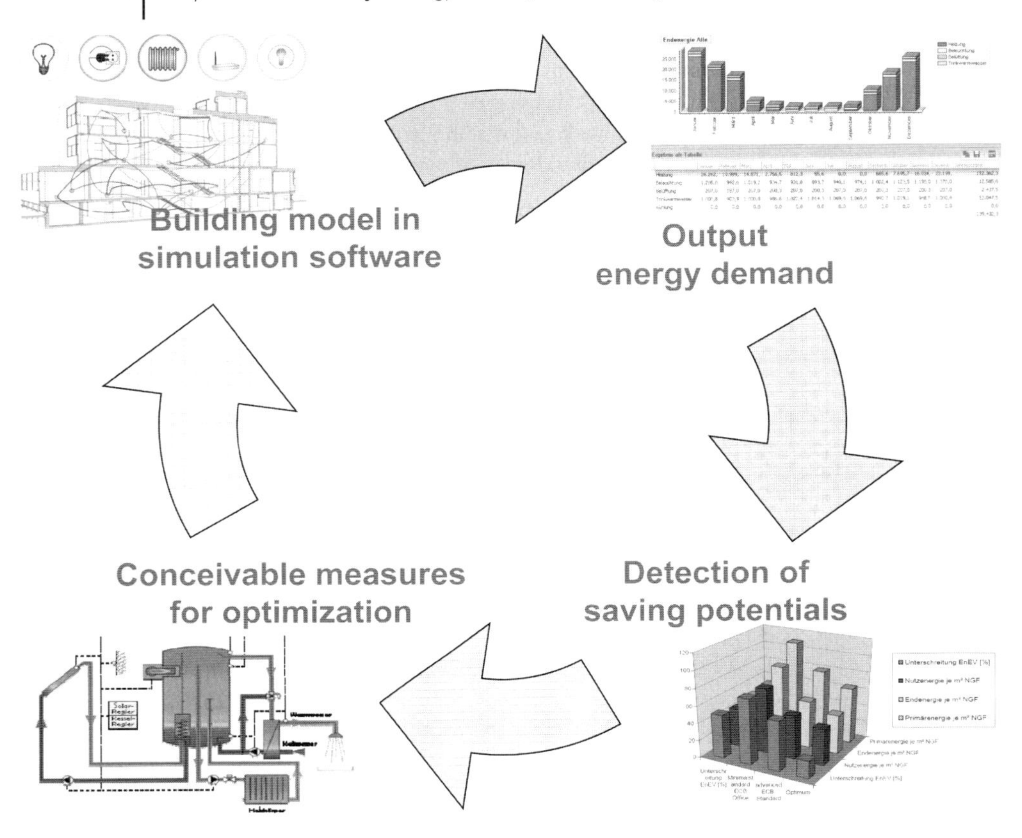

Figure 4.10 Workflow for the generation of saving ideas in buildings.

for savings of more than 30% with a payback time less than half a year. The definition of the profitability will be discussed in more detail in Section 4.5.3.

4.4.9 Energy and Utility Systems

An energy saving project is incomplete if it does not consider the energy and utility generation and distribution. In our workflow we regard the generation and distribution of the following utilities:

- steam (different pressure levels);
- electrical power (different voltages);
- water (for industrial use, demineralized water, cooling towers, chilled water, brine, condensate reuse, ultra pure water);
- air (for industrial use, instrumentation);
- nitrogen (pressurized).

A detailed view on this topic is given in Chapter 10 of this book, so here we will focus just on certain aspects we often observe in our saving projects.

Steam and electricity are generated in a power house which is operated by the customer himself or by an energy provider. In the case where the customer operates the power house, we will perform a technical analysis of the efficiency of the energy generation. In situations where steam and electricity are delivered by an external provider, we normally analyze the contracts in order to quantify the relation between consumption and price. In both cases (own powerhouse or provider) we often see a discontinuous relation between consumption and price. Turbines or generators can be operated in a certain range and deliver certain power. If the consumption goes beyond this range (because of additional demand or less consumption because of saving measures) the specific energy price will change stepwise.

An example: A customer wanted to increase the capacity of a plant, which was provided by its own power house. The proportional calculation of the energy demand with the capacity led to the investment in an additional gas turbine. We performed an energy efficiency project in parallel to the de-bottlenecking project. The result was that the plant with extended capacity could be provided with steam and electricity by the existing turbines. We had to design some modifications in the power house, but the savings in comparison with a new turbine were enormous.

Also the utility distribution, for instance of steam, has to be investigated: a predictive maintenance program for steam pipes and steam traps are just a few measures that can save a lot of energy, reduce CO_2 emissions and costs.

Some remarks concerning usage of water: normally the costs for water and therefore the driving forces for saving measures are lower than for steam or electricity (an exception is demineralized water or in particular ultra pure water in pharmaceutical plants). Nevertheless we analyze the consumption for instance of cooling water. In many cases there is no control valve on the cooling water side of heat exchangers. The hand valve instead is in position 'open'. Here the installation of a process control in some cases makes sense, especially for closed cooling water circuits with cooling towers. If the customer wants to realize a capacity extension then cooling water savings can help to avoid investment in the cooling tower system.

For compressed air or nitrogen, last but not least, we detect leaks in the pipe system or open valves by special ultrasonic leak detectors that detect the gas leakages. The test is not costly and can result in easy savings.

4.4.10
Milestone 2

At the end of the idea generation phase we come up with a complete list of generated measures. Here we reach the Milestone 2 of the project. The main sources for the ideas are, as described, the creativity of the operators, the knowledge of our

No	Title	Description	Affected utility
1	Optimize Dryer Operation/cycle time	Optimize Dryer Operation: Time Cycle, Temp. Requirements	Steam / Electricity
2	Distillation parameter optimization	Do you have constant reflux, no reflux, or stepwise increasing reflux?	Steam
3	Use of reduced amount of Solvent	Reduce amount of solvent acetonitrile	Solvent Reduction
4	Use of reduced amount of Solvent	Reduce amount of solvent xylene	Solvent Reduction
5	Reduction of Heat Losses due to better Insulation	Dryer RVD 120 -Overall: Reduce heat losses, insulation?	Steam / Hot Water
6	Provision of Interlock	Interlock in CHWS Line of Pump 505 to be provided .	Eletricity / Chilled Water.
7	Provision of Interlock	Interlock in CWS / CHWS Line of Condenser, and Dryer 120°C with Vessel Temp.	Eletricity / Chilled Water / Cooling Water.
8	Provision of Interlock	Interlock with CWS / CHWS in condenser of 831B with Oxa Temp.	Eletricity / Chilled Water.
9	Provision of Axuator Valve	Timer in N2 Line of Filter - 119°C - to be provided with actuator valve.	Nitrogen / Electricity
10	Provision of Interlock	In CW/CHW supply Line of condenser 825, valve to be contolled with Temp of boiling point (Oxa).	Eletricity / Chilled Water / Cooling Water.
11	Cyclization Process Time reduction	Reduction in Cyclization Time Cycle with Trial in PID	Electricity
12	Reduction of Pumping Cost by increasing ΔT	Increase in ΔT of cooling medium to reduced pumping cost.	cooling water / Electricity

Figure 4.11 Example for a list of potential measures from the idea generation phase.

unit operations experts and the expertise of the energy consultants. Figure 4.11 shows as an example an excerpt of such a list.

At this stage of the project the ideas were not yet evaluated nor prioritized. In this example we see a mixture of equipment optimization ideas (regarding dryer and distillation, points 1 and 2), of operational parameter changes (solvent concentration, points 3 and 4), good housekeeping (insulation, point 5), automation improvements (points 6 to 10) and other more speculative ideas (points 11 and 12).

In the Milestone 2 meeting the list of saving ideas will be discussed with the customer. Not all ideas can be evaluated in the next project phase. Therefore a pre-evaluation with the customer is necessary in order to reduce project costs. Normally some ideas coming from the operators are already in operation and can be neglected in the next phase. In some cases there are constraints, for example due to safety reasons that impede the implementation of measures. These will of course not be evaluated in the next phase. All these ideas are documented for later use.

All other ideas now will be evaluated in terms of technical feasibility and profitability. This is the task for the next phase of the project.

4.5 Idea Evaluation

4.5.1 Technical Feasibility

All potential measures for saving energy and reducing CO_2 emissions were documented during the idea generation phase in a list shown as an example in Figure 4.11. Some of them go into the idea evaluation phase after Milestone 2. The task during this phase is to assess the ideas regarding two different dimensions: the technical feasibility and the financial profitability of the ideas. This evaluation eventually leads to a prioritization of the potential measures and an implementation plan, which documents the planned actions for the next years on the plant or the site.

The technical feasibility is a soft or qualitative measure. Nevertheless we try to indicate a mark that describes the feasibility of an idea. This mark which lies between 1 (unfeasible) and 10 (feasible) expresses the result of a discussion between consultant, plant manager or operator and experts in unit operations. There are three general outcomes from this discussion in the project team: an idea can be feasible or not feasible or a further evaluation may be necessary. Some questions discussed in the team are:

- Is there a proven technology to implement the measure?
- Are there any reference cases where the measure was implemented successfully?
- Is the feasibility of the measure supported by appropriate calculations based on a simulation model of the plant?
- Will the operators be able to operate the plant after implementation of the measure?
- Are any additional studies required before the implementation?
- Does the effect of the measure justify the risk of implementation?
- Can offline experiments (for instance in the lab) minimize the risk of implementation?
- Are there any side effects that have to be considered?

Depending on the answers to these questions the team will rank a measure with a feasibility parameter between 1 and 10. This parameter will be used for a graphical comparison of all measures in the savings portfolio (see Section 4.5.3). If we need further investigations to determine the feasibility of a measure, then this will be documented accordingly.

4.5.2
Profitability

The profitability of an idea is a quantitative measure. We take the 'return on investment' (ROI) and compare it with the specific ROI target that our customer has defined for investments. A typical value for a target ROI is two to three years. Profitable measures show an ROI below the target ROI, measures with an ROI above the customer's target ROI are marked as not profitable. The ROI itself is calculated as the quotient of the investment for the implementation of a measure (in € or $) and energy saving gained by the measure (in € per year or $ per year). The investment will be estimated roughly in order to keep the effort low. That means we estimate the hardware costs and the customer contributes a so called 'multiplier' which takes into account costs for the implementation of the hardware in the plant. During the detailed design the calculation will be repeated with higher accuracy. The saving measured in € or $ per year will be calculated based on the data collected in the previous steps. It depends not only on the energy saving in kWh per year, but also on the energy price and the use of the plant. Therefore measures that are not profitable today can become profitable in the future when energy prices rise. Documentation of all measures and their profitability calculation is therefore a must.

Further investigation to estimate the investment or the savings derived from an individual measure must also be documented. This example may clarify the procedure:

- **Project title:** Use low pressure steam for the generation of electricity.
- **Current situation:** The plant produces low pressure steam in the range of 20–50 $t\,h^{-1}$ depending on the load of the reaction step. As a basis for the potential calculation is medium value of 30 $t\,h^{-1}$ is taken. For the time being the low pressure steam with 2.5 bar (absolute) is condensed in four air coolers and fed into the condensate vessel.
- **Project idea:** Use of low pressure steam to generate electricity in a saturated steam turbine is proposed. The electricity can be used directly in the plant.
- **Project description:** A turbine should use 30 $t\,h^{-1}$ low pressure steam to generate 2334 kW. The condensation of the flash steam can be achieved at 0.35 bar (absolute) by the existing air coolers (approved by the provider). More than 30 $t\,h^{-1}$ steam cannot be condensed by the existing air coolers so that 30 $t\,h^{-1}$ is the design point. Minimum load of the turbine is 13 $t\,h^{-1}$ low pressure steam which should always be guaranteed. Load variation should not be a problem (information by the provider).
- **Investment calculation:** A turbine is offered by a supplier for 1.1 M€ containing equipment as stated in the proposal (which is included in the documentation). The total investment is estimated with a multiplier of 3 to be 3.3 M€.
- **Project economics:** The savings are calculated to be 1.06 M€ per year and 7960 t per year CO_2. Investment and the savings result in a ROI of 3.1 years.
- **Project rating:** A detailed investigation of the project is necessary. Therefore the rating is B.

4.5.3 Savings Portfolio

We use the feasibility parameter (see Section 4.5.1) and the profitability expressed in years of ROI (see Section 4.5.2) to form a savings portfolio where we insert all considered saving measures. The portfolio is an easy tool to show and compare all executed assessments and to give a quick overview over all measures. A typical savings portfolio is shown in Figure 4.12.

On the *x*-axis we plot the technical feasibility of a measure. The differentiation between difficult and easy realization is the result of the discussion in the project

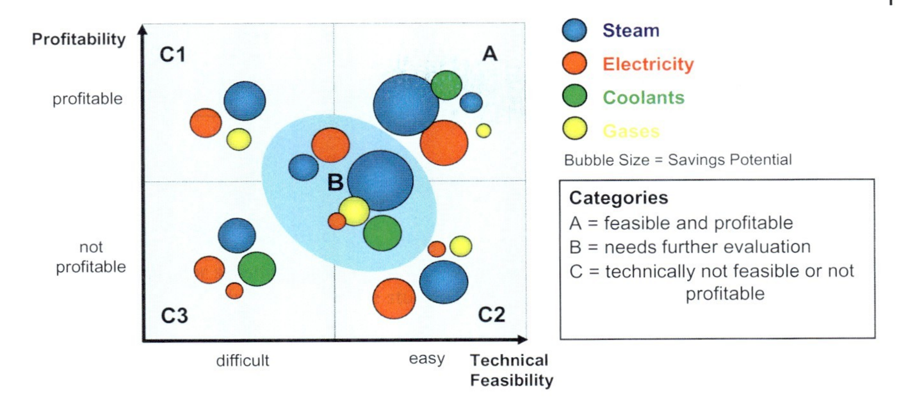

Figure 4.12 Typical savings portfolio.

team described in Section 4.5.1. On the y-axis we plot the profitability of a measure expressed as reciprocal years of ROI. The difference between profitable and not profitable is given by the investment policy of the customer and may also change with time. The limit is normally around three years. The color code describes the utility the measure is referred to (here: steam, electricity, coolants and gases). The bubble size represents the saving potential which was estimated during the evaluation phase.

As shown in Figure 4.12 we can classify the measures in five categories:

- **A measures:** These measures are technically feasible and profitable. Therefore they have a high priority for realization and will be included in the implementation plan (see Section 4.6.1).
- **B measures:** These measures are likely feasible and profitable. Further investigations and evaluations are needed in order to decide whether the measure will be included in the implementation list or not. It can for instance be necessary to perform experiments on the plant or in the lab to assess the feasibility of the idea, or a more detailed design of the measure is necessary in order to estimate the investment costs (and from that the profitability) with the sufficient accuracy.
- **C1 measures:** These measures are profitable but currently not feasible. Progress in technology can improve the feasibility in the future. The measure are documented, but not included in the implementation list.
- **C2 measures:** These measures are technically feasible but currently not profitable. Changes in energy prices or in the customer's investment policy can make them profitable in the future. The measures are documented, but not included in the implementation list.
- **C3 measure:** These measures are neither profitable nor feasible. The measures are documented, but not included in the implementation list.

Typical documentation list

Energy-Efficiency-Check for Process / Plant: Add new Project Sheet | Enumerate | Selected curre Euro | Selected customer:

Plant A / Site Europe 1

Active?	Project No.	Title	Unit/AKZ	Rating	Variable Cost Saving Potential	Fixed Cost Saving Potential	Total Investment	Payback Time	Carbon Dioxide Savings	Affected Product
[-]	[-]	[-]	[-]	[-]	[€/a]	[€/a]	[€]	[yr]	[T/a]	[-]
Y	1	Project 1	Unit/AKZ	N	0	0	0	n.a.	0	Product A
Y	2	Project 2	Unit/AKZ	N	0	0	0	n.a.	0	Product A
Y	3	Project 3	Unit/AKZ	N	0	0	0	n.a.	0	Product A
Y	4	Project 4	Unit/AKZ	N	0	0	0	n.a.	0	Product A
Y	5	Project 5	4W0213	?	8325	0	12000	1,44	63	Product A
Y	6	Project 6	5W0513	?	8325	0	12000	1,44	63	Product A
Y	7	Project 7	4W0215	C3	0	0	n.a.	n.a.	0	Product A
Y	8	Project 8	5W0115	C3	0	0	n.a.	n.a.	0	Product A
Y	9	Project 9	6K1119	C2	11100	0	48000	4,32	84	Product A
Y	10	Project 10	6K1101	A	50875	0	0	0,00	387	Product A
Y	11	Project 11	6K1403	A	10020	0	0	0,00	76	Product A
Y	12	Project 12	6K1423	A	6703	0	0	0,00	51	Product A
Y	13	Project 13	5K0207	A	6947	0	0	0,00	53	Product A
Y	14	Project 14	4K0307	A	7716	0	0	0,00	59	Product A
Y	15	Project 15	6K0514	A	42735	0	0	0,00	325	Product A
Y	16	Project 16	6K1119	A	9250	0	0	0,00	70	Product A
Y	17	Project 17	6K1102	A	57720	0	0	0,00	439	Product A
Y	18	Project 18	6K1102	A	61975	0	75000	1,21	471	Product A
Y	19	Project 19	6K1119/6K1102	A	95682	0	112000	1,17	703	Product A
Y	20	Project 20	7P0104/7P0105	?	0	0	n.a.	n.a.	0	Product A
Y	21	Project 21	Piping	?	0	0	n.a.	n.a.	0	Product A
Y	22	Project 22	7P1730, 7P1725	?	11777	0	n.a.	n.a.	131	Product A
Y	23	Project 23	4W0408	B	122640	0	n.a.	n.a.	832	Product A
Y	24	Project 24	Labor	B	0	0	0	n.a.	0	Product A
Y	25	Project 25	4W0410	N	0	0	0	n.a.	0	Product A
N	26	Project 26	00.00	B	434380	0	n.a.	n.a.	3423	Product A
Y	27	Project 27	6B1412	?	0	0	0	n.a.	0	Product A
Y	28	Project 28	Chiller A	?	0	0	n.a.	n.a.	0	Product A
Y	29	Project 29	7P0518, 7P0511	A	5499	0	0	0,00	61	Product A
Y	30	Project 30	-	B	0	0	n.a.	n.a.	0	Product A
Y	31	Project 31	6C0210, C60211	?	0	0	n.a.	n.a.	0	Product A
Y	32	Project 32	4W0408	C2	0	0	0	n.a.	0	Product A
Y	33	Project 33	7W0704	A	160950	0	135000	0,84	1223	Product A
Y	34	Project 34	7B0701	A	57350	0	n.a.	n.a.	436	Product A
N	35	Project 35	4W0215 / 5W0115	N	2708	0	n.a.	n.a.	0	Product A
Y	36	Project 36	D821	B	14083	0	n.a.	n.a.	107	Product A
N	37	Project 37	D821	N	0	0	4500	n.a.	0	Product A
N	38	Project 38	6W103/203/303/253	A	48563	0	n.a.	n.a.	158	Product A
Y	39	Project 39	6W0501/8	B	91390	0	0,00	0,00	695	Product A
Y	40	Project 40	6WX	B	70219	0	n.a.	n.a.	1492	Product A
N	41	Project 41	4W0213 / 5W0113	A	222000	0	n.a.	n.a.	1687	Product A
N	42	Project 42	6W103/203/303/253	A	683760	0	n.a.	n.a.	4640	Product A
Y	43	Project 43	7P0404/06/08	N	0	0	0,00	n.a.	0	Product A
Y	44	Project 44	Cooling Tower water	N	0	0	0,00	n.a.	0	Product A
Y	45	Project 45	6W501/6W508	A	783475	0	n.a.	n.a.	5954	Product A
N	46	Project 46	6W501/6W508	A	222000	0	108000	0,49	1687	Product A
Y	47	Project 47	6W501/6W508	A	24346	0	15000	0,62	270	Product A
Y	48	Project 48	4W0213/5W0113	A	364820	0	465000	1,27	1406	Product A
Y	49	Project 49	6W0802	A	101750	0	156000	1,53	0	Product A

Figure 4.13 Typical documentation sheet for project results.

The savings portfolio gives a quick view of the assessed measures, therefore it is valuable for discussions with the customer's representatives and for presentations. For detailed discussion of the measures and as documentation of the project results we prepare an excel list of all measures with all information available. A typical documentation sheet is shown in Figure 4.13.

For the plant A on the site 'Europe 1' the team found 49 ideas. (For the purpose of confidentiality, project titles and affected product have been erased.) The list shows for every project idea the equipment or unit, the rating of the idea, cost savings and necessary investment, ROI or payback time and the CO_2 savings. In this example we see some quick wins (savings without investment), profitable measures with investment (rated as A), some ideas that need further investigation (rated as B) and finally measures that are rated as C (not profitable and/or not feasible). For every project idea we have a detailed description and a profitability calculation.

The savings portfolio and the documentation list, together with detailed information about the proposed measures and their evaluation, form the final documentation of an energy efficiency project. The customer can stop here and manage the implementation of measures by himself. We do not recommend an interruption of the project at this point in time. In our experience a project at this stage has a strong momentum that should be used for a successful implementation of profitable measures in the plants. Therefore the project team should stay present in the plants, implement the measures following a jointly developed implementation plan, and work on an energy conscious culture in the plants.

4.6 Sustainable Implementation

4.6.1 Implementation Plan

In most projects we typically identify 20 to 40 measures per plant marked as 'A' or 'B' (see Section 4.5.3), that means they are likely profitable and feasible. B measures need further evaluation in order to finally categorize them as 'A' or 'C'. These investigations should be initiated now in dedicated projects, supported by members of the project team.

All 'A' measures will be then prioritized on the time scale. There are 'quick wins', measures without or with very low investment like change of operational parameters (see Section 4.4.4). These will be realized as soon as possible in order to derive the benefit rapidly. Other measures like process modifications (see Section 4.4.5) or additional equipment are more costly and have to be adapted to the customer's capital expenditure planning. The discussion on the customer's side supported by the consultant on the basis of the documentation sheet, will result in an implementation plan shown in Figure 4.14.

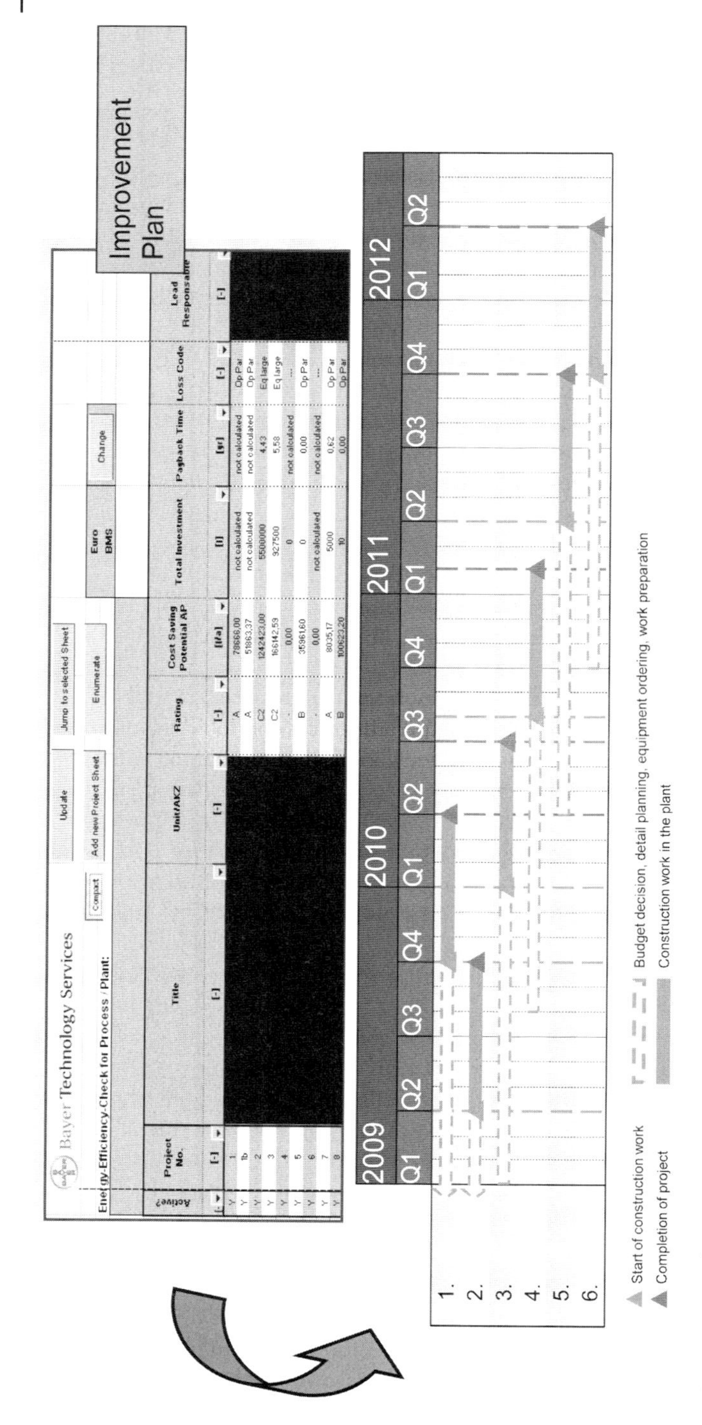

Figure 4.14 Implementation plan.

The implementation plan is a roadmap for energy relevant investments for the next years. Every measure from this plan is a project on its own and will be realized following the project approval and execution procedure of the customer. That means for instance basic design, cost estimation, prove of profitability, project application followed by the project approval. After approval the project will continue with a safety study, detail design, procurement and construction. The consultant should then participate in the start-up of the modified plant in order to support the operators with his knowledge of the designed modification.

It is self-evident but important to mention that the implementation of measures has to be decided by the customer; nevertheless the consultant will support this process by his input.

4.6.2 Monitoring and Controlling

In our experience specific energy consumption in a plant varies widely with time, as illustrated in in Figure 4.15.

The graph shows firstly that the specific energy consumption decreases with the load. That is not surprising because there are a number of energy consumers independent of the actual load, for instance trace heating, fixed refluxes in columns, or coolants that are not controlled. Secondly, we see that the energy consumption scatters in a wide range. Every dot in the graph is a measured energy consumption value at a certain point in time. There is often no reason why the consumption should differ from one point to the other or from one shift to another. Furthermore, there is a best practice curve of the energy consumption that is a function of the load. This curve represents the best values achieved in the plant. The goal now will be to reduce the inconsistencies in the consumption values (for instance on the average curve shown in the graph) and to move the real consumption as close to the best practice curve as possible. A prerequisite for

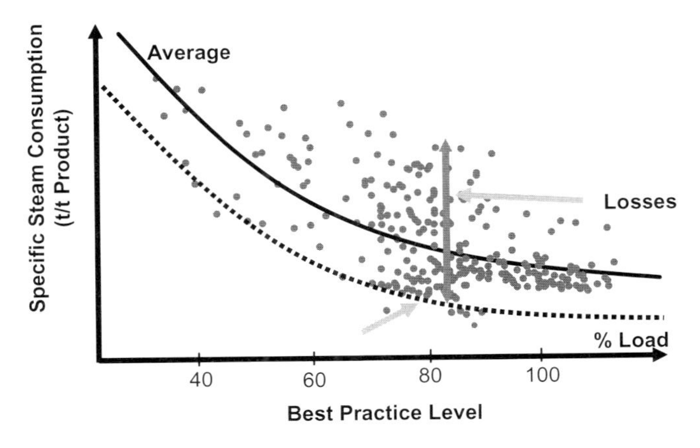

Figure 4.15 Specific energy consumption.

reaching the best practice curve is to give the operators the ability to influence the energy consumption by showing them the actual consumption in the process control system and to compare that value with the best practice.

In order to do so, we implement performance monitoring in the plant. The idea is to depict the energy consumption or energy influencing parameters like concentrations or mass streams online and quantify the difference from calculated or best practice values. Thereby the operators will be given access to the data regarding actual energy consumption of their plant and to react accordingly, when the actual energy consumption is higher than expected.

The following elements are necessary for an online performance monitoring system:

- A distributed control system (DCS).
- Measuring devices for utilities like steam or electricity or measuring devices for energy influencing parameters; alternatively, some parameters can be calculated from a model.
- A mass and energy balance calculation (see Section 4.3.5) for the generation of calculated specific energy consumption values.
- Documented best practice values for energy consumption.

Now the consultant has to discuss the design of the performance monitoring screen in the DCS with the plant staff. An example of an implemented monitoring is shown in Figure 4.16.

It is possible to illustrate the actual energy consumption qualitatively by a traffic light (red: highly above calculated value, yellow: above calculated value and green: near calculated value) or quantitatively by numbers (for instance expressed in kW or $ per hour above calculated value) or in form of a speedometer. The plant representatives and the consultant will discuss and design the interface of the DCS screen. Acceptance by the operators is crucial for the success of the monitoring tool. However, for reporting reasons (see Section 4.6.3) it is important to use common definitions of the energy consumption and its unit for the whole company ('$ per year' or 'MWh primary energy per year' or 'emitted tons of CO_2 per year'). By doing so, it is possible to condense the consumption values for plant, site, business unit and company.

4.6.3
Reporting and Target Setting

A common definition and a common measurement of the energy consumption in a company open up the opportunity of consolidating the gathered energy information within the company from the plant level up to company level. That can be used in two directions:

Firstly for reporting: we recommend energy consumption reporting for all operational units of the company. This means that all hierarchy levels are provided with given consumption information for their units. The plant manager will get

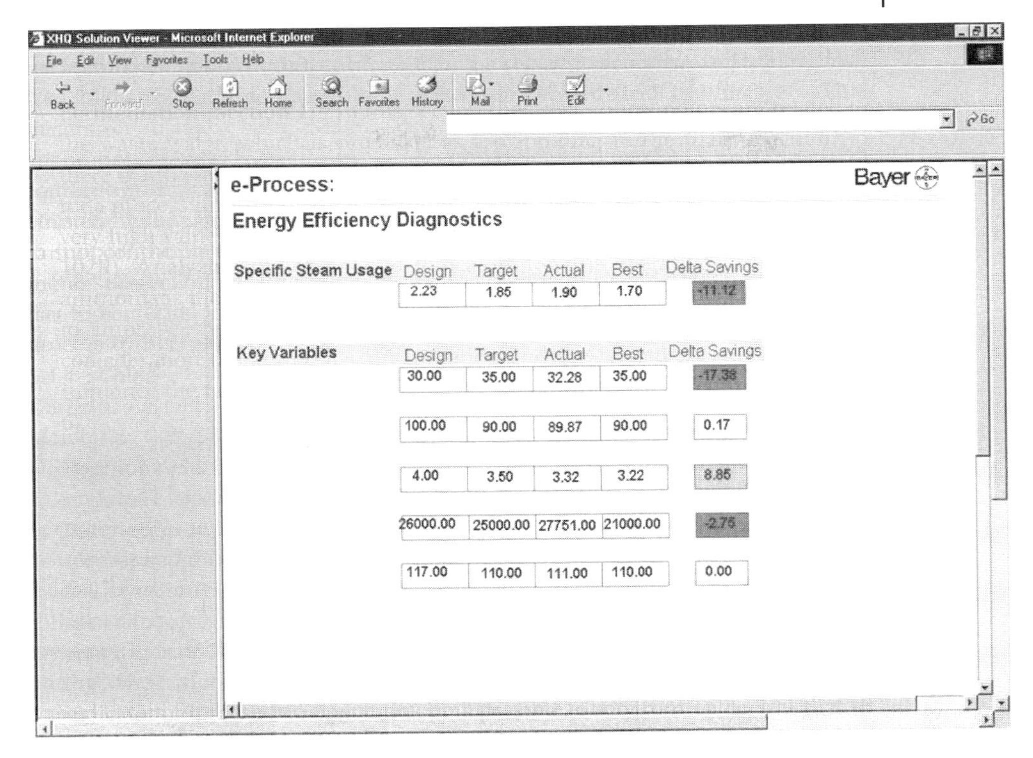

Figure 4.16 Performance monitoring screenshot from DCS.

the numbers for his plant, the site manager for the whole site, the business unit head for his business unit and the CEO finally receives a consumption report for the operations unit of the company. This reporting is available on a daily, weekly and monthly basis. It is possible to provide absolute and specific energy consumption data.

Secondly, for the purpose of target setting: the reporting tool provides unbiased data about the energy consumption in all units of the company, gathered by a common and agreed method. Therefore it can serve as the basis to formulate yearly energy consumption targets for every specific unit and to control the achievement of the targets. Plant managers, site managers or business unit heads are enabled to discuss energy consumption data with their supervisors or to control the effect of efficiency measures with the help of the reporting tool. The result is a transparent target discussion in the company on the basis of neutral numbers.

It is a disadvantage of the reporting tool described above that the energy consumption of different productions is not comparable. Obviously a bulk chemical production normally needs less energy per kilogram of product than chlorine electrolysis. The reporting tool can be used to set targets for the plant managers of the bulk chemicals and the chlorine plant, but it cannot decide which of the two plants is more energy efficient. In order to compare different production facilities we expanded the concept of energy reporting by defining a comparative scale. This concept will be presented in Section 4.6.4.

4.6.4 Energy Loss Cascade

The idea of the energy loss cascade is to compare the actual energy consumption of a production plant with a Plant Energy Optimum defined for that specific plant. The Plant Energy Optimum reflects the minimal energy consumption in kWh primary energy per ton of product which is achievable for that plant at an optimal configuration. Figure 4.17 shows the principle of the cascade. It was developed for Bayer Material Science under the name STRUCTese.

The difference between the minimum and the actual consumption is divided in different loss categories (in our standard approach we consider 10 loss categories). These categories reflect static losses and time dependent, dynamic losses.

Static losses are dependent on the asset configuration and cannot be influenced in the daily business. Examples for potential savings in this category are the implementation of heat integration (for instance preheater) or the replacement of not optimal equipment types.

Dynamic losses include different effects like partial load or a specific product mix which may lead to higher specific energy consumption. All other non-specific losses are summarized under the category 'suboptimal operation'. Examples for potential savings in this category are alternative cleaning operations or increased product concentrations, which both can lead to significantly reduced steam consumption.

Energy saving targets will now be presented to different groups of employees. The static losses can be influenced by investment projects initiated and approved by the upper or middle management. Marketing and Sales department is respon-

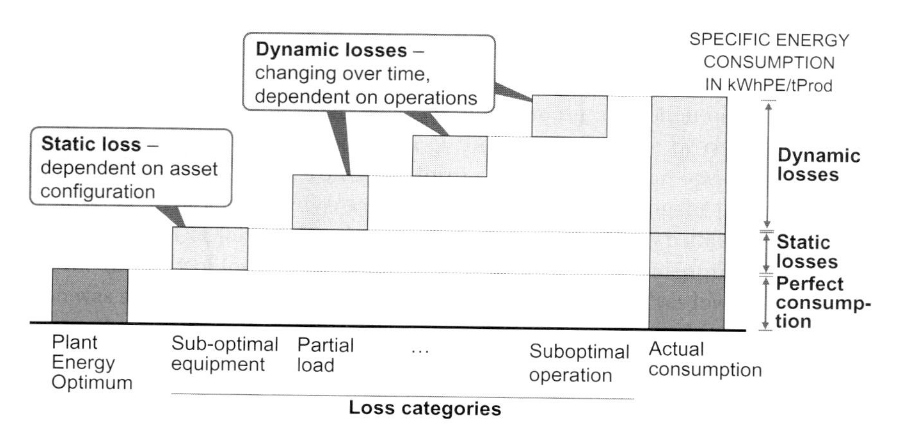

Figure 4.17 Energy loss cascade (STRUCTese).

sible for the reduction of partial load losses. The plant manager will have the target to reduce operational losses. All these groups are committed to saving targets in different loss categories.

Furthermore the energy loss cascade makes it possible to compare the energy efficiency of production plants for different products. The absolute or relative difference between the actual consumption and Plant Energy Optimum is a measure of the absolute energy efficiency of a plant. Plants with high energy awareness will produce near their optimum while plants, which did not focus on energy efficiency in the past, will show a larger difference between the actual and the optimum consumption.

Company-wide definitions are necessary to determine the Plant Energy Optimum and the different loss categories. It is crucial for the acceptance of the program that the values and categories are defined in a uniform way for every plant and the whole company. In a handbook all these definitions and calculation methods were documented in order to achieve a consistent and comprehensive approach.

4.6.5 Energy Management System and Benchmarking

All the measures for sustainable energy efficiency presented above are integrated in a comprehensive energy management tool. It consists of the following core components:

- **Energy loss cascade:** Analysis of the energy efficiency, with technological, operational and production parameters being assigned to specific energy use to enable energy losses to be precisely pinpointed.
- **Real-Time Energy Efficiency Monitor:** Monitoring system which enables energy efficiency to be increased through operational intervention.
- **Energy Efficiency** Key Performance Indicators **(**KPI**):** Quantified energy efficiency indicators that represent efficiency achievements and their temporal development.
- **Energy Efficiency Benchmark:** Standardization of the energy efficiency values and KPIs for the purposes of cross-plant comparability.
- **Sustained Energy Efficiency Improvements:** Target-setting and sustainable tracking of both energy efficiency and the implementation of measures to increase energy efficiency and reduce climate impact.

The management system follows a holistic approach to analyze, evaluate and document technological, production-related and operational changes in energy efficiency, with the aim of facilitating the operation of production processes with an optimal level of efficiency and its continuous and sustainable improvement. The system helps operating personnel to operate the facility in an optimal way and management to develop and implement the best possible production strategies.

The implementation of a management system reduces energy use per metric ton of product. Greenhouse gas (GHG) emissions are lowered and thereby the carbon footprint reduced. Sustainable increases in energy efficiency are assured through KPI-controlled objectives, data aggregation and internal and inter-plant benchmarking. To this end, parameters that have significant impact on energy use are identified, continuously monitored and optimized. In addition, the system provides economic benefits by reducing production costs.

4.6.6
Energy Awareness in Plants

We mentioned before that one of the main targets of an energy savings project is the sustainability and permanency of the savings. Therefore, it is important that the implemented measures will be accepted by the plant staff. We have to avoid for instance that process control loops implemented during a project will be operated manually after a while or that the performance monitoring implemented in the DCS will lose the attention of the plant staff. In order to assure this sustainability it is of high importance to accompany an energy project by a change project that assures that the operators take an active part in the process. Energy awareness in the plants is a key for sustainable savings. This task has been discussed in more detail in Chapter 3; some highlights will be noted here.

One is the participation of the plant staff in the energy savings process. We already mentioned that ideas brought in by the operators have a high value and should be implemented if possible and feasible. Measures on the basis of operators' ideas will not lose their attention.

Energy has to be integrated into the daily workflow in the plant. A daily energy protocol, created by our reporting tool (see Section 4.6.3), shows the 24-hour average energy consumption of selected equipment compared with the target level as defined by the unit supervision. It is a tool for the unit supervision to push for improvements in every morning meeting and keep energy saving in the operators mind.

Awards like 'idea of the week' can increase the motivation of the operators. We see however that the acceptance of such measures depends very much on the local attitudes and cannot be rolled out globally. An adjustment that takes into account the local mentality is necessary.

Training of operators is necessary to assure that measures will be understood and included into the daily work adequately. An important task of the consultant in this phase is to develop a training concept and training documents which describe the implemented technical measures. The consultant will help the nominated representative of the customer to take over the lessons following the 'train the trainer' concept.

Creation of energy awareness is, like every change process, a very long-term process. Therefore, the plant manager and the consultants have to accompany the process for at least 1 to 2 years permanently. The continuous realization of measures following the implementation plan (see Section 4.6.1) is a good tool to maintain energy awareness on the plant and to keep energy savings in mind of the operators.

4.6.7
Repeated Checks

From the perspective of the customer it makes sense to repeat the energy project every 3 to 5 years. There are different reasons to do that:

- Energy prices change (normally rise) and former C measure from our savings portfolio (see Section 4.5.3) can become profitable A measures considering the new prices.
- Plants will be modified and modernized, new equipment for higher efficiency or higher capacity will be installed; the energy efficiency of the modified process is to be checked.
- The specific energy consumption will decrease with time if the energy savings project was successful; this development has to be proved by a repeated energy check.

The effort for a repeated energy check will be of course much lower than for the initial check. All the measures were documented during the energy savings project, the data now only has to be reassessed. New equipment and its interaction with the plant will be considered. The condition of the plant (for instance insulation or possible air or nitrogen leaks) will be checked. The actual process parameters and the status of the control loops will be compared with the values of the initial energy check under consideration of the implemented measures.

Since the effort is low, the profitability of a repeated energy check can normally be easily justified.

4.7
Case Study: The Bayer Climate Check

4.7.1
Situation before the Bayer Climate Check

In the last 20 years Bayer, like other chemical companies, has shown a strong commitment to reduce the GHG emissions of its production facilities. As a result, between 1990 and 2007 the company managed to reduce its direct and indirect GHG emissions by 37.2% or 4.5 million tons of CO_2 equivalents. The total energy consumption of the Bayer group in 2007 equaled 91.7 Petajoule (that means 25.5 Terawatthours).

In order to continue this development the board decided to initiate the Bayer Climate Program in 2007. This program is a long-term engagement of the company.

4.7.2
Goal and Concept of the Bayer Climate Program

The Bayer Climate Program tackles two issues: setting ambitious targets for the own production and development of climate friendly products and production processes.

Based on the GHG emissions in 2005 Bayer MaterialScience plans to reduce the specific GHG emissions per ton of product by 25% until 2020. At Bayer HealthCare and Bayer CropScience the target reductions amount to 5% and 15% of the absolute emissions, respectively. For these targets, the direct and indirect emissions of GHG were incorporated.

In order to achieve these goals, Bayer initiated (among other measures) the Bayer Climate Check as one of the lighthouse projects in the climate program. Within the Bayer Climate Check, which was developed by Bayer Technology Services, we systematically check in a detailed and comprehensive approach the energy efficiency of more than 100 plants at Bayer sites worldwide. These plants are responsible for more than 85% of the total energy consumption and the GHG emissions of all Bayer manufacturing plants. Above that, the new instrument is used to assess the climate impact of new investments that enable the sustainability targets to be achieved.

4.7.3
Realization and Results

The Bayer Climate Check follows the systematic approach described in the first parts of this chapter (Chapter 4).

It incorporates two essential elements, that is, quantification of the climate relevance of the manufactured product by means of the Climate Footprint and identification of savings potential through an 'Energy Efficiency Check'. The value of Bayer Climate Check derives from combining sustainability goals with the corporate-wide motivation to increase its efficiency, reduce costs, and implement innovative solutions.

During the Energy Efficiency Check many potential measures to reduce energy consumption as well as CO_2 emissions are identified in a structured and holistic approach. This involves optimization of utilities' consumption such as electricity, steam, cooling water, chilled water, fuels and compressed air as well as usage of raw materials and auxiliary materials. The energy efficiency of plant facilities, energy generation units and buildings are analyzed in detail.

The project was initiated at the end of 2007 with a pilot phase. In this phase the energy efficiency of four plants in the Bayer subgroups MaterialScience, HealthCare and CropScience was evaluated. By the end of 2009 Bayer Technology Services has checked the energy efficiency of more than 100 plants in the Bayer sites worldwide. The climate checks performed indicated an energy saving potential through increased energy efficiency of more than 10% (A and B measures).

Two examples illustrate the effectiveness of the Bayer Climate Check as an instrument for the identification of energy and CO_2 reduction potentials. We describe how the approach was applied and which measures have already been implemented in the plants.

Example 1 is a production plant for polymer intermediates on a German Bayer site. We began with a detailed analysis of the actual energy consumption. Then we moderated brainstorming sessions with the plant staff and unit operation and equipment experts from Bayer Technology Services. In these sessions the team elaborated an extensive list of saving ideas. The saving potential of likely profitable and feasible measures identified during the Climate Check was 12% of the total energy costs. The proposed measures were divided between those without investment (so-called quick wins) and those where an investment is necessary. The following measures, which were implemented in the short term, resulted in significant savings shortly after implementation:

- The process control strategies of different big columns, for instance the reflux ratios in partial load operation, were optimized by process simulations and plant experiments. Thereby the energy consumption was reduced significantly.
- The energy consuming drive of a big pump was optimized by implementation of a frequency control. Some permanently running redundant pumps were switched off.
- In a filtration apparatus the temperature of a scrubbing solution was reduced. Because the scrubbing liquid had to be heated with external energy, the necessary energy for heating could be reduced.
- The optimization of heated pipes and heat traces resulted in a reduction of energy losses and therefore in a decreased steam consumption.
- The implementation of an innovative monitoring tool for the actual value of the energy consumption resulted in a higher energy consciousness of the operators and a behavioral change in the plant. This change was a main reason for a sustainable reduction of the energy consumption in the plant. Because of the success of the monitoring tool we plan the transfer of the monitoring tool to other plants.

In 2005 the plant consumed for its yearly production 40 647 MWh of electric energy and 320 134 MWh of thermal energy. After realization of the Bayer Climate Check the plant consumed 36 576 MWh of electric and 281 185 MWh of thermal energy in 2007, and that at a comparable production capacity. The year 2006 is not useful for comparison, because the production capacity differed considerably. Taking into account the respective production volumes the savings from 2005 to 2007 sum up to 10% of electric and to 12% of thermal energy. These savings relate to cost savings of 900 000€. Supported by the company-wide energy efficiency management program other measures elaborated during the Climate Check are being implemented. Additional savings of 200 000€ are expected. The energy supply comes from a coal-fired cogeneration power plant.

Example 2 is a production plant for basic materials for the plastic industry, also located in Germany. The saving potential identified during the Bayer Climate Check was 26% of the energy consumption. We found, for instance, saving potentials in the steam distribution net. Large amounts of excess steam, generated from waste heat streams, could thereby be exported into the steam net of the energy provider. As another example the analysis showed the saving potential of a predictive maintenance: if for instance defect steam of cooling water valves as well as steam traps are detected and changed betimes then the waste of energy will be minimized. In 2007 the plant exported 24 979 MWh of thermal energy in the form of steam into the steam net of the provider. After quick implementation of the proposed measures this amount reached 39 991 MWh in 2008. This means an increase of 60%. The customer invested 50 000€ for the implementation of the measures. Cost savings in the first year after implementation were 290 000€ per year. The customer plans to implement further measures supported by the company-wide energy efficiency management program. The plant is provided with energy by a gas and steam power station.

In the framework of the Bayer Climate Program the Bayer Climate Check was established as an important tool for the identification of potentials for energy saving and CO_2 emission reduction. The application of the Bayer Climate Check revealed significant saving potentials which have already been partly implemented. The examples show that quick win measures can result in significant savings at low investment costs. The implementation of further measures is planned and will be supported by a company-wide energy efficiency management program. The Bayer Climate Check is therefore an important element for the achievement of the ambitious climate related targets of Bayer.

References

1 Modig, K., and Johansson, O. (1997) *Six Sigma Guidebook*, TW Tryckeri I Ludvika, Ludvika.

2 Eckes, G. (2001) *The Six Sigma Revolution*, John Wiley & Sons, Inc., New York.

3 Jupke, A., Leimkühler, H.J., and Wolf, M. (2005) 85 Ideen zum Sparen – Energie-Audit identifiziert Energiesparmaßnahmen bei der Kautschuk-Produktion. *Energy*, **1**, 16–17.

4 Linnhoff, B., and Hindmarsh, E. (1983) The pinch design method for heat exchanger networks. *Chemical Engineering*, **38**, 745–763.

5
Sustainable Chemical Process Design

Rafiqul Gani, Henrique A. Matos and Ana Isabel Cerqueira de Sousa Gouveia Carvalho

5.1
Introduction

Economic and industrial activities are continuously increasing, which means that energy consumption, depletion of raw materials and the environmental impact are also growing in their importance to modern society. One of the major concerns, related to the environmental issues, is the increase in CO_2 emissions. The substantial increase of CO_2 concentration in the atmosphere may be attributed as the cause of the greenhouse effect, which in turn, may be responsible for changes all over the world that might not be reversible. Due to these factors the new and the old chemicals-based industries need to maintain a balance between the negative impacts caused by their activity and the benefits to human life provided by their production. New alternatives need to be designed for batch and continuous processes in order to achieve this equilibrium. The object of this chapter is to describe a systematic methodology for sustainable process design (this means less energy consumption, less CO_2 emissions, etc) and the *SustainPro* software, through which this methodology may be applied to generate, screen and then identify sustainable alternatives in any chemical process by locating the operational, environmental, economical, and safety related bottlenecks inherent in the process.

5.2
Definition of Concepts

5.2.1
Process Retrofit

As defined in [1], process retrofitting is the redesign of a chemical process to find new configuration and operating parameters that will adapt the plant to changing conditions to maintain its optimal performance. Retrofitting is an optimization of real plant that includes not only operating variables but also structural characteristics related to the process topology. It is also similar to the design of new plants

Managing CO_2 Emissions in the Chemical Industry. Edited by Leimkühler

ISBN: 978-3-527-32659-4

since it includes both a systematic procedure to develop process alternatives and a procedure to select the optimal configuration. The main objects of process retrofits are to increase the production capacity, to efficiently process the raw materials, to reduce environmental impact, to increase the safety of the process, and to reduce operating costs.

In recent decades different methodologies have been proposed in order to determine the retrofit potential of a chemical process with respect to improvements in cost-efficiency. Some methodologies using heuristic rules for the generation of the new design alternatives have been proposed [2]. Other methodologies based on mathematical concepts and optimization methods, such as mixed integer nonlinear programming (MINLP) have also been proposed by [3] and [4]. Methodologies based on the resynthesis of the entire process by incorporating operating units with enhanced performance attributes have been proposed [5]. Moreover, Lange [6] has proposed a methodology that directly relates the process design alternatives to improvements in the sustainability of the processes, and claims that the optimal solution is a trade-off between the different performance criteria, and therefore, the improvement of sustainability in a process is a balancing of the three areas that most influence it, such as, the environmental, the society and the economical aspects of the process.

5.2.2
Sustainability

The concept of sustainability first emerged in 1970, but it only became an important issue in the modern society with the Brundtland Report, (1987), in which sustainable development is defined as:

> 'Sustainable development is a development that meets the needs of the present without compromising the ability of future generations to meet their own needs'.

The impact of industry on sustainability can be summarized in terms of three criteria – environmental responsibility, economic return (wealth creation), and social development. For industry to guide its activities towards greater sustainability, more engineers need to have the tools to assess the operations with which they are concerned.

The Institute of Chemical Engineers defined a set of sustainability metrics [7]. These metrics will help engineers to address the issue of sustainable development. They will also enable companies to set targets and develop standards for internal benchmarking, and to monitor progress year-on-year.

5.2.3
Safety

It is required that the safety of a process plant fulfills a certain required level, due to general legal requirements, the company image, as well as economic reasons, since an unsafe plant cannot be profitable on account of losses of production and capital. Therefore, safety should influence design decisions from the first moment of the design project. The safety evaluation is usually done by safety analysis methods. Safety analysis is a systematic examination of the structure and functions of the process system aimed at identifying potential accident contributors, evaluating the risk presented by them and finding risk-reducing measures.

An inherently safer design is one that avoids hazards instead of controlling them, particularly by removing or reducing the amount of hazardous material in the plant or the number of hazardous operation. Heikkilä [8] developed a method to study the inherently safety of a given process.

5.3 Methodology for Sustainable Process Design

A methodology to generate new sustainable design alternatives has been developed and can be applied in any chemical and biochemical industrial process.

This tool is particularly useful for retrofitting studies. The methodology identifies the critical points in the process related to energy consumption, raw materials consumption, water consumption, environmental impact reduction and safety. Once those points are identified, sustainable design targets are set and new process (retrofit) alternatives that match the targets are generated and evaluated to identify the optimal more sustainable process design.

The activity–flow diagram for this methodology is shown in Figure 5.1. There are two working process options: continuous operation and batch operation.

5.3.1 Methodology – Continuous Mode

5.3.1.1 Step 1: Data Collection

The object of this step is to collect the mass and energy data corresponding to the steady-state operation of the process under investigation. This data can be provided as steady-state simulation results (for example, from commercial simulators such as *Pro II, Aspen, HYSYS,* or ICAS simulator [9]) or operational data collected from the plant. Therefore, the methodology can be applied to any chemical process, for which steady-state operational data is available. Also the cost data related to prices of chemicals, utilities, etc., need to be specified in this step.

5.3.1.2 Step 2: Flowsheet Decomposition

The object of this step is to identify all the mass and energy flow-paths in the process by decomposing the process into open- and closed- paths for each compound. The closed-paths (CP) represent process recycles from which specific amounts of materials (chemicals) and energy do not leave the process. In other words, they are the flow paths, which start and end in the same unit of the process. An open-path (OP) represents a path for material and energy with an entry- point to the process and an exit-point from the process. The entrance of the compound

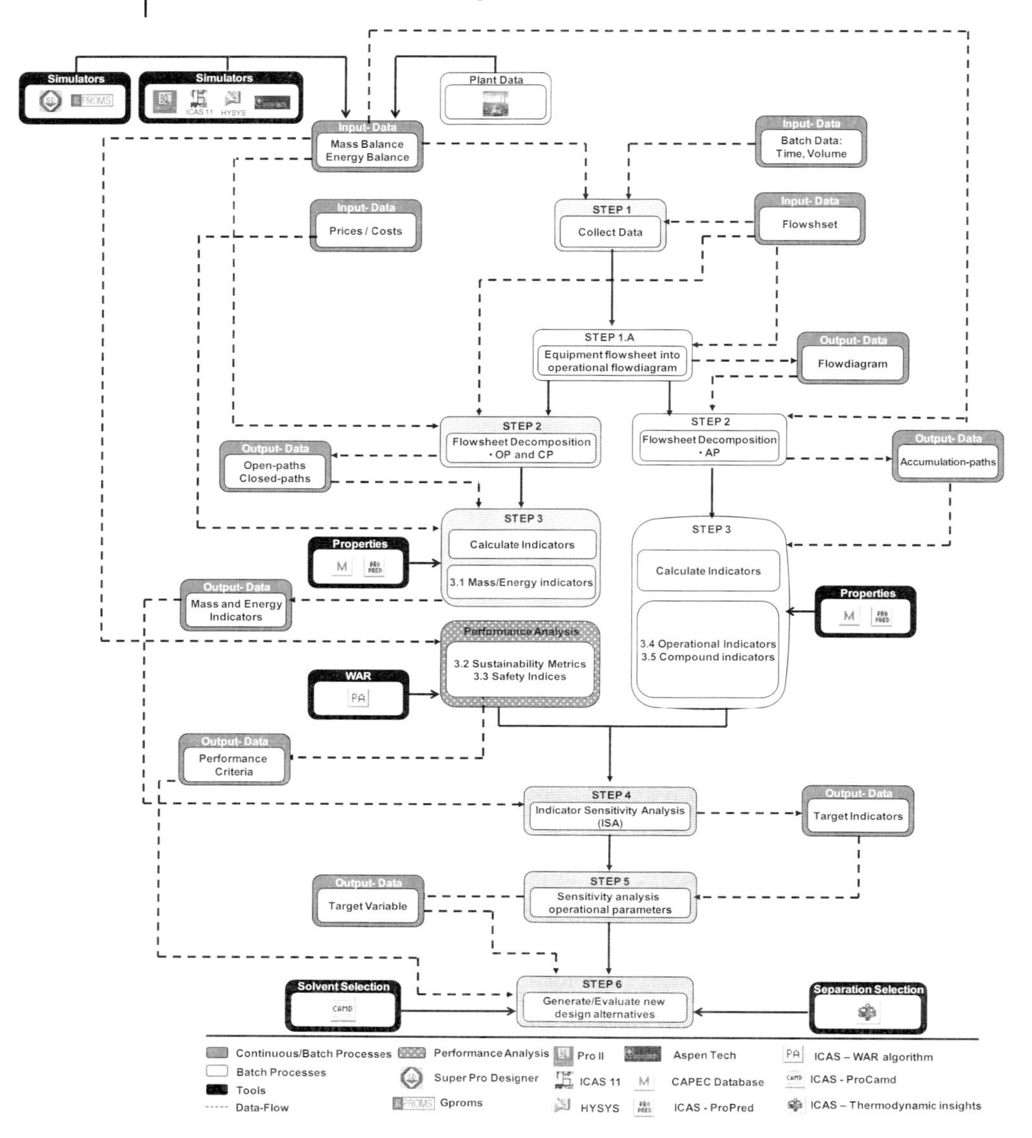

Figure 5.1 Activity-flow of the indicator-based methodology.

in the system can be due to its entrance through a feed stream or by its production in a reactor unit. The exit of the respective compound can be due to a 'demand' (exit) stream or by its reaction in a reactor unit. Figure 5.2 provides an illustration of the open- and closed-paths.

In Figure 5.2 one open-path, which includes streams {1,2,3,4,5} and a closed-path involving streams {2,3,4,6} are highlighted. More details can be found in [10],

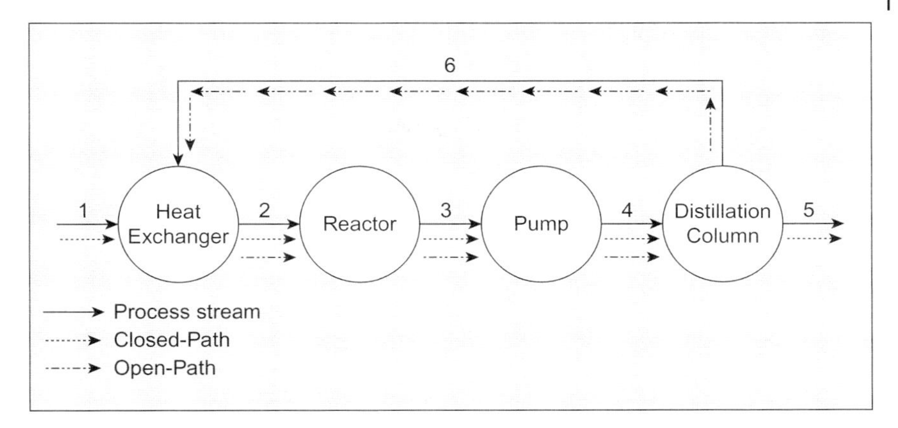

Figure 5.2 Example of closed- and open-paths.

where the algorithms for flowsheet decomposition to identify the open- and closed-paths are explained.

5.3.1.3 Step 3: Calculation of Indicators

In this step the mass/energy indicators, the safety index and the sustainability metrics are calculated.

Calculation of Mass and Energy Indicators There are five mass indicators and three energy indicators. A brief description of each of these indicators is given below. The equations to calculate the indicators can be found in [10].

Material Value Added (MVA) This indicator gives the value added between the entrance and the exit of all compounds present in the system. In other words, this means the value generated between the start and the end point of the path. Consequently, this indicator is only applied to open-paths. To calculate this indicator it is necessary to know the purchase price or the costs related to the production of a given compound as well as its sale price. This indicator is calculated as the difference between the sale price and the purchase price of the corresponding chemical in each open-path. *Negative values of this indicator indicate that the compound has lost its value in this open-path and consequently point to potentials for improvements. Positive values of this indicator show that the compound has win value through the process and consequently that path is a benefit for the process.* The indicator value is given in terms of monetary units per year, which indicates the money lost across that open-path.

Two examples are given to assist understanding:

1) **MVA is negative for a raw material:** This means that a raw material looses its value as it passes through the unit operations in the corresponding open-path and exits the process. That is, the chemical (raw material) at exit has a lower

value corresponding to its value at the entrance of the process. This will point to design alternatives, such as recycling of the raw material, improvements in separation processes, improvements in the reaction conditions, etc., that will make the MVA less negative for the targeted raw material.

2) **MVA is negative for the main product:** This means that the main product is being wasted in the corresponding open-path and the separation unit responsible for its recovery needs to be improved in order to improve the value of this product while reducing the waste.

The selected MVA indicator is the one having the biggest monetary loss and consequently is the one that should be selected in a further methodology step.

Energy and Waste Cost (EWC) This indicator is applied to both open- and closed-paths. The value of EWC represents the maximum theoretical amount of energy that can be saved in each path within the process and consequently this value is always positive. *High values of this indicator point to high consumption of energy and consequently changes in the associated paths should be considered in order to reduce this indicator value.* This can be done in two ways, reducing the path flowrate or changing the conditions in order to decrease the heat duties. The EWC value is given in terms of monetary units per year.

Two examples are given to assist understanding:

1) **EWC is high for a solvent (a chemical found in the system) in a closed-path:** This means that this solvent in a recycle-loop is consuming a lot of energy. The unit operation within the closed-path might be replaced in order to optimize the heat transfer and consequently reduced the energy consumption. Another option is to reduce the solvent flowrate by selecting a new solvent, if it does not compromise the separation efficiency.

2) **EWC is high for an inert compound in an open-path:** This means that the inert compound is entering into the process and consuming a lot of energy. This might be reduced by inserting a separation unit to remove the inert compound before it enters the process. The other option is to improve the post-reaction separations related to the removal of the inert compound.

The selected EWC is the one which has the highest value because it then indicates the biggest monetary loss and consequently it should be selected as the target for further improvement.

Reaction Quality (RQ) This indicator measures the influence of a given path with respect to process productivity. This indicator is applied to both closed- and open-paths. *Positive values of RQ indicate a benefit of this path with respect to process productivity, while, negative values point to a decrease in the process productivity.* This indicator helps in the selection of MVA and EWC indicators. If a path shows a very negative MVA or a high EWC, RQ is used to check if that path is a benefit for the process (positive values) or a bottleneck (negative values).

Two examples are given to assist understanding:

1) **EWC value is high for a raw material in a closed-path and RQ positive:** This means that a raw material in a recycle is consuming a lot of energy even though RQ is positive. Consequently, another EWC should be considered as target for improvement.

2) **EWC value is high for one of the products in a closed-path and RQ negative:** This means that the corresponding product in a recycle is consuming a lot of energy. RQ is negative, which indicates that the recycling of the product is not advantageous for the process. Consequently, EWC should be considered as target for improvement.

The RQ indicators having negative values should be selected for further investigation.

Accumulation Factor (AF) This indicator determines the accumulative behavior of the compounds in the closed-paths. This corresponds to the amount that is recycled relative to the input to the process and not the inventory. This indicator can only have positive values. *High values of this indicator point to high potentials for improvements, due to the high accumulation.*

Two examples are given to assist understanding:

1) **AF is high for a solvent in a closed-path:** This means that the solvent has a large accumulation within in the system. If the analysis is being done for a new process, it indicates that the selection of another solvent which requires a lower flowrate and consequently smaller equipments could be a good option.

2) **AF is high for a raw material and for by-products in a closed-path:** This means that the raw material has a large accumulation and consequently, the by-products might be responsible for the raw material accumulation. Therefore, these by-products should be removed from the recycle by, for example, improving the separations, inserting new separations, or, improving the reaction conditions.

The AF indicator having high values should be selected for further investigation.

Total Value Added (TVA) This indicator describes the economic influence of a compound in a given path. TVA joins two of the previous indicators EWC and MVA and the following equation is used to calculate its value:

$$TVA = MVA - EWC \tag{5.1}$$

Negative values of this indicator point to high potential for improvements in terms of decrease in the variable costs. The values of this indicator should be analyzed carefully, because MVA can have a high positive value and consequently hide the problems in EWC. Therefore, even if the TVA value does not present a very negative value, the values of EWC and MVA should be analyzed separately in order to confirm that they really do not show any problem.

Energy Accumulation Factor (EAF) This indicator determines the accumulative behavior of energy in the closed-paths. This means, that high recycles of energy correspond to high energy integration. *Low values of this indicator highlight the potential for saving energy consumption in the system using heat integration.*

One example is given to assist understanding:

EAF is low for an energy closed-path: This means that a lot of energy is being wasted and consequently heat integration might be considered inside the loop in order to reduce the energy consumption.

The EAF indicators having low values should be selected for further investigation.

Total Demand Cost (TDC) This indicator is applied only to open-paths and traces the energy flows across the process. For each *energy demand* in the process, the sum of all demand costs, which pass through it, are calculated. *High values of this indicator identify the demands that consume the largest values of energy, so these are the process parts, which are more adapted to heat integration.*

One example is given to assist understanding:

TDC is high for an energy open-path: This means that a lot of energy is being released in that stream/unit and consequently units using heating systems might be integrated within this open-path to reduce the energy consumption.

The indicators, presenting high values of TDC, should be selected in a further methodology step.

Calculation of Safety Indices The safety of the process is another important parameter that should be taken into account. In this methodology, the inherent safety index, developed by Heikkilä [8], has been implemented. In order to determine the inherent safety index, values for a set of sub-indices need to be calculated. These sub-indices can be divided into two groups:

1) **Chemical inherent safety:** Heat of the main reaction, heat of the side reaction, chemical interactions, flammability, explosiveness, toxicity, corrosivity.
2) **Process inherent safety:** Inventory, process temperature, process pressure, process layout.

Heikkilä [8] defined a scale of scores for each sub-index. These scales are based on the values of safety parameters, such as the explosiveness, the toxicity, the pressure of the process and so on. The sum of all the sub-index scores is the Inherent Safety Index value; this parameter has the maximum value of 53. Note that the higher the value of this Inherent Safety Index, the more unsafe is the process. *Therefore, the aim in all the design alternatives is to try to reduce the value of this index, if possible.*

Calculation of Sustainability Metrics In this methodology, the sustainability metrics defined by the Institution of Chemical Engineers (see [7]) have been used. Azapagic has defined 49 metrics divided into three main areas: environmental, social and economic. Out of the 49 defined metrics, the methodology uses 23 of them. *The use of the sustainability metrics follows the simple rule that the lower the value of the metric the more sustainable is the process.* A lower value of the metric indicates that either the impact of the process is less or the output of the process is more.

The metrics take into consideration energy consumption, material consumption or water consumption per kg of product or per value added, economic factors such as profit, etc.

For the metrics related to environmental impact , instead of using the definition of Azapagic *et al.* [7] the definition proposed by Cabezas [11] has been used. Cabezas *et al.* proposed the waste reduction algorithm (WAR) in order to calculate the environmental impacts from a chemical process. This algorithm has been implemented as part of the indicator based methodology. To calculate these metrics, the flowrates for each compound coming into the process and leaving the process are needed as known information, that is, the steady-state process stream (simulation) data is needed.

Summarizing, the indicators are applied to the entire set of open- and closed-paths. With their values the critical points of the process and the areas that should be improved in the process are determined. The sustainability metrics and the safety index are calculated using the steady-state data for the global process and they are used to measure the impact of the process in its surroundings. They will be used as performance criteria in the evaluation of the new suggested design alternatives.

5.3.1.4 Step 4: Indicator Sensitivity Analysis Algorithm

In this step the target indicators are determined using the indicator sensitivity analysis (ISA) algorithm (see [10]). To apply this algorithm the indicators having the highest potential for improvements are identified first. Then an objective function such as the gross profit or the process total cost is specified. A sensitivity analysis is then performed to determine the indicators that allow the largest positive (for profit) or negative (for cost) change in the objective function. The most sensitive indicators are selected as targets for improvements.

5.3.1.5 Step 5: Sensitivity Analysis of Operational Parameters

With the target indicators and their variables identified in Step 4 (Section 5.3.1.4) the next task is to determine the process-operational variables that cause the biggest changes in the target indicators for smallest changes in their values. This analysis is done by checking the influence of increments of 5, 10 and 15% in all the operational variables that influence the selected target indicator and the consequent effect in the target. The analysis is done using the following equation (*OPV* is the operational value):

$$\Delta OPV = \frac{OPV_{final} - OPV_{initial}}{OPV_{initial}} \quad (5.2)$$

Through this analysis, it is possible to determine the highest improvement in the indicator value. This value corresponds to the maximum theoretical of improvement that can be achieved in the target indicators. The results determine the operational variables that can cause the highest improvements in the process and consequently the variables that must be targeted to generate more sustainable design alternatives.

5.3.1.6 Step 6: Generation of New Sustainable Design Alternatives

The steps involved in the generation of sustainable design alternatives are highlighted in Figure 5.3. The flow diagram in Figure 5.3 shows four categories where the operational variables are involved.

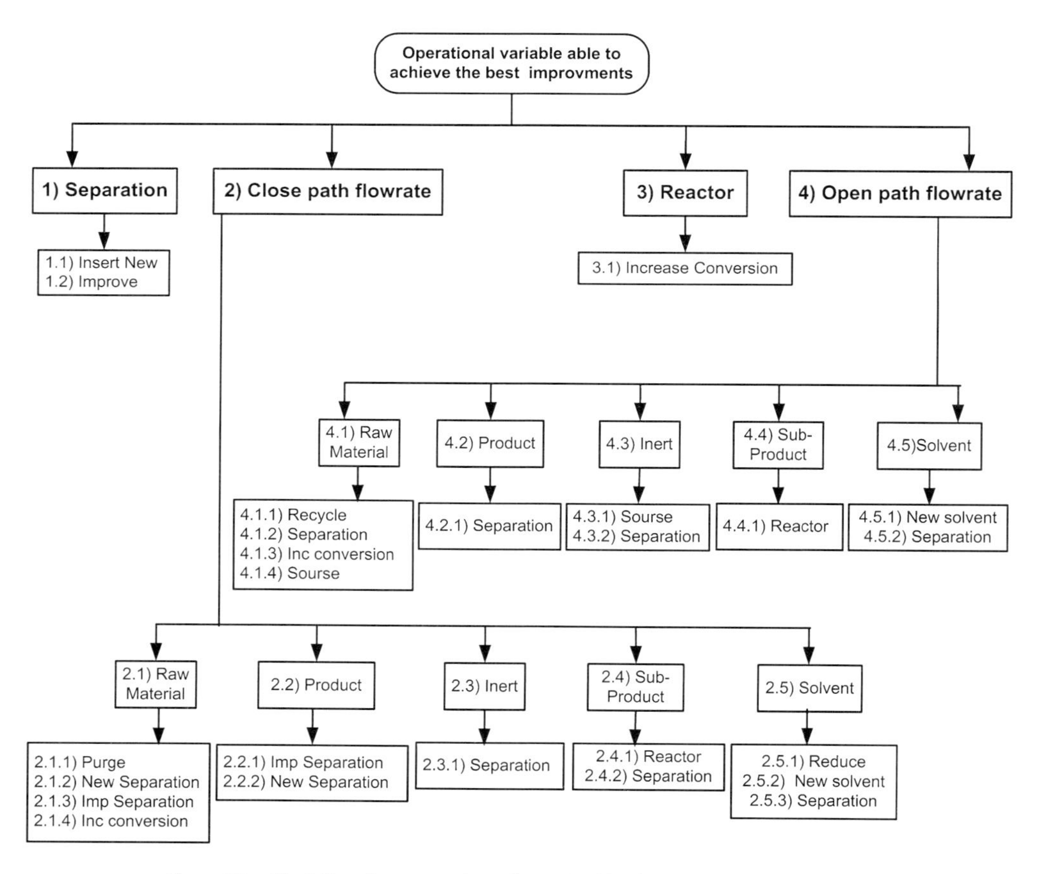

Figure 5.3 Work-flow for generation of sustainable alternatives (Step 6).

- **Category 1:** Operational variables associated with a separation;
- **Category 2:** Operational variables associated with flowrate reduction in a closed-path;
- **Category 3:** Operational variables associated with a reaction;
- **Category 4:** Operational variables associated with flowrate reduction in an open-path.

Once the categories have been identified the corresponding synthesis algorithm is employed to generate more sustainable alternatives. The following synthesis algorithms are recommended for specific targeted problems.

- **Separation synthesis:** Apply the algorithm of Jaksland *et al.* [12];
- **Improvement in a separation unit:** Apply the driving force based reverse design algorithm of D'Anterroches and Gani [13];
- **Improvement in a reactive unit:** Apply the attainable region based reverse design algorithm of D'Anterroches and Gani [13];
- **Selection/substitution of solvents:** Apply the algorithm of Harper [14];

The proposed new alternatives are simulated using the new flowsheet configuration or the new operational conditions. With this new data, the performance criteria are calculated again and a comparison between the new alternatives is done taking into account the following criteria. 'An alternative is considered more sustainable if and only if it improves the indicator targets without compromising the performance criteria'. From the proposed alternatives the one with the better results will be the one selected.

5.3.2 Methodology – Batch Mode

5.3.2.1 Step 1: Data Collection

For the batch case, data on the time of each operation, the equipment volume, the initial and the final mass for each compound in each operation, the mass entering and leaving each batch operation during the operation time and the energy used in each step are required. The purchase and sale prices for each chemical are also needed. All these data can be collected from the real plant and/or generated through model-based simulations.

5.3.2.2 Step 1A: Transform Equipment Flowsheet into an Operational Flow Diagram

For continuous processes the flowsheet diagram is a sequence of different equipments where in each equipment, a specific operation takes place. When the process is operating in the batch mode, the individual equipments may present a sequence of operations. In this methodology the batch process will be treated as a 'continuous' process in terms of the material and energy (data) flow from operation to operation (instead of equipment to equipment). Thus, the equipment-based flowsheet is transformed to an operational flow diagram (illustrated in Figure 5.4).

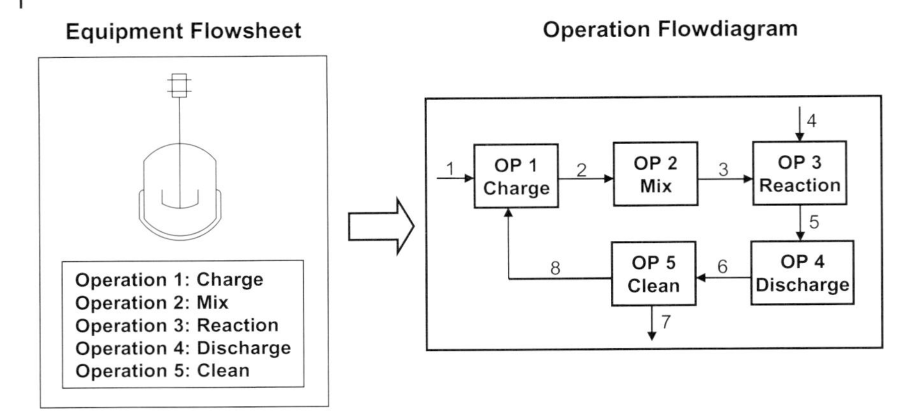

Figure 5.4 Example for the transformation of equipment flowsheet in an operational flow diagram.

5.3.2.3 Step 2: Flow Diagram Decomposition

For a batch process, it is possible to identify all the open- and closed-paths for each compound as in the continuous mode once the operation flow diagram has been generated. However, for a batch operation flow diagram, a new path related to the accumulation of mass and energy is introduced. This new path is called accumulation-path (AP) and corresponds to the accumulation in a given operation. This path represents an average of the mass for each compound during the operation time. More details are given by [15].

5.3.2.4 Step 3: Calculation of Indicators

For batch processes the indicators presented in Section 5.3.1.3 are also applicable. For each type of batch operation, two new indicators are proposed: *operation indicator* (compares the performance of the operation) and the *compound indicator* (indicates for each operation, the compound most likely to cause operational problems). These new indicators provide important information about the batch processes in terms of which operation of a process flowsheet has comparatively more potential for improvements than the others.

Operation Indicators There are three operation indicators, the total free volume factor (TFVF), the operation time factor (OTF) and the operation energy factor (OEF). With these indicators it is possible to have an analysis of the performance of the batch operations in terms of time, volume and energy. In the text below, the operational (batch) indicators are explained in more details.

Total Free Volume Factor (TFVF) This indicator gives the percentage of free volume compared to the total volume of the equipment.

$$TFVF_j = \frac{V_{eq}^j - \sum_c^C \frac{M_{AP,c}}{p_c}}{V_{eq}^j} \tag{5.3}$$

In Eq. (5.3), V_{eq}^j is the equipment volume in operation j, ρ_c is the density of compound c, C is the total number of compounds present in operation j.

High values of this indicator indicate that the equipment volume is not filled to a high level and consequently points to a potential for improvements. Knowing where the equipment is not being fully occupied, there is a good chance of changing the material disposition among the operations in order to improve the performance of the sequence of operations. The indicator value is given as a fraction.

Operation Time Factor (OTF) This indicator points to the fraction of time that a given operation spends compared to the total time taken by the whole sequence of operations.

$$OTF_j = \frac{t_j}{\sum_j^J t_j} \tag{5.4}$$

In Eq. (5.4), t is the time spent in operation j.

High values of this indicator show that a given operation is taking too much time and consequently this operation can be seen as the bottleneck in the operations flow diagram. This is also the limiting operation with respect to time. This indicator value is given as a fraction and it should be reduced in order to improve the process.

Operation Energy Factor (OEF) This indicator gives the percentage of energy used in a given operation compared to the total amount of energy consumed.

$$OTF_j = \frac{E_j}{\sum_j^J E_j} \tag{5.5}$$

Where E_j is the energy consumed in operation j.

High values of this indicator point to an operation consuming too much energy when compared to others. This indicator also helps to identify opportunities for heat integration and to trace the heat integration possibilities among different equipments/operations. This indicator should be reduced to improve the process and its value is given as a fraction.

Compound Indicators A set of compound indicators, which allow the identification of the compound causing a bottleneck in a given operation, have been developed. There are three different compound indicators the free volume factor (FVF), the time factor (TF) and the energy factor (EF). The TF and the EF are applied for each accumulation-path and their calculations are dependent on the type of operation, such as, mixing, reacting and separating operations.

High values of these indicators point to the compounds responsible for the identified problems in the operations. These values should be reduced.

Three examples are given to assist understanding:

1) **TFVF is high for a given operation:** This means that the equipment is not fully occupied and consequently a change in the material allocation among the operations can be made. Also, for a new plant, the size of the equipment, which is going to be acquired, could be made smaller than what was originally designed.

2) **OTF is high for a separation operation:** This means that this separation operation has a bottleneck with respect to time. TF has been calculated for each compound. Supposing that recovering compound A is the object of this separation and that TF has the highest value for compound B. This means that compound B is the responsible for the slow separation. This indicates the need for an improvement in the current separation or to insert a new one that makes the separation of A from B easier.

3) **OEF is high for a reactor:** This means that this operation is consuming a lot of energy. This points to investigation with respect to heat integration, between this operation and some other operation/unit. Supposing that EF has been calculated to all the compounds in the reactor and assuming that the inert compound has the highest EF and, it means that this compound is the one responsible for the high energy consumption in the reactor and options to remove it before the separation process should be investigated.

5.3.2.5 Step 4: Indicator Sensitivity Analysis Algorithm

In this step the target indicators are determined using the ISA algorithm (see Section 5.3.1.4). To apply this algorithm the indicators having the highest potential for improvements are identified first. Then an objective function such as the gross profit or the process total cost is specified. A sensitivity analysis is then performed to determine the indicators that allow the largest positive (for profit) or negative (for cost) change in the objective function. The most sensitive indicators are selected as targets for improvements.

5.3.2.6 Step 5: Sensitivity Analysis of Operational Parameters

A sensitivity analysis with respect to the operational (parameters) variables, which influence the target indicators, is performed. The analysis identifies the operational variables that need to be changed to improve the process in the desired direction (See Section 5.3.1.5).

5.3.2.7 Step 6: Generation of New Sustainable Design Alternatives

Synthesis algorithms are applied to generate new design alternatives (See Section 5.3.1.6).

5.4 *SustainPro* Software

5.4.1 Introduction

Based on the above methodology, a software product (*SustainPro*) has been developed to allow easy application of the methodology to generate more sustainable design alternatives in batch and continuous processes. *SustainPro* is an *Excel* based software, divided into 21 different *Excel* sheets, where two of the *Excel* sheets are Principal Menus, one with options for importing and exporting data and another to guide the user through the methodology steps. The remaining *Excel* sheets represent the different steps and sub-steps of the methodology presented in Figure 5.1. The inputs for *SustainPro* are the mass and the energy balance data as well as the prices of the compounds present in the process. *SustainPro* follows all the steps of the methodology, allowing thereby the creation/evaluation of a new alternatives strategy to any chemical process.

5.4.2 *SustainPro* Architecture

SustainPro architecture is highlighted in Figure 5.5. The main interface of *SustainPro* is divided into three parts:

Part I – Indicator analysis;
Part II – Evaluation;
Part III – Generation and comparison of new alternatives.

To solve a sustainable design problem the user should perform Parts I, II and III sequentially. The built-in color code system guides the user through the different steps of the work-flow (see Figure 5.1). The user must follow the button highlighted in *orange*, which is the next step to be followed. The *light blue* button represents the steps already performed and the *dark blue* buttons indicate the steps that have not yet been performed.

5.4.3 Supporting Tools

Some supporting tools are used by *SustainPro* in each part of the analysis – a summary of the supporting tools is given in Table 5.1.

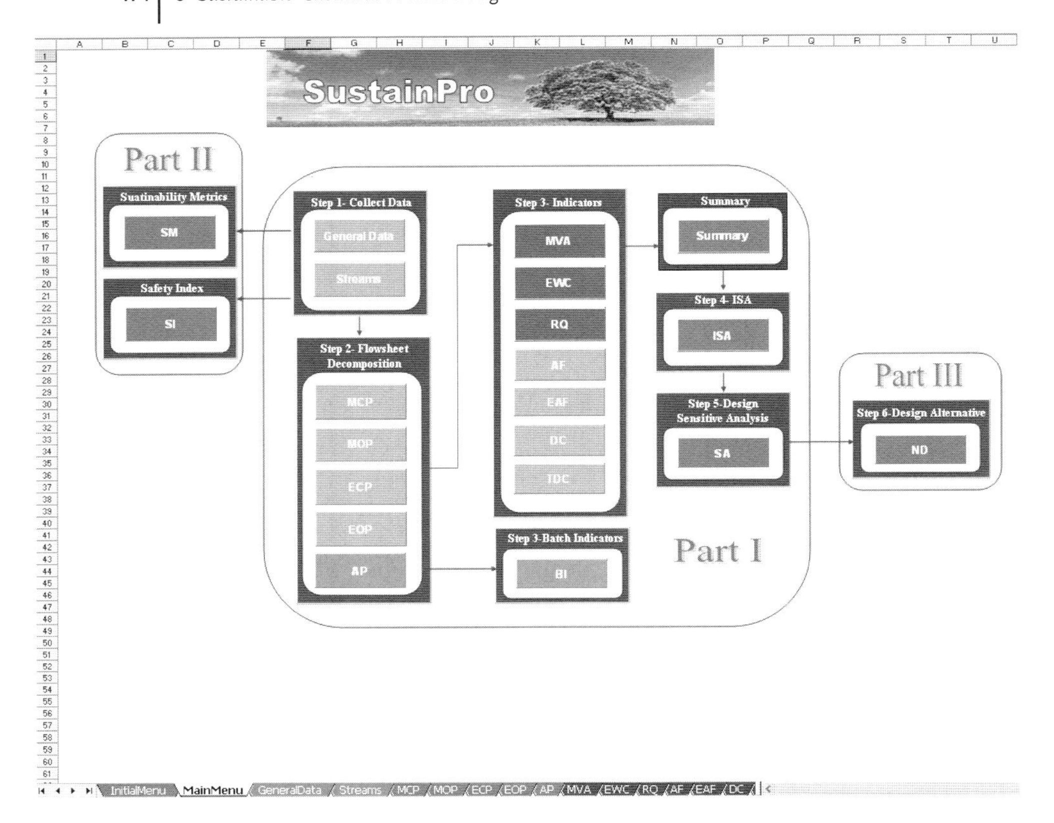

Figure 5.5 *SustainPro* architecture.

5.5
Case Studies

5.5.1
Continuous Processes: Biodiesel Production

A simplified version of the flowsheet for biodiesel production is represented in Figure 5.6. The feedstock (*Jatropha oil*) is first heated before entering the reactor. The unreacted methanol is recovered with a distillation. A liquid–liquid extractor is used in order to separate the glycerol (heavy phase) from the biodiesel (light phase). To finish, both of those products are purified in distillation columns. There is no oil recovery, nor has energy integration been made.

The methodology and *SustainPro* have been applied to this case study.

5.5.1.1 Step 1: Collect the Steady-state Data

The required detailed process data for the biodiesel production plant was taken from a simulation of the process in PRO/II.

Table 5.1 Summary of the supporting tools used by *SustainPro*.

Tools	Purpose	Interaction with *SustainPro*
Simulators	Generate mass and energy balances	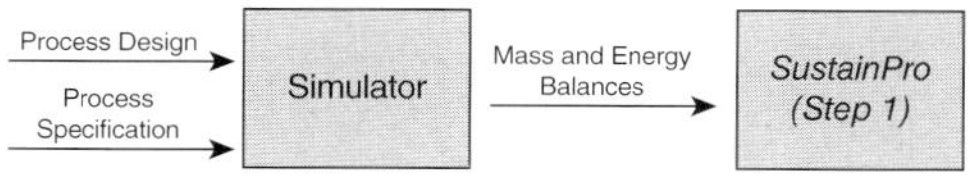 Process Design, Process Specification → Simulator → Mass and Energy Balances → *SustainPro (Step 1)*
CAPEC database	Compound properties	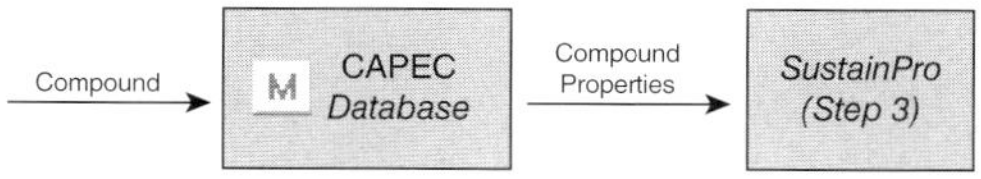 Compound → CAPEC *Database* → Compound Properties → *SustainPro (Step 3)*
ProPred	Property prediction	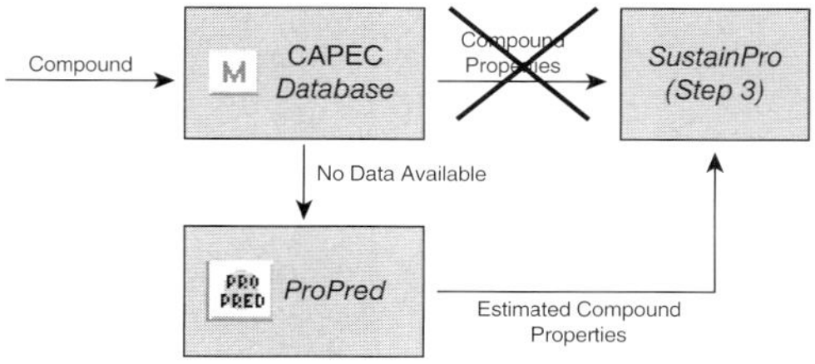 Compound → CAPEC *Database* → ~~Compound Properties~~ → *SustainPro (Step 3)*; No Data Available → *ProPred* → Estimated Compound Properties → *SustainPro (Step 3)*
PA-WAR algorithm	Environmental parameters	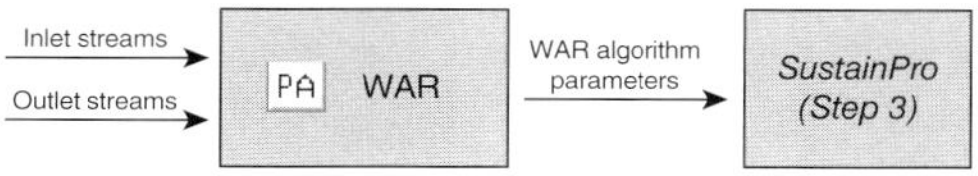 Inlet streams, Outlet streams → PA WAR → WAR algorithm parameters → *SustainPro (Step 3)*
ProCamd	Solvent selection	 *SustainPro (Step 5)* → Operational Conditions → *ProCamd* → Solvent Suggestion → *SustainPro (Step 6)*
CAPSS	Separation technique selection	 *SustainPro (Step 5)* → Operational Conditions → *CAPSS* → Separation Suggestion → *SustainPro (Step 6)*

5.5.1.2 Step 2: Flowsheet Decomposition

For this case study, the flowsheet decomposition generated 7 closed-paths, 34 open-paths.

5.5.1.3 Step 3: Calculate the Indicators, the Sustainability and the Safety Metrics

For the entire set of flow-paths, the full-set of indicators was calculated. The most sensitive mass indicators were identified and they are listed for open- and closed-paths in Tables 5.2 and 5.3.

For this case study the most sensitive indicators are the MVA – material value added, for the open-paths OP28, OP12 and OP2. They have very negative values, which means that a lot of money is wasted from the time the materials (compounds) enter the system to the time they exit. OP 16, OP28 and CP5 show high EWC values and consequently high energy consumption that should be reduced.

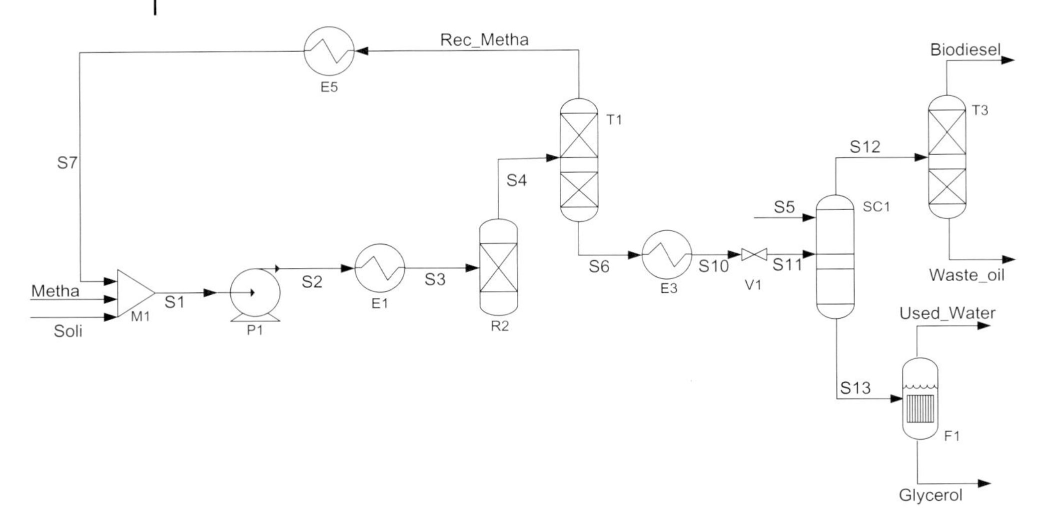

Figure 5.6 Flowsheet for biodiesel production from Jatropha oil.

Table 5.2 Important open-path and the associated indicators.

Path	MVA	Probability	Path	EWC	Probability	Path	TVA	Probability
OP 28	−269.3	High	OP 16	27.8	Medium	OP 28	−293.7	High
OP 12	−266.6	High	OP 28	24.5	Medium	OP 12	−271.6	High
OP 2	−71.2	High	OP 6	19.3	Medium	OP 2	−71.2	High
OP 29	−25.1	High	OP 11	14.0	Medium	OP 29	−25.6	High
OP 1	−9.8	High	OP 5	6.4	Low	OP 1	−10.0	High
OP 33	−1.1	High	OP 12	5.0	Medium	OP 5	−6.4	Low
OP 34	−1.1	High	OP 2	1.0	High	OP 33	−1.9	High

Table 5.3 Important open-path and the associated indicators.

Path	EWC	Probability	Path	AF	Probability
C5	4.11	Check AF	C5	1.04	High
C6	0.00	High	C6	0.00	Medium
C4	0.00	High	C4	0.00	Medium

In Table 5.4 the most important energy indicators are listed.

The high values of TDC show that there is a high potential for improvements and consequently that energy integration can be considered.

The sustainability metrics and the safety index were also calculated (their values are listed in Table 5.6 presented after Step 6).

Table 5.4 Important energy indicators.

Demand	TDC ($/h)
Biodiesel	63
C E3	6

Table 5.5 Indicator sensitivity analysis (ISA) algorithm results for the biodiesel case study.

Path	Indicator	Scores
OP 28	MVA	21
OP 12	MVA	21
OP 2	**MVA**	**29**
OP 16	EWC	14
OP 28	EWC	21
CP5	EWC	18
CP5	AF	18

5.5.1.4 Step 4: Indicator Sensitivity Analysis (ISA) Algorithm

To apply the ISA algorithm the indicators MVA for OP28, OP12 and OP2, EWC for OP 16, OP 28 and CP 5 were selected as possible target indicators. After applying the ISA algorithm, scores were given to the selected indicators (see Table 5.5).

From Table 5.5 it is seen that from the selected indicators, the MVA indicator related to OP 2 for the Jatropha oil is the most sensitive (highest score). MVA values for OP 12 and OP 28 also present high scores. Consequently, these indicators are considered the target indicators for improvements.

5.5.1.5 Step 5: Process Sensitivity Analysis

From a sensitivity analysis of the operational parameters influencing the target indicators (MVA- OP 2; OP 12; OP 28) it was found that the most significant operational variables are the flowrates of the respective open-paths.

5.5.1.6 Step 6: Generation of New Design Alternatives

To generate a new sustainable design alternative, the first thing to do is to verify in which category the selected parameter is included (see Step 6 of the Methodology Section 5.3.1.6). For OP 2, it was found that the operational variable is related to the reduction of an open-path flowrate of a raw material. This pointed to a reduction of the OP 2 flowrate by considering, the recycling of the Jatropha oil. To recycle Jatropha oil, a purge has been considered to avoid the build-up of undesired compounds. Regarding OP 12, it was found that this operational variable is related to the increase of an open-path flowrate of a product, which means that the separation process to recovery biodiesel needs to be improved. The distillation column separation has been optimized in terms of number of trays and reflux ratio. For OP 28 it was found that this compound can be sold out if a required purity is achieved. Consequently, the liquid–liquid extraction separation process needs to be improved. The temperature has been optimized and the required purity has been achieved. Energy integration has been considered. The streams leaving the distillation columns were used to heat the streams entering the distillation column. Energy has been saved with this approach.

Using this information the process was simulated again with all the suggested improvements in order to validate the new design alternative.

For the new sustainable design alternative, which consists of the recycling of Jatropha oil, improved distillation process and improved liquid–liquid extraction process, the following improvements were achieved. The profit increased by 27%, the water metric was improved by 39% and 44%, the energy metric was improved by 71% and 74%, and the material metric was improved by 10% and 17%. The environmental impact has been improved by 11%. The rest of the performance criteria parameters have remained constant (confirming a non-trade-off solution). All the values for the performance criteria are listed in Table 5.6. The improvements in the target indicators are listed in Tables 5.7 and 5.8.

Table 5.6 Comparison of the performance criteria between the 'reference' design and the new sustainable design alternative for the biodiesel production case study.

Sustainability Metrics	Base case	New design	Improvement
Total net primary energy usage rate ($GJ\,y^{-1}$)	31 281	9 991	68%
% Total net primary energy sourced from renewables	0.9999	0.9997	0.02%
Total net primary energy usage per kg product ($kJ\,kg^{-1}$)	1459	417	71%
Total net primary energy usage per unit value added ($kJ\,\$^{-1}$)	1.2	0.3	74%
Total raw materials used per kg product ($kg\,kg^{-1}$)	0.41	0.37	10%
Total raw materials used per unit value added ($kg\,\$^{-1}$)	0.000 34	0.000 28	17%
Net water consumed per unit mass of product ($kg\,kg^{-1}$)	1.7	1.0	39%
Net water consumed per unit value added ($kg\,\$^{-1}$)	0.0014	0.0008	44%
Safety index	22	22	0%
WAR	114	102	11%
Profit ($\$\,y^{-1}$)	3 005 427	3 827 084	27%

Table 5.7 Mass target indicators improvements.

MVA ($/h)	Base-Case	New Design	Improvements
OP 2	−8.7	−1.8	79%
OP12	−33	−1	97%
OP28	−33.7	−31	8%

Table 5.8 Energy target indicators improvements.

TDC ($/h)	Base case	New design	Improvements
Biodiesel	63	11	83%
C E3	6	0	100%

Table 5.9 CO_2 emission for the base case and for the new sustainable design alternative for the biodiesel production case study.

CO2 emission (kg/y)	Initial	New	Improvement
Fuel oil	2 418 021	772 304	68%
Natural gas	2 005 112	640 423	68%

The energy reduction will of course reduce the CO_2 emission. For this case study it is assumed that the furnace uses fuel oil or natural gas to produce steam. The CO_2 emission has been determined for the base case and for the new design alternative, taking into consideration the two established scenarios (see Table 5.9)

Table 5.9 shows that an improvement of 68% in the CO_2 emission has been achieved with the new proposed alternative. These results show that a more sustainable design alternative has been found.

Detailed simulation and process data can be obtained from the corresponding author.

5.5.2 Batch Processes: Insulin Case Study

The insulin process [16] is divided into four sections:

1) **Fermentation:** Here the *E.coli* cells are used to produce the Trp-LE'-MET-proinsulin precursor of insulin, which is retained in the cellular biomass. Fermentation takes place in order to achieve the desired biomass.

2) **Primary Recovery:** In this section a high pressure homogenizer is used to break the cells and release the inclusion bodies. Then with a set of centrifuges and solvents the inclusion bodies are recovered with a higher purity.

3) **Reactions:** In this part of the process there is a sequence of reactions until the production of insulin.

4) **Final Purification:** Finally, a purification sequence based on multimodal chromatography, which exploits differences in molecular charge, size, and

hydrophobicity, is used to isolate biosynthetic human insulin. The crystallization of insulin is the last step of the process.

The flowsheet for the insulin production process is shown in Figure 5.7.

5.5.2.1 Step 1: Collect the Steady-state Data

The required detailed process data for the insulin synthesis plant was taken from a simulation available on SuperPro Designer (2008) software package. The prices and costs were taken from [16], where the insulin production simulation is described in detail.

5.5.2.2 Step 1A: Transform Equipment Flowsheet in an Operational Flowsheet

The equipment flowsheet consists of 31 units, which can be seen in Figure 5.7 (some equipments are represented more than once in the flowsheet; they have however, the same name). Taking into account the sequence of operations, the operational flow diagram is determined. The operational flow diagram has 92 operations, 169 streams and 38 compounds.

5.5.2.3 Step 2: Flowsheet Decomposition

For this case study the operations flow diagram decomposition generated 418 closed-paths, 1022 open-paths and 3344 accumulation-paths.

5.5.2.4 Step 3: Calculate the Indicators, the Sustainability and the Safety Metrics

For the entire set of flow-paths, the full set of indicators was calculated, except for some batch compound indicators whose data were not available. Due to the large size of the flowsheet it is not be possible to present or discuss all the modifications to improve the whole process. Therefore, in the remaining steps, only section 1 and section 3 are highlighted with respect to improvement of their mass indicators and batch indicators.

The most sensitive mass indicators for those sections were selected and they are listed for the selected sections in Table 5.10.

For this case study the most sensitive indicators are the MVA – material value added, for the open-paths listed in Table 5.10. They have very negative values, which means that a lot of money is wasted from the time the materials (compounds) enter the system to the time they exit. The energy consumption and the recycling in the process do not allow very high potential for improvements when compared with the very high values of MVA (see also the EWC values in Table 5.10).

The most sensitive batch indicators were selected and they are listed for each section in Table 5.11.

Operations V-102R, V-103(P8)R, V-105R and V-111R present high values of OTF when compared with the other operations, which means that these operations are spending too much time to execute their respective process operation. V-102R and DS-101(P9) have high values of OEF when compared with the other operations. This indicates that these two operations have high energy consumption. These

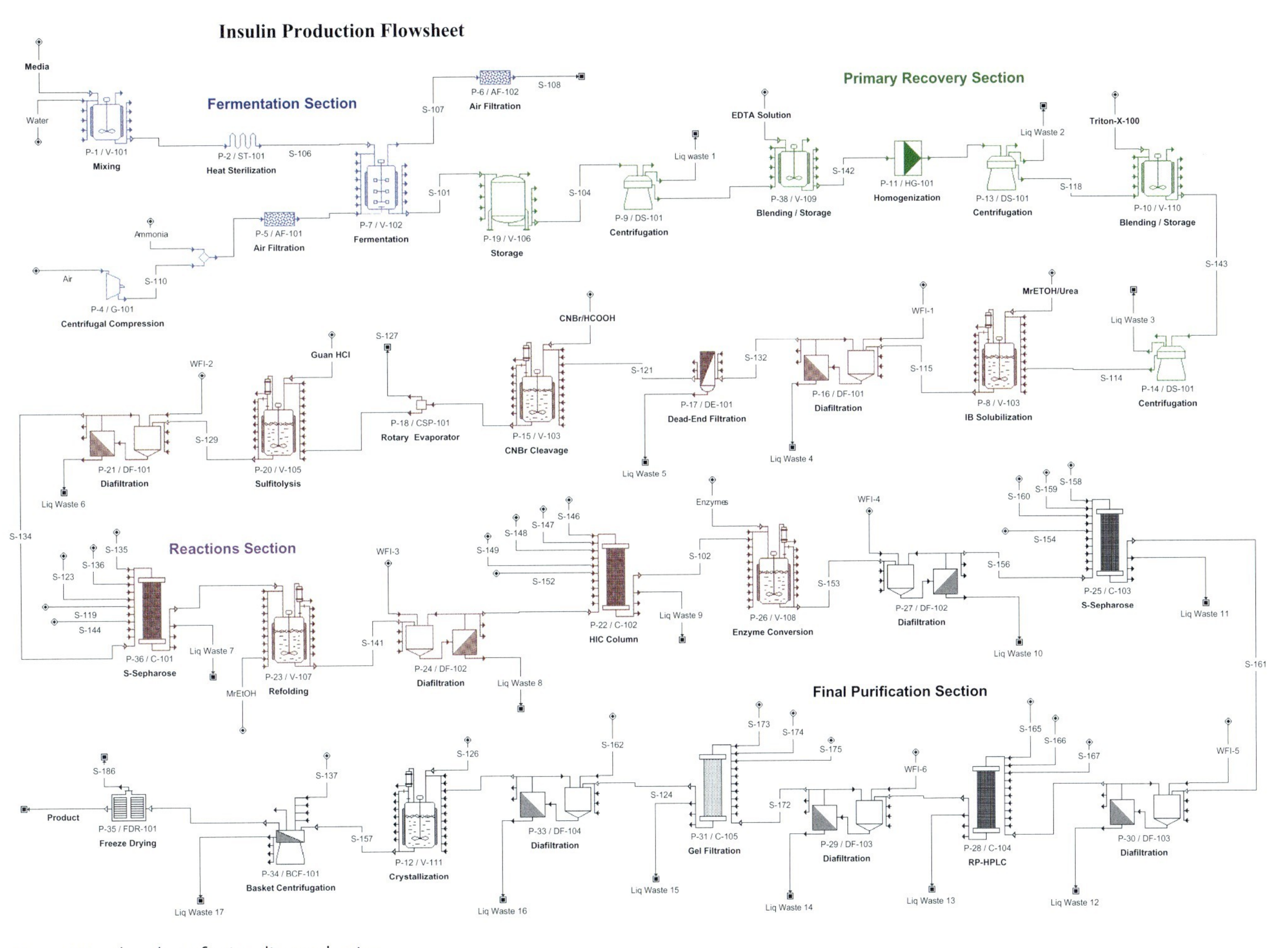

Figure 5.7 Flowsheet for insulin production.

Table 5.10 Mass indicators and their calculated values for the insulin production case study.

Section	OP	Path	Component	MVA (10^3\$/y)	EWC (10^3\$/y)	TVA (10^3\$/y)
Fermentation (Section 1)	OP 37	S4-S26	Water	−22 560	68.90	−22 629
Reactions (Section 3)	OP 620	S79-S80	Urea	−205 917	0.00	−205 917
	OP 591	S54-S60	Formic acid	−137 334	0.34	−137 334
	OP 613	S77-S80	Urea	−90 375	0.00	−90 375
	OP 657	S62-S69	HCl	−74 542	0.01	−74 542
	OP 659	S62-S80	HCl	−51 070	0.01	−51 070
	OP 598	S43-S49	Urea	−43 552	0.01	−43 552
	OP 615	S77-S92	Urea	−37 885	0.02	−37 885
	OP 316	S85-S92	WFI	−21 519	17.82	−21 537
	OP 313	S79-S80	WFI	−19 808	0.00	−19 808
	OP 173	S50-S49	WFI	−17 663	0.00	−17 663
	OP 403	S103-S104	WFI	−16 328	0.00	−16 328
	OP 335	S91-S92	WFI	−13 144	0.00	−13 144
	OP 292	S77-S80	WFI	−10 561	0.00	−10 561
	OP 721	S103-S104	NaCl	−12 210	0.00	−12 210

Table 5.11 Batch indicators and their calculated values for the insulin production case study.

Section	Operation	TFVF	OTF	OEF	AP	Compound	TF	EF
Fermentation (Section 1)	V-102R	0.83	0.07	0.33	230	Oxygen	27 058	0.06
					231	Glucose	9 884	0.07
					232	Salts	63 021	0.00
					233	Water	66 039	3.06
					234	Biomass	Non-defined	Non-defined
					235	Ammonia	66 244	0.00
					236	CO_2	44 668	0.10
Reaction (Section 3)	V-103(P8) R	0.87	0.03	0.01	943	Cont Proteins	20 410.07	Not available
					944	IBs	90 061.83	Not available
					945	Trp-proinsulin	Non-defined	Not available
	V-105 R	0.74	0.04	0.002	1388	$NaSO_3$	23.49	Not available
					1389	$Na_2O_6S_4$	100.90	Not available
					1402	Denatured protein	660 080.16	Not available
					1403	Proinsulin-SSO_3	Non-defined	Not available

indicators show high potential for improvements and their values should be reduced. The options for improvements in the selected section are analyzed below.

- **Fermentation (Section 1):** In this section the most critical points are related to the waste water, which is produced as a by-product in the main reaction, so there is little likelihood of reducing it. Consequently this is not the best choice for a process improvement. Regarding the batch indicators it can be seen that a very high value of OTF has been calculated for the fermentation operation (V-102R). Analyzing the compound indicators for this operation it is seen that ammonia is the compound which is limiting the operation time. Consequently to improve the fermentation process it would be necessary to take into consideration the ammonia concentration and the related parameters, which influence the rate of the reaction (this point is further discussed in steps 5 and 6).
- **Reactions (Section 3):** This section involves many solvents (urea, WFI, formic acid, HCl, NaCl) which are not recovered and recycled within the process. The best option to improve these indicators, and consequently the process, is to recover and recycle the solvents. For some of them it might not be economically feasible. Some waste solvents, however, may be sold to other users. For example, urea can be further processed and used as nitrogen fertilizer [16]. Here, two operations, V-103 R and V-105 R, indicate high values of OTF, which point out that their operation time should be reduced. Regarding the compound indicators for these two operations, it is possible to see from Table 5.11 that IB and denatured proteins are the compounds causing the high time consumption. In order to decrease the time factor, it is necessary to analyze the rate of reaction conditions.

The sustainability metrics and the safety index were also calculated (their values are listed in Table 5.13 presented after Step 6).

5.5.2.5 Step 4: Indicator Sensitivity Analysis (ISA) Algorithm

To apply the ISA algorithm the indicators listed in section 3 of Table 5.10 were selected as possible target indicators. After applying the ISA algorithm it is seen that from the selected indicators, the MVA indicator related to OP591 for formic acid is the most sensitive. Consequently, this indicator is considered the target indicator for improvements (see Table 5.12, row highlighted with bold letters).

For batch indicators, the most sensitive indicator in section 1 is the TF of ammonia in the fermentation operation (V-102R).

5.5.2.6 Step 5: Process Sensitivity Analysis

From a sensitivity analysis of the operational parameters influencing the target indicator (MVA–OP591) it was found that the most significant operational parameter is the flowrate of OP591.

The fermentation process time is mainly dependent on the specific cell growth rate, which is represented by the following equation ([17])

Table 5.12 Indicator sensitivity analysis (ISA) algorithm results for the insulin production case study.

Path	Indicator	Compounds	Scores
OP 721	MVA	NaCl	2
OP 620	MVA	Urea	3
OP 613	MVA	Urea	2
OP 403	MVA	WFI	2
OP 335	MVA	WFI	8
OP 313	MVA	WFI	5
OP 292	MVA	WFI	3
OP 173	MVA	WFI	4
OP 316	MVA	WFI	10
OP 591	**MVA**	**Formic acid**	**15**
OP 615	MVA	Urea	10
OP 598	MVA	Urea	11
OP 657	MVA	HCl	7
OP 659	MVA	HCl	6

$$\mu_g = \mu_{g\max} \frac{Co_{Glucose}}{(k_{Glucose} + Co_{Glucose})} \frac{Co_{O_2}}{(k_{O_2} + Co_{O_2})} \frac{Co_{NH_3}}{(k_{NH_3} + Co_{NH_3})} \frac{Co_{H_3PO_4}}{(k_{H_3PO_4} + Co_{H_3PO_4})} \tag{5.6}$$

In Eq. (5.6), μ_g is the specific cell growth rate, $\mu_{g\,max}$ is the maximum specific cell growth rate, k is the monod constant for each compound and Co is the concentration.

To analyze the operational parameters that influence the batch target indicator (TF), Eq. (5.6) was used and it was possible to verify that the ammonia (NH_3) concentration is the most significant parameter in order to reduce the time of the reaction.

5.5.2.7 Step 6: Generation of New Design Alternatives

To generate a new sustainable design alternative, the first thing to do is to verify in which category the selected parameter is included (see Section 5.3.1.6). It was found that the operational parameter is related to the reduction of an open-path flowrate. This pointed to a reduction of the OP591 flowrate by considering, the recycle of the formic acid. To recycle formic acid, a separation operation needs to be inserted in order to purify/recover this compound. Applying the process separation algorithm of Jaksland and Gani [12], a set of feasible separation techniques for the recovery of formic acid coming from stream S60 was identified. Pervaporation was selected as the separation operation, because it involves lower operational costs when compared with the other separation techniques and it does not need external compounds for the separation. In the literature, Nakatani *et al.* [18], found that membranes such as aromatic imide polymer asymmetric, are available to purify/recover formic acid from water (which is the mainly impurity compound in S60). To estimate the selectivity of the membrane, it is assumed that this system (membrane to separate) has the same behavior as that of a similar mixture considered by Huang *et al.* [19]. Using this information the process was simulated again in order to validate the new design alternative. To reduce the fermentation time the concentration of ammonia needs to be increased. The concentration was increased by 2%, and 0.2% of fermentation time reduction was achieved, which is not a significant improvement. This indicates that the fermentation process is already optimized and nothing could be done to improve it. Also, the fermentation operation has more constraints that cannot be violated without changing the enzyme.

For the new sustainable design alternative, which consists of the recycling of formic acid, the following improvements were achieved. The profit increased by1.98%, the water and the energy metrics per value added improved by 2%. The material metrics improved by 2% and 4% respectively per kg of final product and per value added. Finally, the environmental impact output was improved by 31.7%. The rest of the performance criteria parameters have remained constant. All the values for the performance criteria are listed in Table 5.13. The target indicator was improved by 99.9% (MVA–OP591 Initial = -1.37×10^{8} \$/y, Final = -169×10^{3} \$/y). Note that MVA should be positive (or less negative). Clearly, the new design has made this target indicator less negative.

In this case there is no energy reduction. For this case study the supplied material has been reduced. However the CO_2 emission measured per value added has been reduced. This means that for a greater profit it was not necessary to increase the CO_2 emission. The CO_2 emission has been determined for the base case and for the new design alternative, taking into consideration the two established scenarios (see Table 5.14).

Table 5.13 Comparison of the performance criteria between the 'reference' design and the new sustainable design alternative for the insulin production case study.

Metrics	Initial	Final	Improvement
Total net primary energy usage rate (GJ/y)	26 727	26 727	0%
% Total net primary energy sourced from renewables	0.72	0.72	0%
Total net primary energy usage per kg product (kJ kg^{-1})	292 397	292 397	0%
Total net primary energy usage per unit value added (kJ/\$)	4.55×10^{-4}	4.46×10^{-4}	1.94%
Total raw materials used per kg product (kg/kg)	43 029	42 083	2.20%
Total raw materials used per unit value added (kg/\$)	6.70×10^{-5}	6.42×10^{-5}	4.10%
Fraction of raw materials recycled within company	0	0	0%
Fraction of raw materials recycled from consumers	0	0	0%
Hazardous raw material per kg product (kg/kg)	4 932.35	3 986.22	19.18%
Net water consumed per unit mass of product (kg/kg)	7 162.22	7 162.22	0%
Net water consumed per unit value added (kg/\$)	1.11×10^{-5}	1.09×10^{-5}	1.94%
Safety index	20	20	0%
WAR	23 709	16 188	31.7%
Profit (\$/y)	7.42×10^{9}	7.56×10^{9}	1.98%

Table 5.14 CO_2 emission for the base case and for the new sustainable design alternative for the Insulin Production Case Study.

CO2 emission (kg/value added)	Initial	New	Improvement
Fuel oleo	2.78×10^{-4}	2.73×10^{-4}	2%
Natural gas	2.31×10^{-4}	2.27×10^{-4}	2%

These results show that a more sustainable design alternative is presented, since the material metrics, the energy consumption per value added, the profit and the WAR algorithm improved and the other parameters remained constant.

5.6 Conclusions

The proposed methodology is able to analyze the process, identifying the process bottlenecks, and then generate new sustainable design alternatives. The new design alternatives lead to less energy consumption, less CO_2 emissions, less water and material consumption, which consequently means less environmental impact. The methodology can be applied to processes operating in continuous, semi-continuous and batch mode. The software, *SustainPro*, which applies the described methodology allows a simple, accurate and faster analysis of any chemical process, simple or complex, big or small, batch or continuous and it is integrated with other (needed) external tools (process simulators, process synthesis, tools, etc). The capabilities of the methodology and the use of the software have been presented and they were highlighted through non-trivial case studies (biodiesel production and insulin production).

References

1 Guinand, E.A. (2001) Optimization and network sensitivity analysis for process retrofitting, PhD-thesis, Massachussetts Institute of Technology (MIT), Chemical Engineering Department, Boston.

2 Rapoport, H., Lavie, R., and Kehat, E. (1994) Retrofit design of new units into an existing plant: case study: adding new units to an aromatics plant. *Computers and Chemical Engineering*, **18**, 743–753.

3 Ciric, A.R., and Floudas, C.A. (1989) A retrofit approach for heat exchanger networks. *Computers and Chemical Engineering*, **13**, 703–715.

4 Jackson, J.R., and Grossmann, I.E. (2002) High-level optimization model for the retrofit planning of process networks. *Industrial and Engineering Chemistry Research*, **41**, 3762–3770.

5 Liu, J., Fan, L.T., Seib, P., Friedler, F., and Bertok, B. (2006) Holistic approach

to process retrofitting: application to downstream process for biochemical production of organics. *Industrial and Engineering Chemistry Research*, **45**, 4200–4207.

6 Lange, J.P. (2002) Sustainable development: efficiency and recycling in chemical manufacturing. *Green Chemistry*, **4** (6), 546–550.

7 Azapagic, A. (2002) Sustainable development progress metrics, IChemE Sustainable Development Working Group, IChemE Rugby, UK.

8 Heikkilä, A.-M. (1999) Inherent safety in process plant design – an index-based approach, Ph.D Thesis, VTT Automation, Espoo, Finland.

9 Gani, R., Hytoft, G., Jaksland, C., and Jens, A.K. (1997) An integrated computer aided system for integrated design of chemical processes. *Computers and Chemical Engineering*, **21** (10), 1135–1146.

10 Carvalho, A., Matos, H.A., and Gani, R. (2008) Design of Sustainable Chemical Processes: Systematic Retrofit Analysis Generation and evaluation of alternatives. *Process Safety and Environmental Protection*, **86**, 328–346.

11 Cabezas, H., Bare, J., and Mallick, S. (1999) Pollution prevention with chemical process simulators: the generalized waste reduction (WAR) algorithm. *Computers and Chemical Engineering*, **23** (4–5), 623–634.

12 Jaksland, C., Gani, R., and Lien, K.M. (1996) Separation process design and synthesis based on thermodynamic insights. *Chemical Engineering Science*, **50** (3), 511–530.

13 D'Anterroches, L., and Gani, R. (2005) Group contribution based process flowsheet synthesis, design and modelling. *Fluid Phase Equilibria*, **228–229**, 141–146.

14 Harper, P.M., and Gani, R. (2000) A multi-step and multi-level approach for computer aided molecular design. *Computers and Chemical Engineering*, **24** 2–7, 677–683.

15 Carvalho, A., Matos, H.A., and Gani, R. (2009) Design of batch operations: systematic methodology for generation and analysis of sustainable alternatives. *Computers and Chemical Engineering*, **33** (12), 2075–2090.

16 Petrides, D., Sapidou, E., and Calandranis, J. (1995) Computer-aided process analysis and economic evaluation for biosynthetic human insulin production – a case study. *Biotechnology and Bioengineering*, **48** (5), 529–541.

17 Singh, R., Gernaey, K.V., and Gani, R. (2009) Model-based computer-aided framework for design of process monitoring and analysis systems. *Computers and Chemical Engineering*, **33** (1), 22–42.

18 Nakatani, M., Sumiyama, Y., and Kusuki, Y. (1994) EP0391699. Pervaporation method of selectively separating water from an organic material aqueous solution through aromatic imide polymer asymmetric membrane.

19 Huang, S.C., Ball, I.J., and Kaner, R.B. (1998) Polyaniline membranes for pervaporation of carboxylic acids and water. *Macromolecules*, **31**, 5456–5464.

6
Heat Integration and Pinch Analysis

Zoran Milosevic and Alan Eastwood

6.1
Introduction

Heat integration is a subject that lies within a broader scope of process integration – a comprehensive approach to process design, which considers the system as a whole, and exploits the interactions between different unit operations from the outset, rather than optimizing them separately.

While process integration may include all aspects of process design, such as integration of distillation columns and reactors, heat integration focuses on improving the plant's energy performance, while at the same time reducing the capital investment in the energy recovery equipment. It applies to both new designs and retrofits.

The term *heat integration* means a number of things to different people. It may be applied to a simple heat exchanger that recovers heat from a process product stream, to waste-heat recovery from a gas turbine, to the integration of a number of process units in an oil refinery, or to a complete integration of an industrial complex. In all these cases, the integration makes it possible to identify how a process can use the heat rejected by another process to reduce the overall energy consumption, even if the units are not running at optimum conditions on their own. Such an opportunity would be missed with a unit-by-unit analytical approach, as it would seek to optimize each unit, thereby losing the opportunity to maximize the re-use of heat by looking 'across the fence'.

Among the heat integration methodologies, *pinch analysis* is certainly the most widely used since its inception in the early 1980s [1, 2]. The popularity is due to the simplicity of its underlying concepts, and to the excellent results it has obtained in numerous projects worldwide.

Pinch analysis, which is sometimes synonymous with heat or energy integration, and sometimes referred to as pinch technology, enables calculation of thermodynamically attainable energy targets for a given process, and provides instructions as to how to design them. A key insight is the existence of the 'pinch point' – that being the most constrained point in the integration process.

Managing CO_2 Emissions in the Chemical Industry. Edited by Leimkühler

ISBN: 978-3-527-32659-4

Typically, pinch analysis is employed at the start of a project, for example in a new plant design, to screen out the promising options and optimize the flowsheet. It is often used in conjunction with various simulation and optimization tools that assist in identifying and verifying the opportunities for improved integration and design.

Having once been a manual calculation technique, heat integration and pinch analysis are today greatly assisted by the use of advanced software tools.

6.2 Heat Integration Basics

6.2.1 Why Heat-Integrate for Optimum Heat Recovery?

It should be obvious why the design of process units incorporates some degree of heat recovery. Energy is used to drive process operations, for example by heating up the feed to the distillation column. This energy essentially ends up in the product streams, which are therefore hotter than the feed. As the end products that are sent to storage normally need to be cooled, the heat contained in them is recovered and used to preheat the feed, thereby reducing the duty of the feed heater. Figure 6.1 shows a distillation unit in which the heat contained in the bottom product is recovered, while the side product's heat is sent to air coolers.

Because the heat contained in the side products is wasted, this schematic represents a unit that is 'poorly heat integrated', with a modest feed preheat temperature of only 180 °C.

A much more complex feed preheat system for crude and vacuum distillation units in an oil refinery is shown in Figure 6.2. The crude oil preheat temperature is considerably higher when compared with that in Figure 6.1, however it is achieved at the cost of installing many more exchangers.

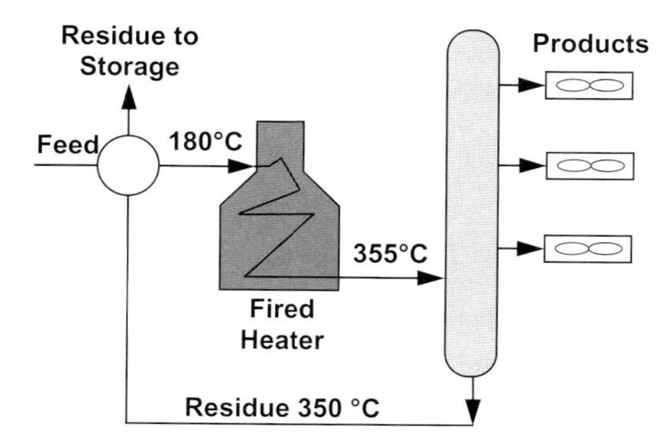

Figure 6.1 Crude distillation unit–simple heat integration.

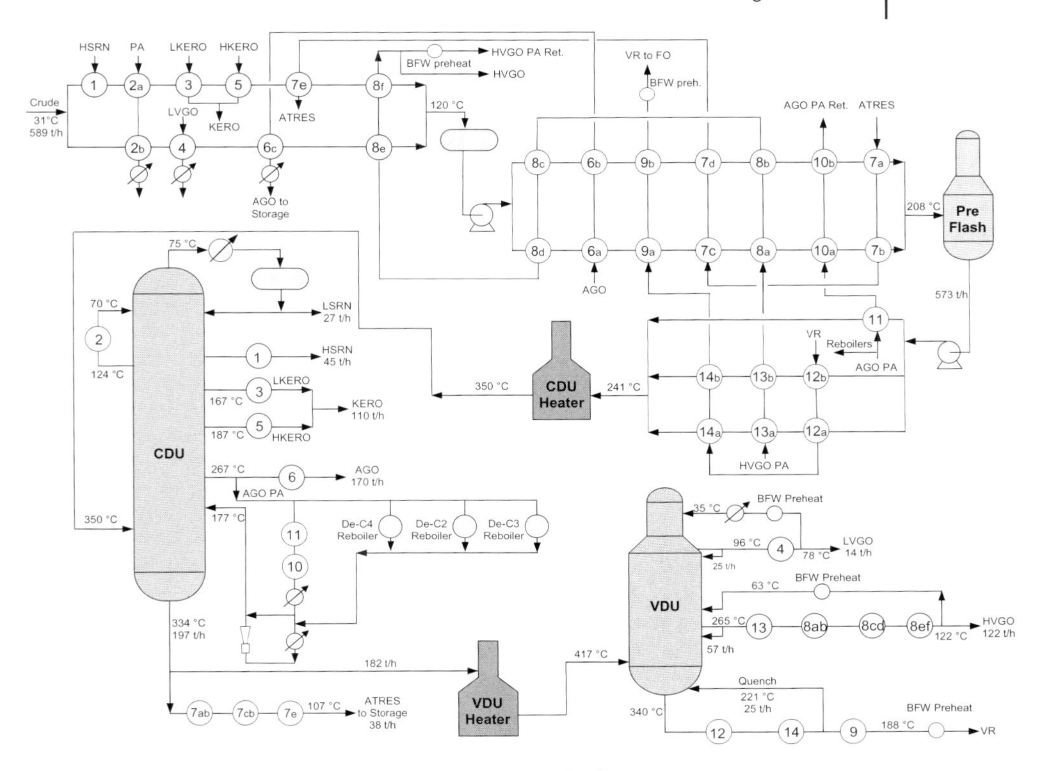

Figure 6.2 Complex heat integration of crude and vacuum distillation units.

It is precisely this trade-off between the value of the increased heat recovery and the increased capital cost that is at the heart of effective heat integration and pinch analysis.

6.2.2
Inter-Unit Heat Integration

Figure 6.2 shows two process units heat integrated, and thereby illustrates what is known as inter-unit heat integration. The arrangement is sometimes referred to as 'hot-feeding'. This means that the product of unit A, which is going to be further processed hot in unit B, is sent there directly, that is, without first recovering its heat in unit A, and then being re-heated in Unit B. The hot products from both units are then used to preheat the stream feeding the unit A. The advantage of this type of integration is that the feed temperature is increased in both units, but more importantly, that the overall exchanger area is considerably reduced relative to the sum of the exchanger areas in the two units when designed as self heat integrated [3]. Heat integration of the crude and vacuum distillation units has become a standard feature of energy efficient oil refinery designs.

A good time to develop the heat integration scheme is when a major investment is being planned and before the process design is fixed. This is because heat integration opportunities may suggest an overall process design that is different to the one that would be developed without considering heat integration. Design features such as column sequencing, reboiling strategy, or column pressures might be chosen differently if heat integration options are considered.

In retrofit projects, improvements in energy efficiency usually require some capital expenditure. In those cases, heat integration and pinch analysis can be specifically directed towards maximizing the return on investment.

6.2.3
Benefits of Heat Integration

Typical savings achieved through the application of pinch analysis in various industries, and expressed as a percentage of the base case energy consumption are as follows (Table 6.1):

A range of savings are shown, reflecting the very site-specific facet of energy conservation. While pinch technology may often identify benefits in excess of those presented in Table 6.1, it is to be noted that sometimes not all of those opportunities get implemented in practice. Some projects may fail to meet the owner's investment criteria, or they may be perceived to cause potential operational problems, such as controllability or product quality. A skillful designer should however be able to spot any such effects, and adjust the design accordingly.

6.2.4
Pinch Analysis

Pinch analysis consists of several parts, characterized by their own specific techniques and methodologies. These main elements can be briefly described as follows:

Table 6.1 Typical savings achieved through pinch analysis.

	%	
Oil refining	10–15	On refinery total consumption. Assumes total site integration
Oil refinery processes	15–35	On the process unit's energy
Petrochemicals	5–10	On plant's total energy
Iron and steel	10–30	On plant's total energy
Chemicals	5–15	On plant's total energy
Food and drink	20–35	On plant's total energy
Pulp and paper	15–30	On plant's total energy

6.2.4.1 Energy Targeting

The targeting step includes the thermodynamic analysis of the process streams, their heat content, and the heat demand pattern, in order to develop optimized energy targets for each process unit, ahead of design. The methodology is based on the use of *composite curves*.

6.2.4.2 Process Modifications

This step considers how the energy targets may be improved by making changes to the core process. By use of a combination of the composite curves and *grand composite curves* the sequence of distillation columns and their operating pressures along with appropriate location of heat pumps can be evaluated before entering the process synthesis step and design of any heat exchanger network.

6.2.4.3 Process Synthesis

In this step, the heat integration scheme is produced. Synthesis may be an iterative process, whereby process design is successively changed to accommodate improved heat integration. The energy versus investment trade-off is considered and optimized during this phase. The 'pinch rules' and a specific network design methodology are employed.

6.2.4.4 Utilities Optimization

In the optimization step the cost of utilities is minimized by maximizing the use of the low cost ones. For example, given the choice, the use of LP steam for process heating is maximized, while the use of HP steam is minimized. The methodology is based on the *grand composite curve*.

6.2.4.5 Total Site Optimization

The *total site pinch technique* is used to optimize the use of utility across a site. It takes into account site steam demand versus site power demand and optimizes the operation of turbines, steam generation from waste heat, and the overall steam balance.

6.3 Introduction to Pinch Technology

6.3.1 The Concept of Quality of Energy

Process operations, like distillation or chemical reaction, are driven by energy. It may be supplied in various forms, such as the heat required for column reboiling, the heat of reaction, or refrigeration in low temperature processes.

The energy that is used by the process is not destroyed, as the first law of thermodynamics postulates. It only passes through the system, and having completed the task of keeping the process in operation, it is eventually rejected to the

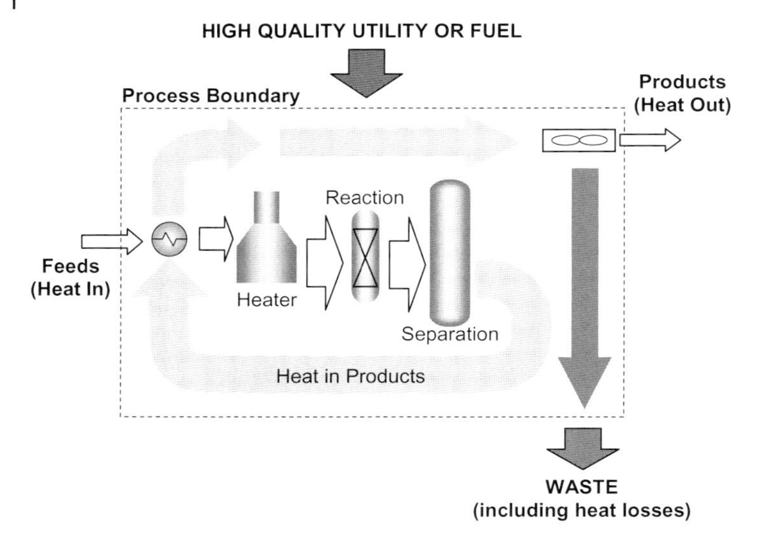

Figure 6.3 Process heat flows.

atmosphere as low grade (i.e., low temperature) heat. In other words, it is not the quantity, but the quality of the energy input that is destroyed by the process.

The process of the degradation and conversion of high quality energy input into low grade rejected heat is illustrated in Figure 6.3. It shows the costly (high quality) fuel being burned in the feed heater to supply energy to the process. A part of that heat is then rejected to the atmosphere at a lower temperature by cooling the reaction products down to their storage temperatures. Cooling water or air can be used as coolants.

It should be noticed that the low grade heat that is sent to waste has no good use. While the useful, high quality and versatile energy contained in the fuel is not destroyed by the process, it is degraded to the level of uselessness – a price to pay for process' feasibility and operability.

As shown in Figure 6.3, the process recovers some of the product's heat internally, in the feed/product exchanger, and only the remaining part is wasted in the product coolers. Assuming, for simplicity, that the temperatures of the feed and of the final products are similar, it follows, from the heat balance around the process boundary, that the quantity of the waste is approximately equal to the energy input! It also follows that if the internal heat recovery in feed/effluent exchanger is improved, the waste would be reduced, and again, from the overall heat balance, that the utility consumption (both heating and cooling) would reduce by precisely the same amount.

A question can be asked 'What is the minimum energy consumption of a super-efficient process?' This could be rephrased as: 'How much can the heat waste be reduced by improving heat recovery?' It seems to make sense for one to think that the waste could be reduced to close to zero, by recovering all product's heat into the feed and minimizing the losses, and that this would define the minimum energy consumption of the process at hand.

Not so! It will be shown later that there is a limit as to how much of the product's heat can be recovered internally, and that some waste heat will almost always be rejected. This constraint is revealed and explained by pinch analysis.

Pinch analysis uses a rigorous technique to suggest how to optimally match the individual demands for heat with the suitable supply. It further seeks to use the heat available in the internal process streams to minimize the use of the heating utility, and minimize the energy waste. In other words, it seeks to match a 'hot' stream (e.g., hot product) with a 'cold' one (e.g., feed) to maximize heat recovery.

The defining parameter in determining the suitability of a match is the 'quality' of its streams – the temperatures of the heat source and the heat sink. Pinch analysis ensures that the high quality of a high-temperature hot stream is not used prematurely to preheat a low-temperature cold stream (an example of this is the hot bottom product of a distillation column preheating the cold feed at the beginning of the preheat train, Figure 6.1). High temperature hot streams should be saved for high temperature services, for example, by using hot residue in the last exchanger of the feed preheat train.

A major advantage of using the pinch analysis over any other integration methodology is that the minimum consumption of the high value utility can be calculated and targeted just by analyzing the stream data, ahead of process design. This knowledge guides the designer through the process of optimizing the trade-off between the extent of heat recovery and the installed cost of exchangers.

6.3.2 Energy Targeting

Energy targeting, as well as many other pinch concepts, is best illustrated in the T–H (temperature–enthalpy) diagram, Figure 6.4. The 'hot' and the 'cold' process streams are shown as more-or-less straight lines. A hot process stream is the one that needs cooling from its supply to its target temperature (the reactor effluent, for example), and conversely, a cold stream is the one that needs heating (e.g., the reactor feed).

The T–H diagram can also be used to illustrate the heat recovery options. Figure 6.5 shows a two-stream heat recovery example. By superimposing the hot and the cold streams, and by then moving them to the left or to the right, the extent of the heat transfer between the streams can be varied. The heat exchanged is proportional to the horizontal length of the shaded area. The temperature differences that drive heat transfer in the exchanger are also shown as vertical distances between the streams.

If the streams are positioned close to each other (Figure 6.5a), the diagram depicts a 'high heat recovery' case, with high exchanger duty and relatively small temperature differences. Because of this the exchanger is going to be large. Conversely, if the streams are pulled apart (Figure 6.5b), the opposite would happen.

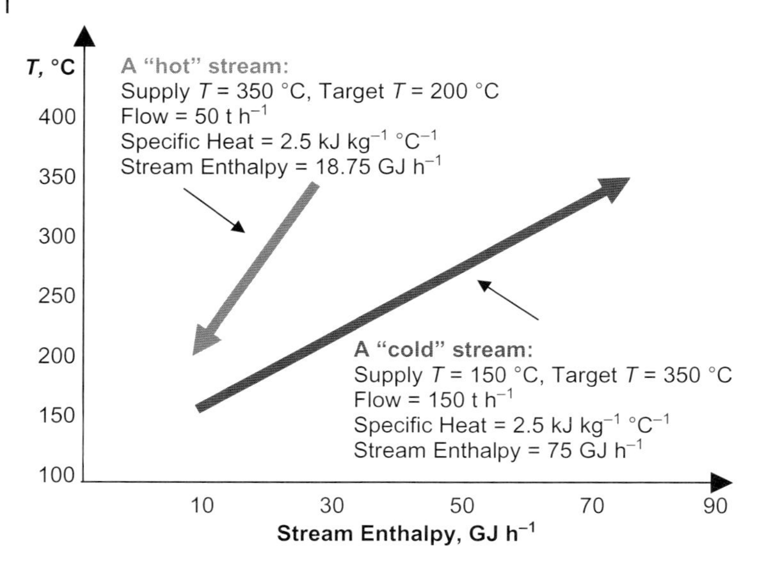

Figure 6.4 Illustrating process streams in *T–H* diagram.

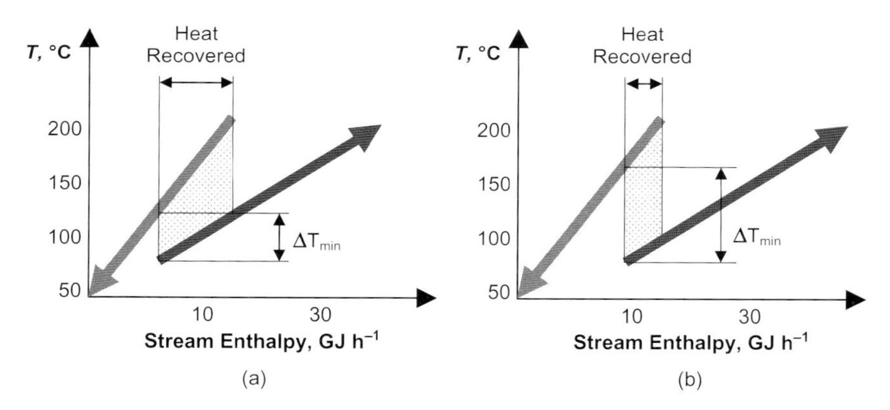

Figure 6.5 Using *T–H* diagram to analyze heat recovery.

In this simple example the designer will only have to decide on the size of one exchanger. This decision will be based on a trade off between the value of the heat recovered and the equipment cost. Essentially, the designer will determine the $\Delta T_{\min}$ that satisfies the desired return on investment, and the $\Delta T_{\min}$ so chosen would set the exchanger size.

If the two lines in Figure 6.5 are moved towards each other so that they touch, a maximum theoretical heat recovery can be determined. Of course, $\Delta T_{\min}$ would in that case be equal to zero, and the exchanger would be of infinite size, so to have a feasible heat recovery target, some positive $\Delta T_{\min}$ has to be maintained.

It is to be noted that some of the heat available in the hot stream cannot be recovered, because it is available at a temperature that is too low to be used against any part of the cold stream. This, low grade heat has to be wasted to air or water coolers, because it has no 'home' in the particular system at hand. It may, however, be used elsewhere, as part of the total site integration described later.

6.3.3 Composite Curves

The above two-stream example is simple. Many process units have multiple cold and hot streams, in which case the problem becomes more complex. A large number of combinations of hot and cold streams coupling in a network of exchangers may exist. Oil refinery or petrochemical processes provide examples of such complex heat recovery schemes.

The approach to a multi-stream problem, however, would be similar to that shown in Figure 6.5, with the difference that instead of using single streams in the analysis of their heat availability and the heat demand pattern, the streams would be combined in what are known as 'composite curves'.

The construction of a composite curve that combines two hot streams is shown in Figure 6.6, using the data in Table 6.2.

Table 6.2 Data for construction of a composite curve.

Stream	Start temperature (°C)	Target temperature (°C)	Duty (kW)	CP (kW °C^{-1})
1	180	80	2000	20
2	130	40	3600	40

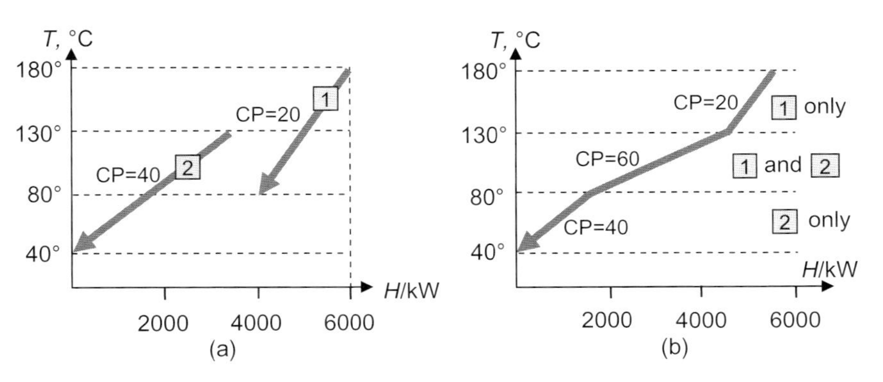

Figure 6.6 Construction of the hot composite curve.

The stream heat content, CP, is defined as (mass flowrate) × (heat capacity). For example, Stream 1 is cooled from 180 °C to 80 °C, releasing 2000 kW of heat, and so has a CP of 20 kW °C^{-1}.

Figure 6.6a shows the two hot streams plotted separately on the *T*–*H* diagram. The hot composite curve in Figure 6.6b is constructed by adding the enthalpy changes of the individual streams within each temperature interval. In the interval 130–180 °C only Stream 1 is present. Therefore, the CP of the composite curve equals the CP of Stream 1, that is, 20 kW °C^{-1}. In the interval 80–130 °C both Streams 1 and 2 are present. The CP of the hot composite here is the sum of the CPs of the two streams, that is, 20 + 40 = 60 kW °C^{-1}. In the interval 40–80 °C only Stream 2 is present, so the CP of the composite is 40 kW °C^{-1}.

The construction of a cold composite curve would be similar to that of the hot composite.

6.3.4 Setting the Energy Targets

To find the minimum energy target for a process, the hot and cold composite curves are placed on top of each other and moved around, as was done with single streams in Figure 6.5.

A pair of composite curves is shown in Figure 6.7. How close the curves can be moved towards each other is set by the minimum allowable temperature difference (ΔT_{min}) at the point of their closest approach. A value of 20 °C has been chosen in the example.

This place of minimum approach is called the pinch point, and its significance is so prominent, and its existence so insightful, that the whole methodology for the analysis and synthesis of heat exchange networks is named after it [2].

The composite curves almost always have that characteristic broken (kinked) shape, which sets a limit to how close together they can be moved. This leaves the cold end of the hot composite and the hot end of the cold composite 'uncovered', necessitating some minimum use of both the hot and the cold utilities. When

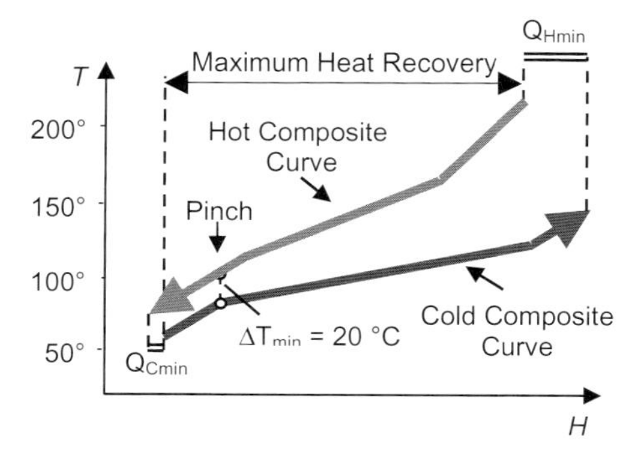

Figure 6.7 Using the composite curves.

analyzing individual processes, there will almost always be some part of the hot composite the heat of which cannot be usefully recovered, and will therefore have to go to waste.

The overlap between the composite curves shows the maximum possible heat recovery. The remaining heating and cooling represent the minimum hot utility requirement (Q_{Hmin}) and the minimum cold utility requirement (Q_{Cmin}) of the process for the chosen ΔT_{min}.

6.3.5
Setting the Area Targets

The minimum heat transfer area of the network, required to achieve the energy targets can be found from the composite curves (Figure 6.8):

$$\text{Network Area, } A_{min} = \sum_i \frac{1}{\Delta T_{LM}} \sum_j \frac{q_j}{h_j}$$

where:

- i denotes ith enthalpy interval
- j denotes jth stream
- ΔT_{LM} is log mean temperature difference in interval
- q_j is enthalpy change of jth stream
- h_j is heat transfer coefficient of jth stream.

This area target is based on the assumption that 'vertical' heat exchange will be adopted between the hot and the cold composite curves across the whole enthalpy range.

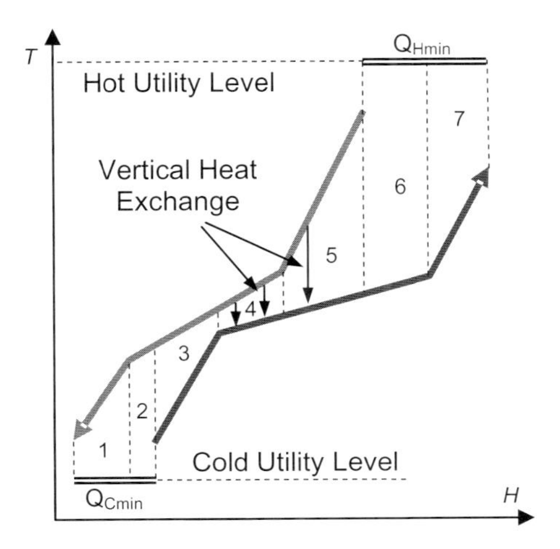

Figure 6.8 Area targeting.

This vertical arrangement, which is equivalent to pure counter-current heat exchange within the overall network, gives a minimum total surface area. In new designs, where there are no existing exchangers, it should be possible to configure a network with the total area that is close to this target.

6.3.6
Capital/Energy Trade-off

As with single streams, the choice of ΔT_{min} is decisive for the project economics. The value is determined as a trade-off between the value of the heat recovered and the investment cost.

Although there are occasions where pinch analysis can lead to savings in both energy *and* capital, saving energy generally implies increased capital spending.

The trade-off can be analyzed by examining the composite curves (Figure 6.9).

As the separation between the hot and the cold composite curves (ΔT_{min}) increases from 10 °C to 20 °C, their overlap decreases, thereby reducing the recovery of heat from the hot into the cold streams, and consequently increasing the utility demands. On the other hand, with the curves further apart, the temperature differences between them and the driving forces for the heat exchange will increase. This reduces the required exchanger area.

The energy and the exchanger area targets can be calculated for different values of ΔT_{min} using the procedures described in Sections 6.3.4 and 6.3.5. Knowing the unit costs of energy and exchanger area, the cost curve can be developed. Figure 6.10 shows the typical curves of the capital and the energy cost as a function of ΔT_{min}. There is an optimum value for ΔT_{min} that gives the lowest total annual cost of the utilities and the capital.

The shape of the composite curves affects the resulting optimum ΔT_{min}. Typically, for curves that are almost parallel a higher value will be chosen than in the cases where they sharply diverge.

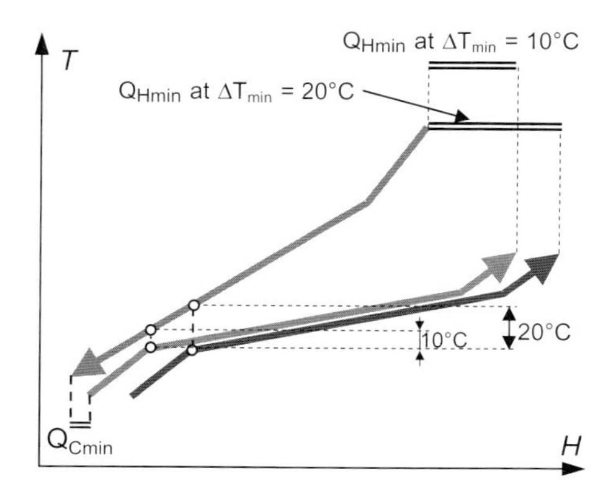

Figure 6.9 The effect of increasing ΔT_{min}.

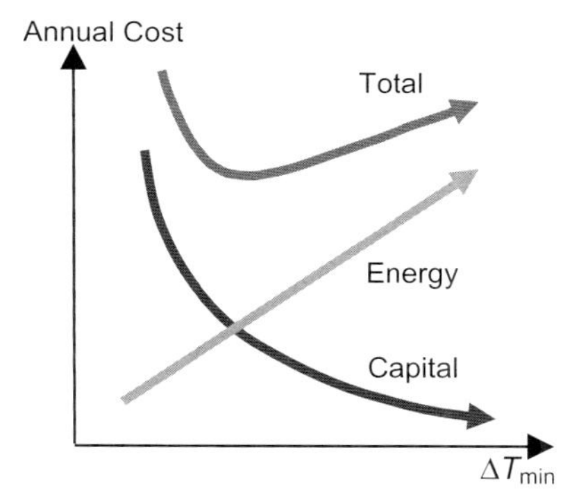

Figure 6.10 Energy/capital trade-off.

For chemical processes the values are typically between 10 °C and 20 °C. However, in systems with high exchanger fouling, or where the heat transfer coefficients are low for other reasons, ΔT_{min} of 30–40 °C may be chosen. Sub-ambient temperature processes using refrigeration are designed with a close approach (3–5 °C) to minimize the load and power demand of refrigeration compressors.

6.4 Minimizing the Cost of Utilities

6.4.1 Utility Costing

Processes use different forms of energy, and some forms are costlier than others. For example, electricity is more expensive than steam per unit of energy delivered, and refrigeration is costlier than cooling carried out by air or water. Any energy cost reduction program will endeavor to replace a high cost energy form with a lower cost one. An example is the use of low pressure (LP) where possible, instead of using high pressure (HP) steam.

Correct energy and utility costing is therefore a prerequisite and an essential element of effective energy management. What is relevant here is the *marginal* (incremental) cost – the net cost of producing (or purchasing) one additional unit of energy.

The marginal costs will need to be calculated for fuel, power and steam at different pressure levels, and sometimes for cooling water and refrigeration. This calculation starts with identifying the marginal mechanism – the way in which the plant responds to an incremental change that is made somewhere in the system. For example, when steam consumption is increased a series of changes occur throughout the utility system: more fuel is burned in a boiler, more boiler feed water is needed, the boiler blowdown increases, the de-aeration steam demand increases, the load on a turbo-generator transferring steam from high pressure to low pressure may increase etc. All these changes will in one way or another affect the operating cost, and they should be correctly depicted and used in the calculation of the incremental cost of steam.

The calculation of the cost of a plant's marginal fuel is quite simple when the marginal fuel is purchased from outside. It is equal to the purchased price. The calculation can, however, be more complex if the fuel is an intermediate product, such as oil refinery vacuum residue, the true value of which will depend on its blending properties as a component of the sales fuel oil.

The marginal mechanism for increased power consumption will normally be the increase in power import. The marginal power cost will be the cost of purchased power. If, however, the site is a power exporter, the mechanism will be the reduction in power exports. If the site is in power balance, the marginal mechanism may be an increase in turbo-generator load (typically a condensing turbine). In each case, the cost of marginal power would be different, but quite straightforward to calculate.

It is the cost of steam, and especially that of the medium and the low pressure steam, that deserves special attention. It is surprising how many operators value their steam incorrectly, which obviously can lead to incorrect operating as well as investment strategies.

To calculate the marginal cost of steam rigorously, it is best to use a simulation model of the steam and power system. The underlying logic of this calculation will be presented here in a simplified form.

A simple steam system is shown in Figure 6.11. It consists of two pressure headers, a turbo-generator between them, a boiler, and a de-aerator. The flow through the turbo-generator controls the LP header pressure. Any variation in the

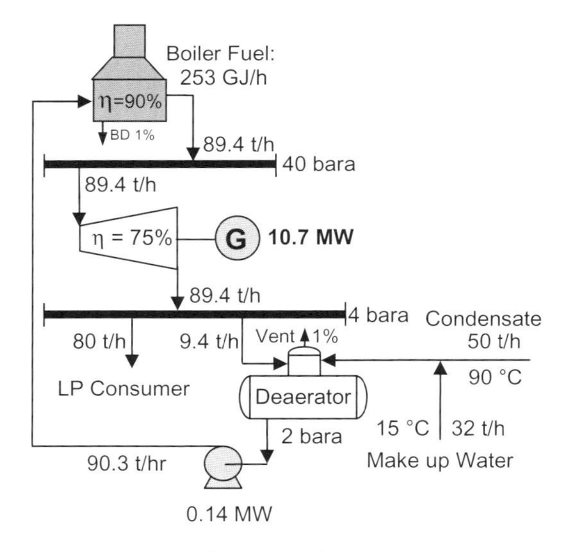

Figure 6.11 A simple steam and power system.

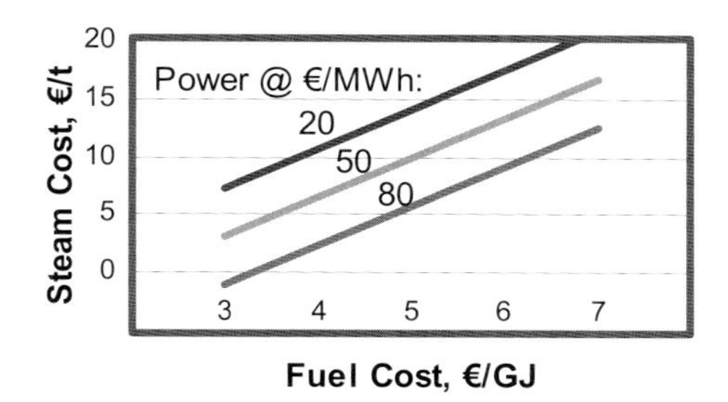

Figure 6.12 Low pressure steam cost (condensate lost).

LP steam demand would therefore change the generator steam flow and the amount of power that is produced. This change in power generation will affect the net cost of the LP steam.

The calculation of the marginal cost of HP steam is straightforward – the changes to its demand do not significantly affect power generation. The cost of HP steam is approximately equal to the value of the extra fuel that is burned to generate an extra ton of steam. Note: it is not *exactly* equal to the fuel cost, because there would still be a small change in power generation due to the increased use of LP steam for de-aeration, which slightly changes the steam flow through the turbo-generator.

The net cost of LP steam will be equal to the cost of HP steam *minus* the value of the power that the steam generates on its way from the HP to the LP headers. The value of the power generated depends on a number of things such as: header pressures, steam temperature, turbine efficiency and the cost of the imported power. The higher the value of the power produced, the lower will be the net cost of marginal LP steam.

For the system shown in Figure 6.11 the cost of marginal LP steam, as a function of fuel and power costs, is shown in Figure 6.12. It is assumed in the example that the marginal condensate is not returned. At high power costs and low fuel costs, the cost of marginal LP steam can be very low, and even negative. The negative cost of LP steam implies that it is profitable to vent the steam in order to maximize power generation.

It is sometimes observed that operators calculate the LP steam cost on an enthalpy basis. This is erroneous if LP steam generates power. The enthalpy-based valuation would in that case overestimate the cost of LP steam.

The values of the utility costs used in the pinch analysis will affect its results. Those costs will therefore need to be calculated and selected with proper care and understanding.

6.4.2
Targeting for Multiple Utilities: The Grand Composite Curve

The use of the composite curves to target process energy consumption in order to achieve the optimum balance between the energy cost and the exchangers' cost was described in Section 6.3. Although the procedure sets the total energy consumption target for a process, it does not specify which energy form, or which kind of hot or cold utility should be used.

Most processes are heated and cooled using several utility levels, for example different steam pressure levels, refrigeration levels, hot oil circuit, furnace flue gas, etc. The costs of these utilities may vary considerably, as discussed in Section 6.4.1. To minimize the overall energy costs, the designer's object would be to maximize the use of the cheaper utility and minimize the use of the more expensive ones. For example, it is preferable to use LP steam instead of HP steam, and cooling water instead of refrigeration when possible. The use of different utilities is another target that can be developed by pinch analysis.

Consider Figure 6.13a – it shows the composite curves for a design case where the heating target is entirely met by using HP steam, available at temperature T_{HP}. At the same time, LP steam is also available, at temperature T_{LP}.

Figure 6.13b illustrates a possible use of LP steam, replacing the dotted section of the hot composite, which is now pushed to the right and thereby 'saved' to exchange heat against the hotter part of the cold composite. If the object is to maximize the use of LP steam, the designer can keep increasing its usage until another pinch, called the 'utility' pinch, is created. This sets the LP steam target.

One can use the composite curves to carry out the 'utility cost minimization' exercise as shown in Figure 6.13, but the proper tool for setting multiple utilities targets is the grand composite curve. This is the plot of process energy deficit (when analyzing the above-the-pinch heat transfer) and the energy surplus (when looking at the below-the-pinch area) as a function of temperature.

The grand composite curve is constructed by plotting the horizontal distances between the hot and cold composite curves at different temperatures (Figure 6.14). It provides a graphical representation of the heat flow through the process: from hot utility to the process (above the pinch), and from the process to the cold utility (below the pinch).

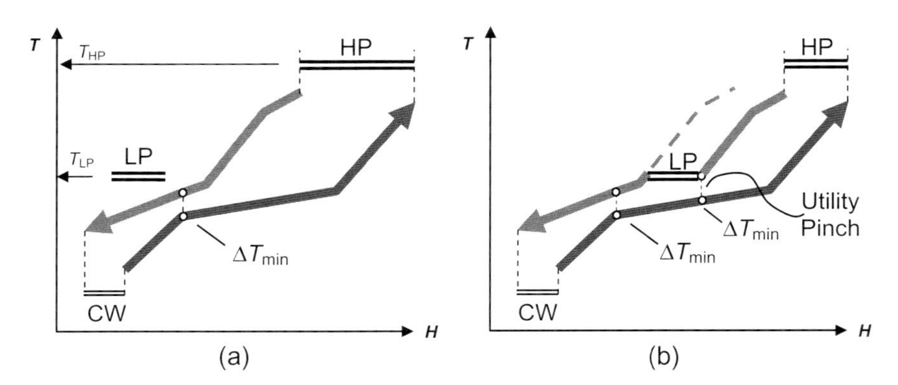

Figure 6.13 Using low pressure steam instead of high pressure.

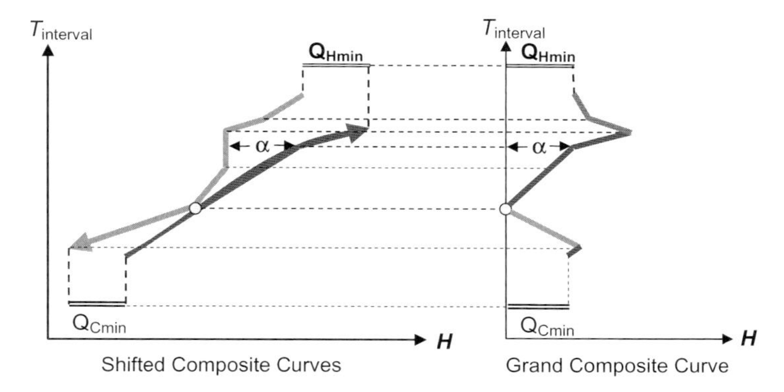

Figure 6.14 Construction of the grand composite curve.

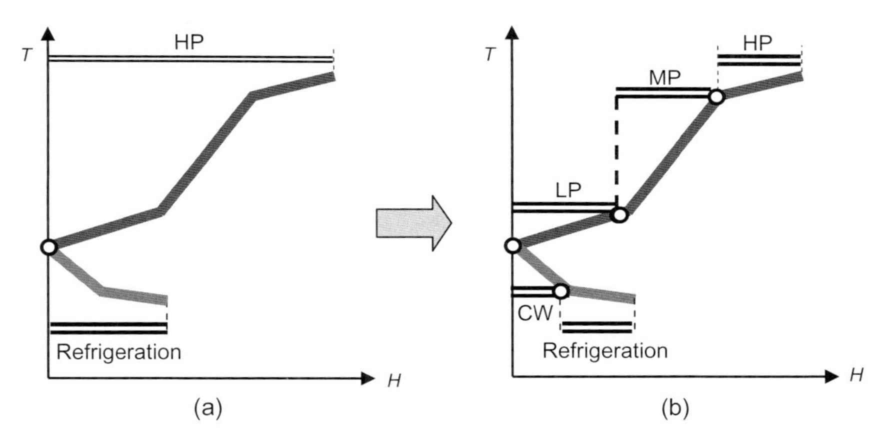

Figure 6.15 The grand composite curve for multiple utilities targeting.

As will be discussed later, when the pinch rules are introduced, an ideal process does not transfer heat through the pinch point. Because of this, the grand composite curve, representing the heat flows, has that characteristic shape of touching the y-axis at the pinch point.

There is a small, mathematical adjustment to make to the composite curves prior to converting them to the grand composite curve. The separate hot and cold composites are 'shifted' by moving them down (hot curve) and up (cold curve), each by half the ΔT_{min} until they touch at the pinch point. The resulting composite curves are referred to as shifted curves and have no real physical meaning but are merely a step in the construction of the grand composite curve. This ensures that the resulting grand composite shows the required zero heat flow at the pinch point.

Figure 6.15a shows a grand composite curve for a case where the HP steam is used for heating and refrigeration is used to cool the process. In order to reduce the utility cost, the intermediate utilities are introduced: LP steam, medium pressure (MP) steam and cooling water (CW).

The target for LP steam (the least cost hot utility) is first set by plotting a horizontal line at the LP steam temperature from the y-axis until it touches the grand composite curve (Figure 6.15b). The MP steam target then follows in a similar way. The remaining heating duty is finally satisfied by HP steam. This minimizes HP consumption in favor of LP and MP steam and thus minimizes total utility cost. A similar construction below the pinch maximizes the use of cooling water ahead of refrigeration.

The points where the LP, MP and CW levels touch the grand composite curve are called 'utility pinches'. Similarly to the process pinch, the existence of the utility pinch sets a limit to how much heat can be exchanged between a particular utility and the process.

6.4.3 Total Site Integration

The overall process design of a chemical plant is a complex task that may start with designing the reactor and the separation system, followed by heat exchanger network and finally the design of the utility systems.

In the 'total site' approach, the designer seeks to take advantage of the interaction between these four parts of the system, particularly in the choice of utilities that would be best suited to the rest of the process systems [4].

Where appropriate, total site analysis can be used to extend the pinch analysis to site-wide integration of a number of processes by means of the utility system.

A direct integration of two or even three processes is common. However, to heat-integrate more units, or across the whole site, while potentially beneficial is in most cases impractical. A wider integration may pose operational problems, and is in most cases infeasible, considering the physical distances between the units. However, an indirect site-wide integration can be achieved by using the utility system as the intermediary. For example steam can be generated in a unit that has heat in excess, and the steam can then be used in a unit that requires heating. Matching these heat sources and heat sinks is central to total site analysis. It is applied as follows:

- In grass-roots designs the choice of utility levels, such as steam header pressures, is almost unlimited. Total site technique is used to match the overall site sources and sinks and determine the most appropriate utility levels. The site profiles are developed by using the grand composite curves of each process. The designer sums up the utility requirements of all processes above the pinch thus creating the site 'sink profile', and below the pinch, creating the site 'source profile' (Figure 6.16). The source and the sink profiles are used to determine how much heat and at what temperature can be exchanged between the processes, by using the steam network as the site's energy transmission system.

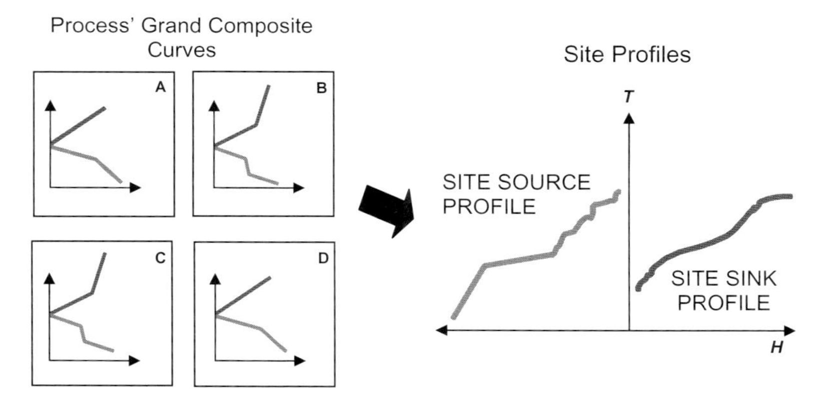

Figure 6.16 Developing the site profiles.

- In revamps – total site techniques identify the inefficiencies in utility use across a site and find opportunities for exploiting the potential that is there. In many cases it is possible to balance steam supply to consumers while taking advantage of existing steam turbines to generate power.

6.4.4 Steam and Power System and Efficient Power Generation

Steam and power systems tend to be large and complex in process industries. They consume substantial amounts of fuel, and are therefore subjected to extensive study and optimization. Good understanding of what improves the steam system's efficiency provides an insight into its use for effective 'total site' integration.

Power can be produced from heat using various cycles and power engines. Due to the very nature of this process, some heat will always be rejected as low grade energy, and will not be converted to power.

Any industrial facility that consumes heat or steam provides a heat sink which enables generation of power at high efficiency. While conventional power plants using a condensing steam turbine cycle (the 'Rankine' cycle) may achieve power generation efficiencies of no more than about 40%, a backpressure turbine would generate power essentially at the boiler efficiency (i.e., 90% or higher). This is because the condensing cycle sends the rejected heat from the turbine outlet to waste, while the steam at the outlet of a backpressure turbine is used by the industrial process (Figure 6.17). Maximizing backpressure power generation would be a clear objective of an energy-conscious process designer.

The steam system of an energy efficient industrial plant would therefore include HP steam boilers, typically >40 bar. The steam demand at lower pressures will be met by passing the HP steam through efficient backpressure turbo-generators. The following set of additional rules will further guide efficient design:

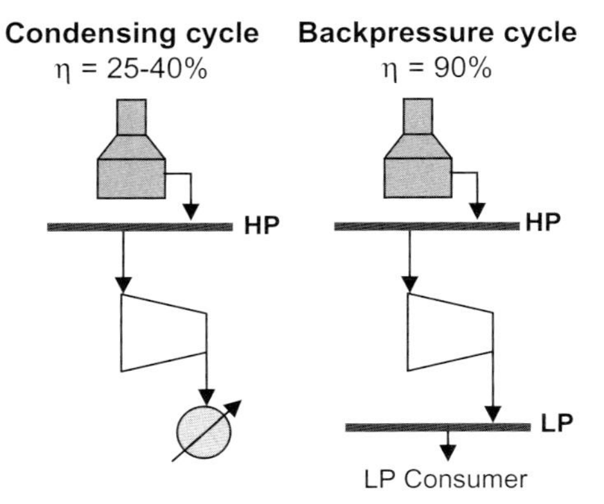

Figure 6.17 Power generation cycles.

- Consume steam at the minimum pressure that the user's temperature profile allows.
- Minimize the use of small process turbine drivers, and maximize the use of electric motors.
- When possible, avoid generating LP steam from waste heat–instead use any waste heat available to preheat feeds to process units, or to preheat furnace combustion air.
- For high power consuming sites consider the use of gas turbines coupled with their waste heat recovery boilers.

The above features of an efficient steam system will also direct the total site integration. For example, the designer will need to ensure that the source profile is used to generate the highest pressure steam possible, because this would maximize power generation. There will be incentive to replace HP steam consumers with the lower pressure ones, and save on the relative costs, etc.

6.4.5 Options for Low Grade Heat Use

With the total site approach the whole plant becomes heat integrated. Whatever the extent of this integration, however, there will almost always be some low grade heat that is in excess and remains unused, that is, wasted to air or cooling water. The amount of the waste may be substantial, and questions naturally arise as to how this 'free' energy could be used. Obviously, an additional useful heat sink has to be found. It can be located inside or outside the facility that produces the excess heat. A 'useful' sink is either an existing consumer whose energy requirement incurs cost (e.g., a low temperature consumer that uses steam), or a new consumer that is unprofitable to run at the actual energy prices (e.g., refrigeration for additional process cooling).

Potential uses of the low grade heat are:

- Install a hot water circuit as a new utility level. It would recover heat from waste heat sources which are too low in temperature to generate LP steam. The hot water could then be used for the following purposes:
  - Heating low temperature heat sinks to save LP steam
  - To drive an absorption refrigeration unit to produce chilled water. This can be used to reduce power demand from conventional refrigeration units or air conditioning units.
- District heating. Exporting heat 'across the fence' for either municipal heating or to a neighboring industrial consumer. This will be further discussed in Section 6.7.1.

In addition, there are two technologies that allow power generation from waste heat: the organic Rankine cycle and the Kalina cycle. However, they both have limited application due to the relatively high investment cost, and very low efficiency (less than 10%) of conversion of heat to power in the range of 110–140 °C, which would be considered for a hot water circuit or LP steam.

6.5
Process Synthesis

The targeting procedure of pinch technology sets the energy consumption targets ahead of design. One can sensibly question weather it is always possible to design the process that meets those predetermined targets. The answer is affirmative – pinch technology provides the methodology for designing heat exchanger networks that would invariably produce configurations featuring the targeted utility consumption.

This process of configuring the heat exchanger network is called 'synthesis'. In the formative days of pinch technology, a manual procedure was used. This is well documented [5, 6] and the interested reader may find it useful to familiarize with the methodology. Today, however, heat exchanger network design is almost fully automated, and the advanced software tools that combine stream analysis, targeting and network design are extensively used.

Either manual or automated, the synthesis methodology is built around several basic pinch rules.

6.5.1
The Pinch Rules

The existence of the pinch point divides the heat recovery region, defined by the composite curves, into the hot and the cold sections. They are in heat balance with their respective utilities.

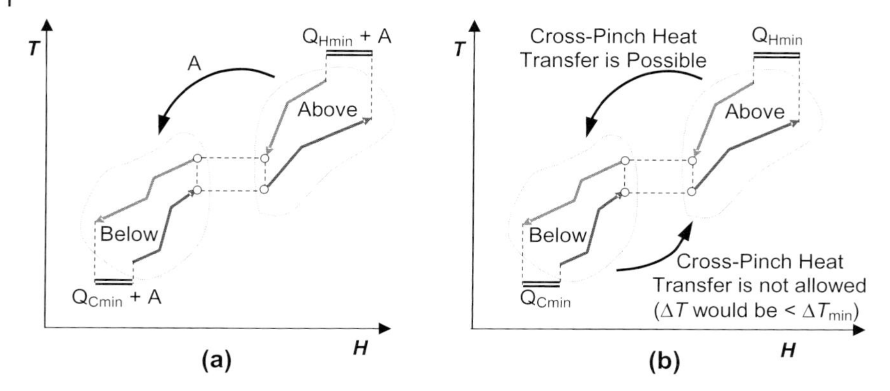

Figure 6.18 Cross-pinch heat transfer.

In principle, the region above the pinch requires only hot utility, and that below the pinch only the cold one. An ideal design would reflect this arrangement.

Following the logic of this separate heat balances above and below the pinch, the design rules instruct that no heat should be transferred from a hot stream placed above the pinch point to a cold stream placed below the pinch point. If such heat is transferred from the hot side of the pinch to its cold side, the hot and cold targets would not be met. Instead, this cross-pinch heat would need to be replaced by an equivalent amount of hot utility above the pinch, and, consequently, the consumption of the cold utility below the pinch (air, cooling water, etc.) would be increased by the same amount (Figure 6.18a).

As shown in Figure 6.18b, it is possible to use a hot stream to perform some low temperature heating, and indeed many sub-optimal designs feature such heat recovery. However, a careful designer would not allow the transfer of heat from the region below the pinch to the region above the pinch, because it would violate the chosen ΔT_{min}.

To meet the hot and cold utility targets, suitable matches for this heat must be found on the hot side of the pinch rather than on the cold side.

There are three rules for achieving the minimum energy target for a process.

- Heat must not be transferred across the pinch;
- There must be no cold utility used above the pinch;
- There must be no hot utility used below the pinch.

Violating any of these rules will lead to 'cross-pinch' heat transfer, which increases the energy requirement beyond the target. The 'pinch equation' defines this rule:

$$A = T + XP$$

Where:

A = Existing (or actual) energy consumption
T = Target energy consumption
XP = Cross pinch

In other words, the difference between current energy use and the target is the sum of all cross-pinch inefficiencies.

Eliminating any cross-pinch heat transfer would be the focal point of many revamping projects of existing exchanger networks.

6.5.2
Network Design

Network designs are usually and conveniently carried out and presented in a 'grid' diagram. The hot streams are shown as horizontal lines running from left to right, and cold stream running from right to left (Figure 6.19).

A heat exchanger transferring heat between two process streams is shown as a vertical line joining the circles that are placed on the two matched streams. Heater and coolers are shown as single circles placed on the relevant streams. The pinch division is clearly shown in the grid diagram, and consequently any cross-pinch heat transfer would be conveniently indicated as an exchanger stretching across the pinch division line.

The network design procedure [5, 6] can briefly be summarized as follows:

- To ensure that no cross-pinch heat transfer occurs, the design starts at the pinch, by placing the 'pinch' exchangers first. These are the exchangers that bring a stream precisely to the pinch temperature.
- After these exchangers have been correctly placed and any cross-pinch is thus prevented, the remaining exchangers are placed according to designer's preferences and convenience.
- Ensure that the heat is properly 'cascaded', meaning that the high temperature hot streams are used against high temperature cold stream, that is, that the heat transfer is 'vertical' when observing it in the composite curves.
  - Criss-crossing, when it happens on the same side of the pinch does not incur energy penalty, but would be wasteful of the temperature driving forces, and thereby of exchanger area.

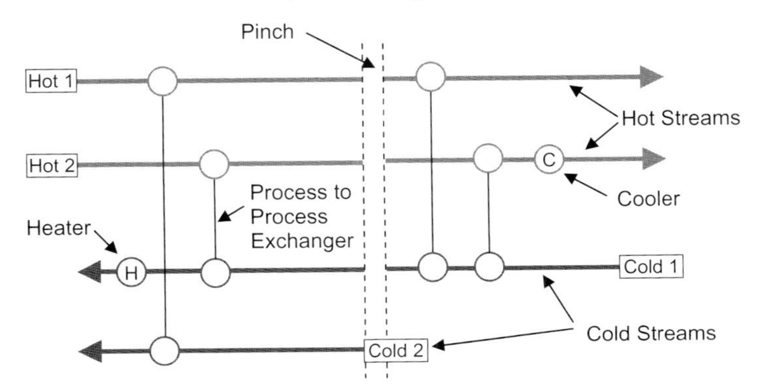

Figure 6.19 Grid diagram.

6.5.3 Network and Process Design Interaction

6.5.3.1 Process Modifications

The minimum energy requirements set by the composite curves are based on a given process heat and material balance. The targets and the optimum heat exchanger configurations identified using the composite and grand composite curves are based on these fixed process conditions. The proposed configuration will not have any effect on the fundamental energy requirements and the process temperatures.

In this context, *process modifications* are changes to these conditions. As the designer modifies the heat and material balance, the composite curves are changed, and consequently the heat recovery. Appropriate process changes can 'bend' the curves in a way that actually improves the heat recovery.

There are several important parameters that the designer may want to review. These include distillation column operating pressures and reflux ratios, feed vaporization pressures, column side cooling (pumparound) duties and flowrates, etc.

For example, increasing the heat removal duty in pumparounds (side coolers) of distillation units 'flattens' the hot composite and increases heat availability above the pinch, thus reducing the use of hot utility.

Figure 6.20 shows a distillation column with four products and two pumparounds (P/A-1 and P/A-2). Increasing the bottom pumparound duty and reducing the overhead condenser duty correspondingly changes the shape of the hot composite and reduces the hot utility target at constant ΔT_{min}.

Pumparound maximization strategy is often employed when designing or revamping distillation columns for improved energy performance. It is to be noted that the increase in pumparound duty reduces the column reflux, which in turn affects the fractionation sharpness. A careful designer will take this into consideration when optimizing the combined yield and the energy performances of the unit.

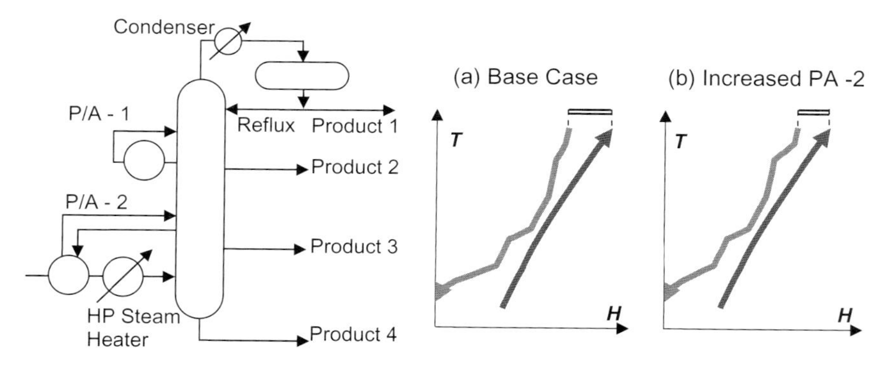

Figure 6.20 Process modification changes the shape of the composite curves.

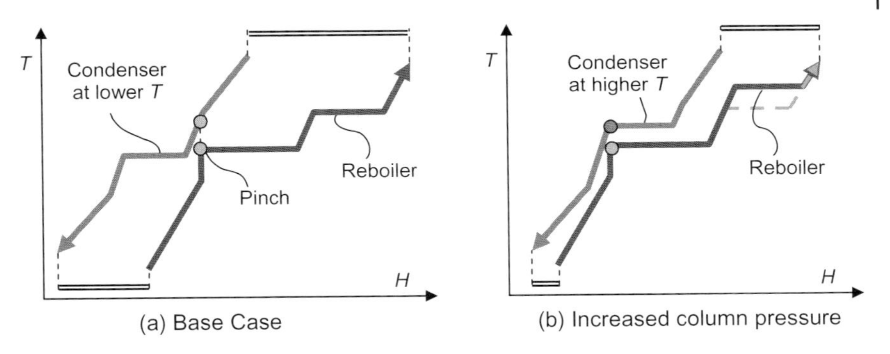

Figure 6.21 Adjustment column pressure to improved heat integration.

Another example of manipulating process design parameters in order to improve heat integration is the adjustment in condenser temperature by changing the column pressure.

The set of composite curves in Figure 6.21 shows a reboiler and a condenser of a distillation column. The cold composite has another flat part in it, representing another vaporizer. If the column pressure is sufficiently increased, its condenser temperature may become high enough to preheat the flat part (the vaporizer) of the cold composite, and the hot utility requirement would be greatly reduced.

6.5.3.2 **The Plus/Minus Principle**

The number of choices and potential modifications is large. An exhaustive search to identify the three or four such parameters that could be changed to the overall benefit of the process would be time consuming. However, thermodynamic rules based on pinch analysis can be applied to identify the key process parameters that can have a favorable impact on energy consumption.

In general, the hot utility target will be reduced by:

- increasing hot stream (heat source) duty above the pinch;
- decreasing cold stream (heat sink) duty above the pinch.

Similarly, the cold utility target will be reduced by:

- decreasing hot stream duty below the pinch;
- increasing cold stream duty below the pinch.

The set of four simple rules, described in the four dot-points above, is termed the '+/−principle' for process modifications [2]. This simple principle provides a reference for any adjustment in process heat duties, and indicates which modifications would be beneficial and which would be detrimental.

As shown in the example Figure 6.21, it is sometimes possible to change temperatures rather than the heat duties. It is clear from the composite curves that temperature changes that are confined to one side of the pinch will not have any effect on energy targets. However, temperature changes *across the pinch* can change them. The pattern for shifting process temperatures can be therefore summarized as follows:

- shift hot stream from below the pinch to above;
- shift cold streams from above the pinch to below.

6.5.3.3 Integration Rules for Various Process Equipment

In terms of their 'appropriate placement' relative to the process pinch point, the following equipment items deserve special attention:

- **Distillation columns:** Although not always possible, the most energy efficient integration of a distillation column is on one side of the pinch, so that either its condenser duty can be used to heat up the background process (if the column is placed above the pinch), or its reboiling duty can be supplied by the process (if the column is placed below the pinch).
- **Heat pumps:** The appropriate placement of a heat pump is across the pinch, meaning that heat is taken from below the pinch and rejected above the pinch. A similar principle applies to refrigeration systems.
- **Gas turbines:** Gas turbines should be integrated above the pinch, so that the exhaust heat is fully used. This is usually the case, as the hot flue gas is at high enough temperature to generate HP steam.

6.6 Revamping Heat Exchanger Networks

Retrofitting existing networks is a much more complex task than designing the new ones.

If an engineer were to start a heat exchanger network design with a blank sheet of paper and a good knowledge of pinch technology, the resulting minimum energy design would, almost always, be appreciably different from the existing design for the same process.

Revamping an existing network to achieve a 'pinch design' can therefore be quite costly, considering the changes that would have to be made to existing equipment. Because of this, different rules have been developed for retrofits to obtain the best energy savings from the fewest modifications to the existing design – a pragmatic approach that recognizes the limits concerning the changes that can be economically made to an existing network. It is also recognized that the final design will rarely be identical to the ideal grassroots design obtained from Pinch principles.

6.6.1 Area Efficiency Method

The minimum heat exchanger network surface area is obtained when all heat flows from hot streams to cold streams are vertical when analyzed against the

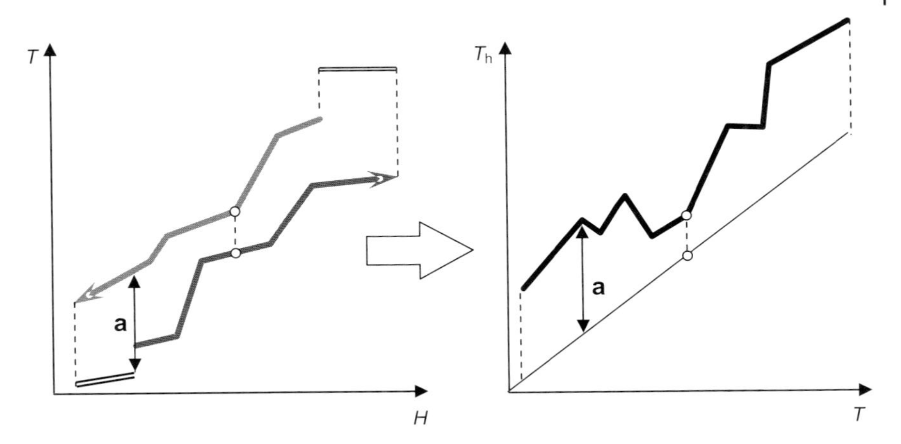

Figure 6.22 The driving force plot.

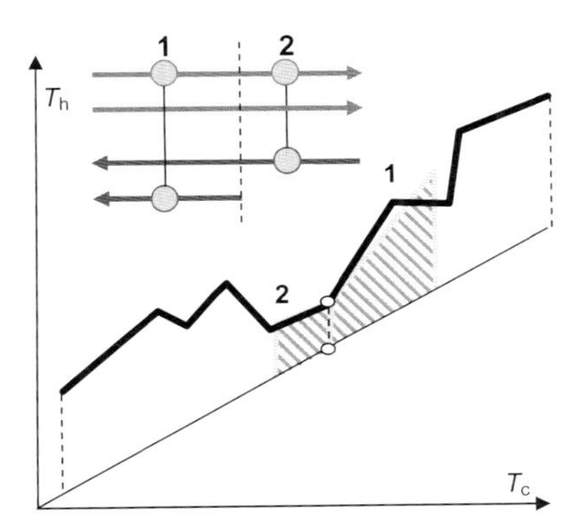

Figure 6.23 Using the DFP to assess exchanger positioning.

composite curves. One approach to retrofitting existing networks is to check the individual exchangers and see if their surface areas are used effectively [7].

This introduces the concept of the 'driving force plot' (DFP), which is a graph of ideal temperature differences (between hot and cold streams) across the whole temperature range. It is derived from the composite curves (Figure 6.22).

Once this diagram is prepared, the 'fit' of individual heat exchangers to the ideal profile can be assessed. If the matches closely follow the DFP, this results in a vertical heat transfer that leads to minimum area designs.

In Figure 6.23 the two highlighted heat exchangers are well-designed from basic pinch principles, and show a good fit with the DFP.

In most existing networks, individual heat exchangers will not follow the ideal temperature profiles (even if there is no cross-pinch heat transfer) and, consequently, the surface area required is more than the minimum required, thermodynamically. This tends to be because exchangers may have a larger than ideal temperature difference at one end, the benefit of which is outweighed by a smaller than ideal temperature difference at the other end. Overall, the heat transfer per unit area is less than ideal. An experienced pinch engineer may use the DFP to ensure that each exchanger in the network fits closely to the ideal plot, and hence contributes to the minimum surface area of the overall network.

6.6.2 Modern Retrofit Techniques

Modern, pragmatic approaches to network improvement seek to squeeze the best performance out of the existing exchangers and minimize the need for new exchangers. Typical retrofits may involve surface area enhancement equipment, such as tube inserts and twisted tube exchangers, and often one new exchanger or exchanger shell, but will avoid extensive changes to the network.

The advanced design techniques include the use of 'loops' and 'paths' within a network (Figure 6.24), because these provide a degree of freedom to adjust the heat flows. Paths are the flow trails within the network. They connect the cold and the hot utilities, and because of this any improvement in the heat recovery along a path can reduce the consumption of both utilities (Figure 6.25). A loop is a closed energy path within the network.

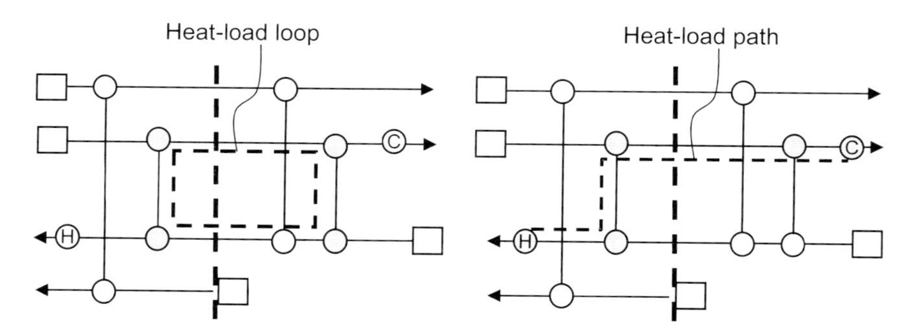

Figure 6.24 Loops and paths.

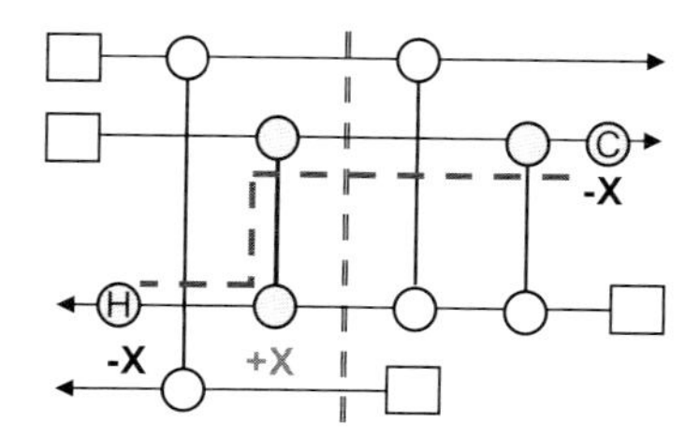

Figure 6.25 Exploiting a path to save energy.

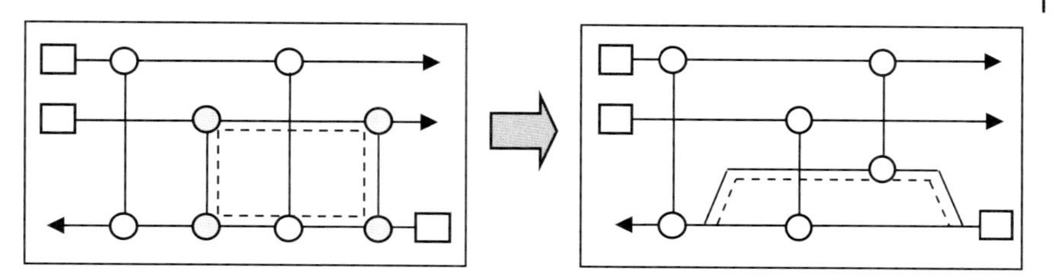

Figure 6.26 Exploiting a loop the reduce the number of exchangers.

In a grassroots design, the skilled engineer would try to eliminate the loops, as they are always caused by having more than the minimum number of heat exchangers and indicate some redundancy in the network. By breaking a loop, the number of heat exchangers, and hence investment cost, will reduce (Figure 6.26).

In a retrofit design, however, paths are more important, and they form the basis of the *network pinch*, a relatively recent development in pinch analysis. Network pinch addresses the additional constraints imposed by the specific configuration of an existing facility. It is aimed specifically at finding the best energy savings for the least investment cost.

6.6.3
The Network Pinch

Existing networks can usually be improved by using paths to shift the loads between exchangers, but eventually a design will be reached from which no further improvement is possible, although it is still far from the pinch target. The starting network configuration imposes a constraint that hinders further improvement.

Network pinch analysis identifies the heat exchanger forming the bottleneck to increasing heat recovery and provides a systematic approach for removing this bottleneck. It is a step-by-step method for implementing energy savings in a series of consecutive projects.

Once the offending exchanger is identified, following four options can be considered for removing the constraint [8].

- re-sequencing – reversing the order of exchangers to improve heat recovery;
- re-piping – changing the matched streams to improve heat recovery;
- adding a new exchanger – to change the load on the offending exchanger;
- stream splitting – to reduce the load on a stream in the offending exchanger.

This is a software-led process, so that all possible paths in the network are explored and new ones identified by, for example, placing one new exchanger. Each path is then tested to see how much energy can be economically squeezed from that path, and the various paths are ranked in terms of their potential energy saving. It has been shown that these paths are independent of each other and a series of paths can be successively analyzed to produce a cumulative saving. The

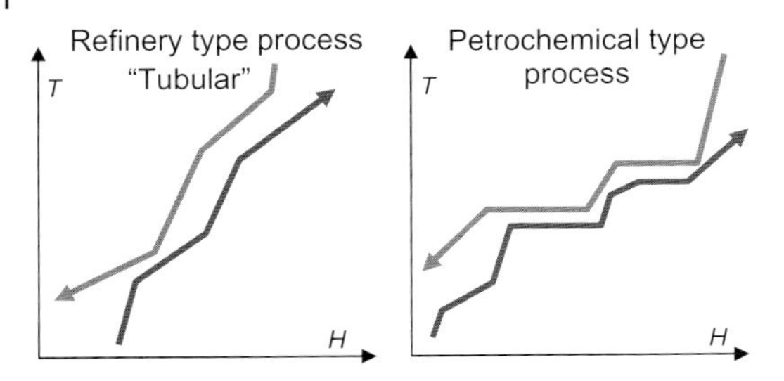

Figure 6.27 Two types of composite curves.

network pinch revamp method is therefore sequential, but it examines various configurations in a systematic way, at the same time allowing the designer to interact with the software-led design procedure.

The network retrofits tend to fall into two different categories because there are, generically, two types of heat exchanger networks: the 'oil refining' type and the 'petrochemical' type (Figure 6.27).

The 'oil refining' type consists of sloping composite curves displaying a long tubular region like an extended pinch point and single hot and cold utilities. Refining networks are usually quite complex.

The 'petrochemical' type displays horizontal segments in the composite curves (change of phase), multiple utilities and fewer exchangers per stream.

In the 'oil refining' type, the paths are used first and the loads are shifted to improve heat recovery, usually by area enhancement on selected exchangers. Once the network pinch is reached, the four options presented above are exploited. The retrofit design often spans a large section of the process, and the projects may affect a number of exchangers.

The analysis of the petrochemical type of network usually involves successive removal of small numbers of exchangers from the network (two or three at a time), and re-targeting each time to identify the potential saving. Petrochemical revamps tend to cover smaller sections of the process and produce more or less independent projects, each affecting few exchangers only.

6.7 Other Applications of Pinch Technology

6.7.1 Area Integration

Previous sections have discussed the application of pinch analysis to individual process units. The technique for extending the analysis to the overall manufacturing site, total site analysis, is also described.

In certain cases, pinch analysis can be applied to whole areas, usually exclusively industrial but sometimes integrated with communities [9].

An area-wide pinch technology analysis was applied to Kashima chemical complex in Japan [10]. This case study showed that although the individual plants in the complex were quite efficient, there was large energy saving potential through energy sharing among the various sites.

A similar study was carried out for the Rotterdam industrial area, with over 60 manufacturing sites, including four oil refineries and several petrochemical, chemical and food processing plants and a municipal incinerator. The study identified around 2000 MW of recoverable waste heat from these companies, of which 400 MW could be shared between them, 500 MW could be used for power generation and 1100 MW exported to heat the nearby greenhouses [11].

Notice that while area-wide integration is technically quite straight forward, its commercial aspects may sometimes be challenging. A detailed pinch analysis would require the sharing of sensitive process information. The success of a project will depend on good selection of data requirements so that sufficient data could be obtained to produce a meaningful analysis without endangering commercial confidences.

6.7.2
Water Pinch

Water pinch is a systematic technique for analyzing water networks and identifying projects to improve the use of water in industrial processes. It identifies improvement opportunities in water re-use, regeneration (the partial treatment of process water that allows its re-use) and effluent treatment.

Like energy, water can be quantified in terms of quality and quantity, and similarly looking composite curves can be produced. Figure 6.28 shows the profile of the available sources arising within the process, and the various sinks or demands

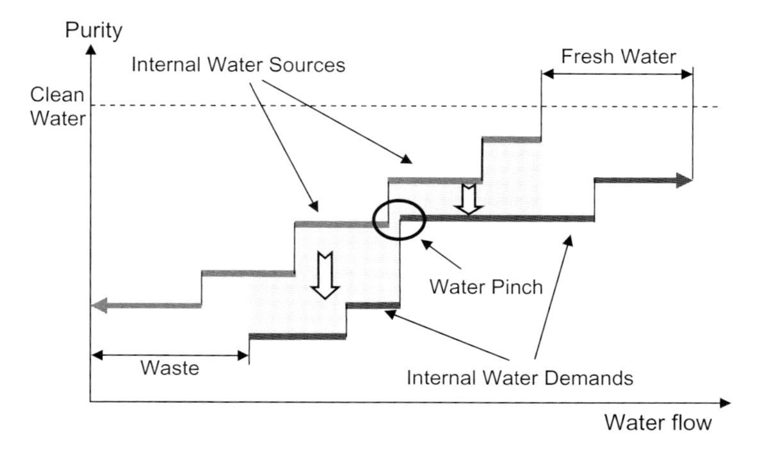

Figure 6.28 The water pinch.

for water. As for energy, the overlap region identifies where the internal wastewater streams can be re-used.

The one major difference between energy and water is that energy has only one dimension of quality and that is temperature. Water, on the other hand, can have several values, one for each contaminant (e.g., conductivity, dissolved solids, organics, etc). This would require that the composite curves for each impurity are developed. If each impurity created the same pinch point within the process, the analysis would not be too difficult to carry out manually. In reality, each contaminant creates a pinch at a different point and water pinch analysis has to rely on mathematical programming to identify the improved designs and optimize the trade-offs.

6.7.3 Hydrogen Pinch

Hydrogen systems can also be analyzed by pinch techniques, again employing the concept of quality versus quantity. It is largely confined to the oil refining industry where the demand for hydrogen is continually increasing due to legislative pressure on low-sulfur products and the processing of increasingly sour crude oils.

The analysis is conceptually similar to the water pinch, as both hydrogen and water networks can be defined in terms of flowrate and purity. Their composite curves are hence alike. The scope for hydrogen re-use is defined by the overlap of the source and the sink composites, which is limited by the pinch point, as before. The target for minimum pure hydrogen make-up (from a hydrogen plant or from import) is given by the horizontal difference between the curves at the high purity end. The minimum purge rate is defined by the horizontal difference between the curves at the low purity end. In the case of hydrogen systems, the purge is however not wasted, but is fed to the site fuel gas system.

Similarly to water systems, hydrogen flows have multiple contaminants, but these are often approximated to a single 'impurity' since the different contaminants do not normally exclude re-use of hydrogen streams in other processes. However there are some factors specific to hydrogen systems:

- Cost – because of the high cost of hydrogen and hydrogen generation equipment, systems are usually highly integrated with significant re-use of purge gas.
- Pressure – as the compression costs are high, pressure is an important parameter in the overall economics of the system.
- Effect of purity on production – hydrogen purity influences the economics of refinery unit operation in terms of throughput, yield or run length.

To analyze hydrogen systems, which is a complex problem, computerized algorithms are used , and they incorporate these additional parameters. The software will look for an overall optimum, maximizing the plant's profitability by simultaneously addressing hydrogen system operating costs and the process benefits, and minimizing the investment required to achieve them.

References

1 Linnhoff, B., and Flower, J.R. (1978) Synthesis of heat exchanger networks: I. Systematic generation of energy optimal networks. *AIChE Journal*, **24** (4), 633–642.

2 Linnhoff, B., and Vredeveld, D.R. (1984) Pinch Technology has come of age. *Chemical Engineering Progress*, **80** (7), 33–34.

3 Milosevic, Z., and Rudman, A. (2009) Energy integrated refinery of the future. *Oil Gas European Magazine*, **35** (2), 86–90.

4 Dhole, V.R., and Linnhoff, B. (1993) Total Site targets for fuel, co-generation, emissions, and cooling. *Computers and Chemical Engineering*, **17** (Suppl.), S101–S109.

5 Linnhoff, B., and Hindmarsh, E. (1983) The Pinch Design method for heat exchanger networks. *Chemical Engineering Science*, **38** (5), 745–763.

6 Kemp, I.C. (2007) *Pinch Analysis and Process Integration. A User Guide on Process Integration for the Efficient Use of Energy*, Elsevier.

7 Tjoe, T.N., and Linnhoff, B. (1986) Using Pinch Technology for process retrofit. *Chemical Engineering*, **93** (8), 47–60.

8 Asante, N.D.K., and Zhu, X.X. (1996) An automated approach for heat exchanger retrofit featuring minimal topology modifications. *Computers and Chemical Engineering*, **20**, S7–S12.

9 Perry, S., Klemes, J., and Bulatov, I. (2008) Integrating waste and renewable energy to reduce the carbon footprint of locally integrated energy sectors. *Energy*, **33** (10), 1489–1497.

10 Matsuda, K., Yoshiichi, H., Hiroyuki, T., and Shire, T. (2009) Applying heat integration Total Site based Pinch Technology to a large industrial area in Japan. *Energy*, **34** (10), 1687–1692.

11 E den Dekker, E., and Keuken, H. (1999) Heat-sharing between companies and within industrial areas. Proceedings of PRES'99 – 2nd Conference on Process Integration, Modelling and Optimisation for Energy Saving and Pollution Reduction, Budapest, Hungary.

Further Reading

Shenoy, U.V. (1995) *Heat Exchanger Network Synthesis*, Gulf Publishing Co, Houston.

7
Energy Efficient Unit Operations and Processes

Andreas Jupke

7.1
Introduction

The cost of energy has risen dramatically in recent years. It now contributes significantly to the total manufacturing cost of chemical products. Consequently, a key lever to increasing profitability is the reduction of the energy costs. Chemical companies contribute to man-made greenhouse gas emissions. Therefore, growing attention has been given to reducing the emission of greenhouse gases in order to combat climate change. Reducing greenhouse gas emissions is required for economic and social reasons. As a conclusion, chemical companies have started initiatives that help to reduce energy costs and, consequently, decrease CO_2 emissions.

Energy efficiency is an issue for energy generation and distribution as well as for energy usage within the production processes. The area of energy generation and distribution is presented in Chapter 10. Energy efficient equipment is discussed in Chapter 8 and heat integration by pinch analysis in Chapter 6. Therefore, this chapter will focus on measures for improving the energy efficiency within chemical production processes, especially on energy-intensive unit operations and on the entire process.

Energy is required in nearly every unit operation in a chemical process. For the purposes of energy efficiency the most energy-intensive operations, which represent the main levers for optimization, must be identified. Heat usage is dominated by the unit operations distillation, evaporation, crystallization and drying. In addition, improvement measures for large electricity-consuming equipment like pumps and blowers are presented. Further electricity-consuming equipment such as compressors is discussed in Chapter 8.

For many chemical processes, changes in the reaction step and modifications in the overall process offer a large potential for reducing energy requirements. The reaction and the entire process are closely related since measures in the reaction usually have a large influence on other parts of the process. Therefore, the reaction and the entire process are discussed together in this chapter.

Managing CO_2 Emissions in the Chemical Industry. Edited by Leimkühler

ISBN: 978-3-527-32659-4

Energy efficiency optimization measures can be divided into three main categories:

1) improved operation of existing plant;
2) improved design of existing plant;
3) new plant with improved design.

Activities falling into the first category are the responsibility of the plant management. One area of measures involved in operating plants efficiently is often described as good housekeeping and will be presented in this chapter. Improvements to existing plants can be achieved by optimized operation as well as by optimized design of the equipment or the entire process. This chapter will focus on both topics. Improved design of new plants involves research and development that is discussed in Chapter 5.

Most of the measures presented in this chapter for improving the energy efficiency of chemical processes are realistic and based on established technologies. They are also often economical. It depends on a detailed technical and economical study of each potential measure whether they should be implemented or not.

7.2 Good Housekeeping

A wide range of maintenance measures, which generally represent good plant management and operation are covered by 'good housekeeping'. Many of these measures can be realized with no or low investment, but they often require time and manpower. In addition, many good housekeeping measures require regular attention.

When performed on a regular basis, good housekeeping generally also leads to extended periods between equipment replacements, and lower production costs, because proper maintenance increases the reliability and durability of the equipment [1].

Good housekeeping for energy saving includes such activities as identification of steam and compressed air leaks, tuning boiler and furnace burners, checking insulation of the pipes, replacing leaking steam traps, shutting off equipment when it is not required, as well as preventing maintenance on heat transfer equipment (i.e., heat exchangers), and on pumps, fans, compressors, measuring devices, and control systems. A list of recommended improvement measures follows [1]:

Steam systems

- detect and eliminate leaks;
- increase condensate recovery or reuse;
- improve operation of steam traps;
- increase recovery of flash steam;

- use lower steam pressure if possible;
- improve maintenance of steam ejectors;
- turn off heat tracing if not needed, for example, in summer;
- install steam monitoring system.

Insulation

- use IR camera to detect inefficiencies;
- improve protection against rain and water;
- check thickness, increase if necessary;
- ensure flanges, valves, fittings and manholes are insulated;
- replace after maintenance.

Compressed air systems

- eliminate leaks;
- use minimum operating pressure;
- use minimum inlet air temperature;
- reduce pressure drops in pipes;
- use appropriate filters/dryers for specified tasks;
- install buffer tanks in case of fluctuating usage to optimize compressor operation.

Cooling water systems

- minimize flow at heat exchanger (maximize temperature increase);
- switch off when not required;
- reduce circulation flow in entire system to required value (maximize temperature increase);
- check 'cascade' flow through units in series.

Vacuum systems

- eliminate leaks;
- replace steam ejectors by more efficient vacuum pumps;
- consider dry vacuum pumps.

Chiller systems

- adjust supply temperature to the highest possible value;
- reduce circulation flow in entire system to required value (maximize temperature increase).

Illumination

- switch off lights if they are not needed, for example, outdoor lighting during daytime;
- clean lamps regularly;
- ensure that walls and floors are light in color;
- install motion sensors in rooms that are lit and not used continuously;
- install time switches for simple switch-off functions.

Heat exchangers

- improve control concept to reduce utility consumption;
- avoid sub-cooling or over-heating if not required;
- use lowest grade of possible heating and cooling media, for example, cooling water instead of chilled water;
- monitor fouling and clean surfaces frequently to ensure best possible heat transfer.

7.3 Centrifugal Pumps and Blowers

Equipment that consumes a large amount of power is pumps and blowers. The energy efficiency of these operations can be improved with the use of more efficient equipment, better matching of equipment capacity to throughput, and more energy efficient flow control such as variable-speed drives. Energy efficient equipment is discussed in detail in Chapter 8

7.3.1 Centrifugal Pumps

Pumps account for a large part of industrial electricity consumption. In addition, the energy costs account for a significant part of the lifecycle costs of a pump. Therefore, optimization of the design is also essential. Since energy losses can be found in each part of a pump system consisting of pipeline, pump, gear, motor and frequency converter as illustrated in Figure 7.1, the whole system needs to be considered to identify improvement measures.

A large part of the energy consumption of a centrifugal pump can be saved by optimization of the control concept. In older plants we usually observe throttle control with fixed-speed motors where the flow is adjusted by a control valve in the pipe. This concept often causes a high energy loss mainly because a part of the hydraulic energy of the pump is converted into heat in the control valve. The

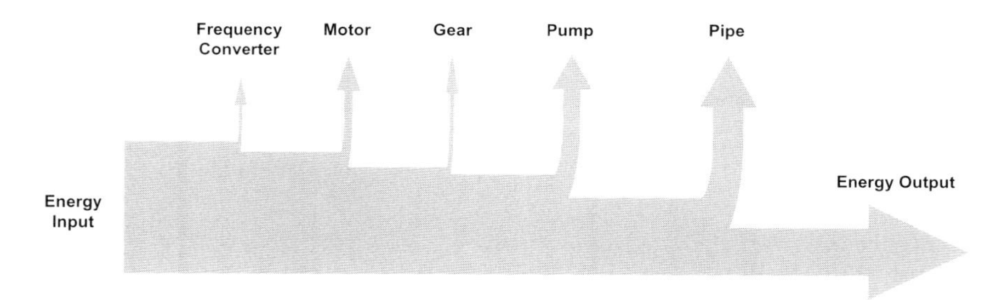

Figure 7.1 Schematic energy losses in a pump system.

throttle control is only recommended for smaller pumps or if there are minor deviations from the operating point or it is necessary for process reasons to have defined pressure conditions downstream of the control valve. The pressure drop at the control valve has to be minimal.

Bypass control is used primarily with displacement pumps where throttling of the volumetric flow is not possible. A further application is the control of minimum quantities with high-pressure centrifugal pumps. An advantage of bypass control is that the pump is always operated at its design point–but this is also the main disadvantage if it is operated for longer periods in bypass mode, since during this period it will require 100% energy without effective work. Bypass control with centrifugal pumps is in most cases less favorable in terms of energy than throttle control. An exception are pumps that have a power consumption which increases with a reduction in the volumetric flow rate, for instance axial pumps. In this case bypass control is more favorable in energy terms than throttle control. This type of control is only used in approximately 5% of pumps to be controlled, for instance small metering pumps.

Speed control influences the hydraulic performance of pumps and therefore the pump characteristic curve. With centrifugal pumps a variation of speed enables control of pressure and volumetric flow. The speed n of a centrifugal pump influences the volumetric flow Q, the pump head H and the power consumption P in the manner shown in the following equations:

$$Q \sim n$$

$$H \sim n^2$$

$$P \sim n^3$$

Based on these relationships, it can be seen that if the flow rate is reduced by 50% through a 50% reduction in speed, the power consumption will be reduced to 12.5% of the value at full speed. On the other hand, it is only possible to reach 25% of the pump head H.

The effect of reducing the speed of a centrifugal pump is comparable to the effect of reducing the impeller wheel diameter. Energy savings become more pronounced for pumps with steep characteristic curves (e.g., axial and semi-axial pumps) then for pumps with flat characteristic curves (e.g., radial pumps).

Frequency control drives offer considerable potential savings for load profiles that fluctuate substantially. The costs of a variable-speed drive will almost always be recouped if the load profile exhibits substantial differences over the operating period. A detailed analysis is not worthwhile for pumps with a power consumption less than approx. 15 kW or if the operation time is less than 2000 h/a.

Figure 7.2 shows a comparison of an old pump system design with a fixed-speed motor and throttle control with a new pump system design with variable speed control and a high efficiency motor. Considering all elements a difference in the system efficiency of approx. 40% is possible.

Old design: Fixed-speed machine, flow adjusted by control valve, system efficiency: app. 30 %

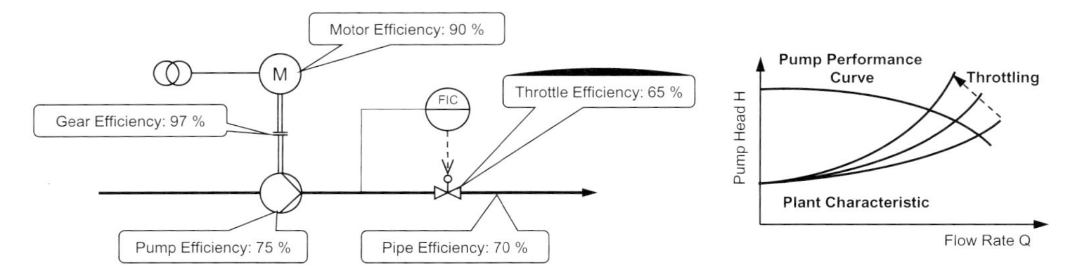

New design: Variable speed control with high efficiency motor, system efficiency: app. 70 %

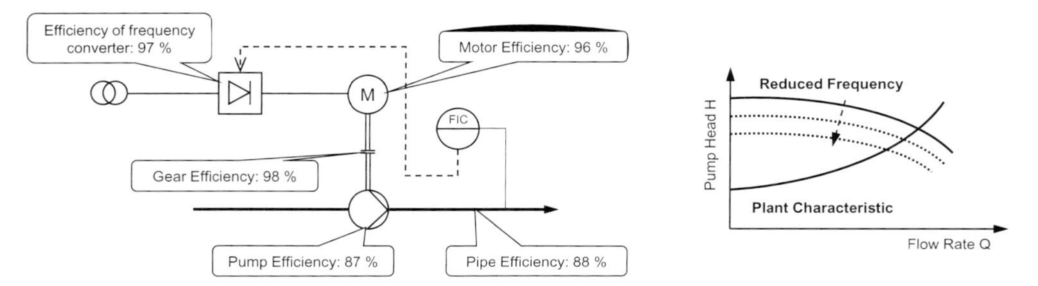

Figure 7.2 Comparison of pump system concepts.

Important improvement measures for pump systems are:

- Consider if flows can be reduced.

  Often low temperature difference in circulation systems for heating or cooling media can be observed due to high circulation flows; reduction of circulation flow saves electrical energy.

- Switch off when not required.
- Adjust impeller wheel diameter.

  A variation in the diameter of the impeller wheel has an effect similar to that of a change in speed. This is a suitable measure for pumps with a constant output. After adjusting the impeller wheel diameter the pressure drop at the control valve should be minimal.

- Reduce pressure drop in piping.
- Consider replacement of undersized lines and valves.
- Optimize motor drive.
- Consider replacement of old motor with low efficiency by new motor with higher efficiency.

- Avoid oversized pumps and motors.

  Pump and motor should have maximal degree of efficiency at required operating conditions; over-sizing increases energy consumption.

- Implement predictive maintenance.

7.3.2 Blowers

Beside pumps, a large part of the industrial electricity consumption is used for blowers. Approx. 90% of the blower lifecycle costs are attributed to the energy costs. In most cases pumps are purchased as manufacturer-configured functional units (housing / impeller wheel / drive and control electronics). With blowers, however, there is substantially greater freedom in the configuration of the components. This enables a high degree of adaptation of the components to the intended use, but at the same time requires the equipment user to devote more time to a consideration of the requirements and options to arrive at the correct specification for the suppliers.

Important improvement measures for blowers are:

- Switch off when not required.
- Consider if flows can be reduced.
- Reduce pressure drop in piping.
- Improve control concept.

  Speed control by frequency converters is the current method for adapting the blower output to the required value in ventilation systems. Components such as throttle valves, pole switching units, bypasses and blade positioning that were previously used no longer satisfy current engineering standards and result in unnecessary energy consumption.

- Replace impeller wheel with one of better flow geometry.
- Replace motor with one that is more efficient (easy with standard motors).
- Replace inefficient drives, for example, belt drives by direct-drive systems.
- Use blowers with high speeds as long as there are no technical or noise protection-related reasons.

7.4 Distillation

The most widely used unit operation in the chemical industry for the separation of mixtures is distillation. In addition, distillation is also one of the most energy-intensive separation techniques in the chemical and petrochemical industries.

7.4.1
Basic Principles

The separation principle of distillation is based on different volatilities of components in a boiling mixture. The methods for decreasing the energy consumption of distillation generally focus on the reduction of the reboiler heat duty, because this represents the major energy requirement or on the recovery of the condensation heat at the condenser [1–3]. Electrical energy, which is necessary for pumping the process streams and cooling medium, is usually of minor importance compared with the heat demand [1].

The reflux ratio is the key parameter that influences the energy consumption of a distillation column. The lower the reflux ratio, the lower the energy consumption of the column. However, to achieve the required purity of the distillate, a minimum reflux ratio is required, which is based on the thermodynamic properties of the mixture [2, 3]. The reflux ratio and consequently the required heat input decrease as the number of theoretical stages in a column increases for a given separation. The boundaries are the minimum reflux ratio ν_{min} with an infinite number of theoretical plates and the minimum number of theoretical plates at total reflux N_{min} [2, 3], as shown in Figure 7.3.

Based on the assumption of constant mass flows, constant relative volatility α and a boiling feed flow, the minimum reflux ratio ν_{min} can be determined by Fenske and Underwood [4, 5] where x_D and x_F are the molar fractions of the low boiling component in distillate and feed, respectively:

$$\nu_{min} = \frac{1}{\alpha - 1}\left[\frac{x_D}{x_F} - \alpha \cdot \frac{1 - x_D}{1 - x_F}\right]$$

The design of distillation columns is usually a trade off between the operating costs and investment cost. Often reflux ratios between 10 to 30% above the minimum are chosen [1, 2].

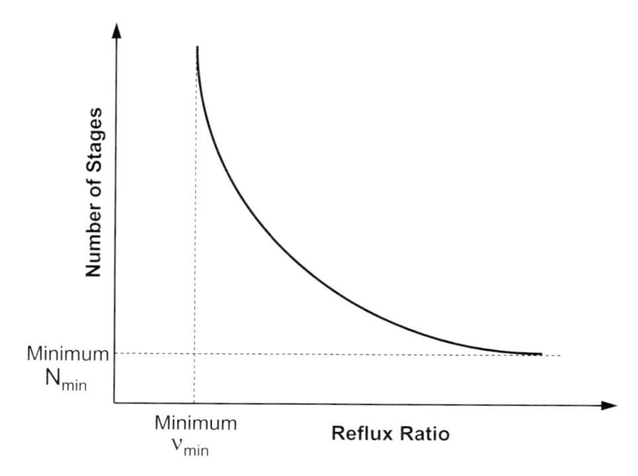

Figure 7.3 Relationship between reflux ratio and number of plates.

However, beside the minimum reflux ratio based on thermodynamic principles the minimum reflux flow based on the fluid dynamics of the columns internals also needs to be considered. Therefore, in some cases an existing column has to be operated at a higher reflux than the minimum reflux ratio defined by the thermodynamics to achieve the required wetting of the column internals.

There are two main ways to increase the energy efficiency of distillation. In general, the first and the cheapest way is to improve operation and control. The second area of measures is improved column design for single and for multiple columns.

7.4.2 Operation and Control

The reflux ratio is the main variable influencing the energy consumption of an existing column. It can be observed that distillation columns are frequently operated at much higher reflux ratios than required. This leads to increased energy consumption. Therefore, a major measure for energy efficiency represents the proper adjustment and control of the reflux ratio. A number of parameters can be improved in order to reduce the reflux.

7.4.2.1 Purity of the Product

Frequently, the purity of the product–distillate or bottom product–is higher than the required specification for further processing or sale, because the reflux ratio is set higher than necessary. Therefore, the reflux rate can be decreased until the desired purity is obtained. However, the minimum reflux flow to achieve the required wetting of the column internals needs to be considered. It is recommended that this procedure should be supported by thermodynamic calculations, like the determination of the minimum reflux ratio, or by the simulation of the distillation column to determine the minimum reflux flow.

Process fluctuations are often the reason for a reflux ratio that is too high. In this case, process fluctuations need first to be reduced, before reducing the reflux ratio, for example, by process control measures.

7.4.2.2 Operating Pressure

For many mixtures, the relative volatility increases with decreasing pressure [2, 3]. As a consequence, lower pressure operation can be used to achieve the same product purity at lower reflux ratios and with lower energy requirements. However, condensation of the overhead vapors with cooling water or air at approx. 35 °C is a limit for the operating pressure decrease. Alternatively other cooling media like chilled water or refrigerants can be used for the condensation, but these are usually more expensive than cooling water. Additionally, the heat transfer conditions of the reboiler and the condenser will change. A pressure decrease will also cause the volume flow of the vapor phase in the column to increase, influencing the capacity of the column. Consequently, detailed considerations and calculations are usually required for a large decrease of the operating pressure. Therefore, for existing columns typically smaller pressure reductions are realized.

On the other hand, operation at increased pressure can enable a more effective recovery of waste heat due to the higher condensation temperature at the condenser, although the column may require more energy.

7.4.2.3 Sub-Cooling of the Reflux Flow

Sub-cooling of the reflux flow far below the condensation temperature of the overhead product causes an increased energy demand. An explanation for sub-cooling of the reflux flow is a too high flow of cooling medium in the condenser, because the cooling medium is often not controlled. If there are no requirements for sub-cooling, the flow of cooling medium to the condenser should be controlled in a suitable manner. However, it is necessary to evaluate if a certain level of sub-cooling is necessary in a particular case, for instance to condense low-boiling components in the overhead vapors of the column.

7.4.2.4 Location of Feed Point

The ideal plate for the feed point should be used. The optimal point, at which the feed should enter the column, is where the composition on the tray closely matches the composition of the feed. Frequently columns are designed with several feed points so that it can easily be relocated.

7.4.2.5 Fouling or Damage of Internals

Separation performance can be reduced after longer periods of operation, due to fouling or damage to the column internals. A lower efficiency of the internals is often compensated by a higher reflux flow leading to increased energy usage. The energy consumption can be reduced through improved maintenance and cleaning of the trays or packing and the heat exchanger. In addition, the liquid distributors need to be mounted properly to exploit the full efficiency of the internals.

7.4.2.6 Process Control

The aim of the control system for a distillation column is to stabilize and to optimize the operation. The control system should also ensure the production of products in specification at the minimum energy consumption. Conventional control systems are often designed for stabilization only. They rely on the intervention of an operator for any optimization.

In Figure 7.4 a conventional control system for a two-component separation is shown [1]. In this conventional control system the product flows are level-controlled and the reflux is set at a fixed flow, which is usually too high. This conventional control system often leads to an operation in a comfort zone with less trouble for the operators but low energy efficiency.

Advanced process control (APC) aspires to establish best practice process conditions and to maintain them as rigorously as possible, regardless of disturbances, grade changes or load changes, while satisfying process constraints and produc-

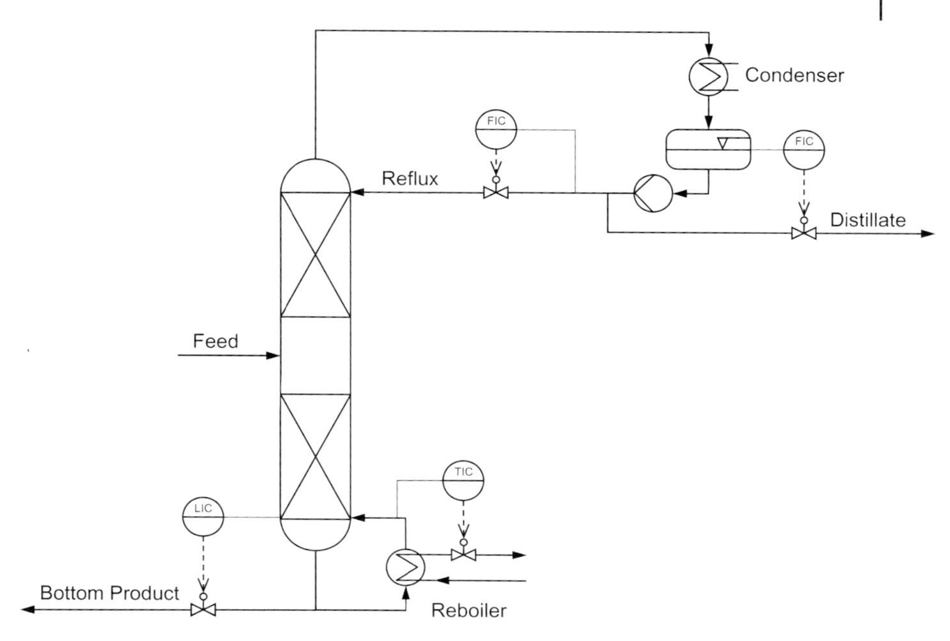

Figure 7.4 Conventional control system for a two-component separation.

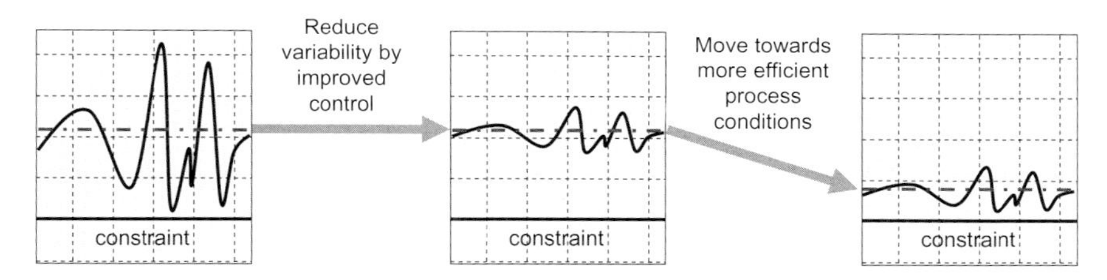

Figure 7.5 Strategy for energy saving by improved process control.

tion goals. The most efficient process conditions are often subject to process constraints, such as the purities at top and bottom, weeping and flooding. Since there is always process variability due to disturbances, a distance is kept from the efficient conditions in order to avoid constraint violations. Examples for external disturbances are feed flow rate and composition, feed temperature, reflux temperature (air coolers) and steam pressure. This motivates a useful strategy for energy saving as shown in Figure 7.5.

Advanced control systems comprise feed-forward control, material-balance control and frequently computers for optimization. These systems control the column at the optimum operating point, provide more stable operation and reduce the need for operator action. Additionally, they offer considerable potential for

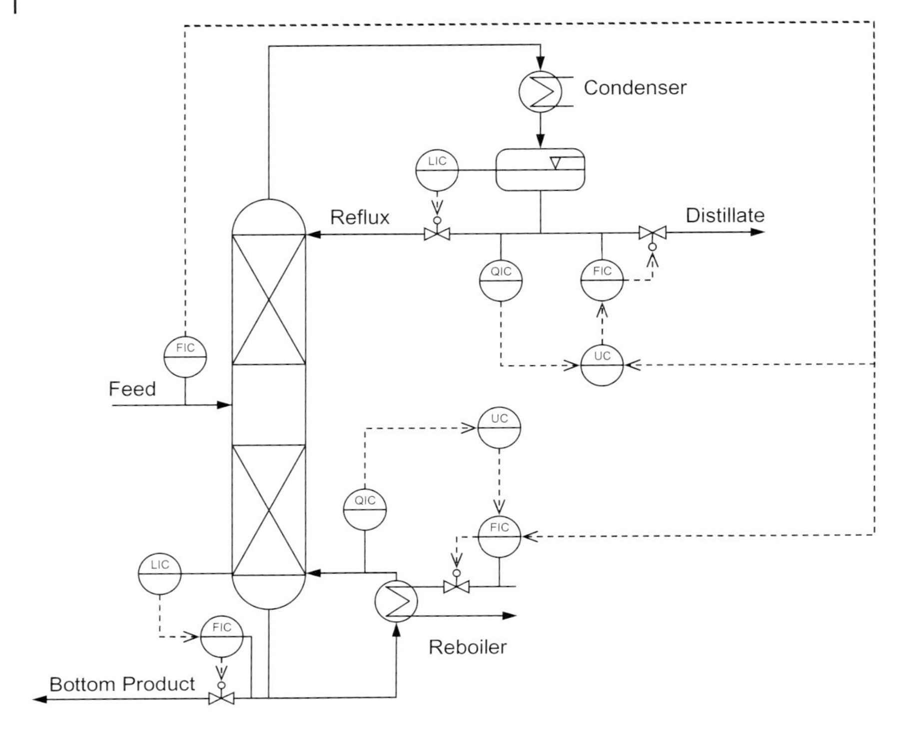

Figure 7.6 Example for an advanced process control system for a two-component separation.

reducing energy requirements and operating costs. An example for an advanced control system for a two-component separation is shown in Figure 7.6.

In addition, online product analyzers to measure the concentration or purity are recommended to allow a tight process operation. Examples for online product analyzers are NIR, UV, VIS, Raman spectroscopy, or gas chromatography [6, 7].

7.4.3 Improved Design for Single Columns

7.4.3.1 Column Internals

It is recommended that for the design of new columns a reflux ratio close to the minimum should be chosen. However, many existing columns were designed for higher reflux ratios. In these cases, energy consumption can be reduced by increasing the number of stages.

The number of stages and the stage efficiency or the total height of a packing and the height of a transfer unit for packed columns influences the performance of a distillation column. For tray columns the number of theoretical stages can be increased by increasing the number of trays, for example, by increasing the column height or reducing the tray spacing. However, increasing column height

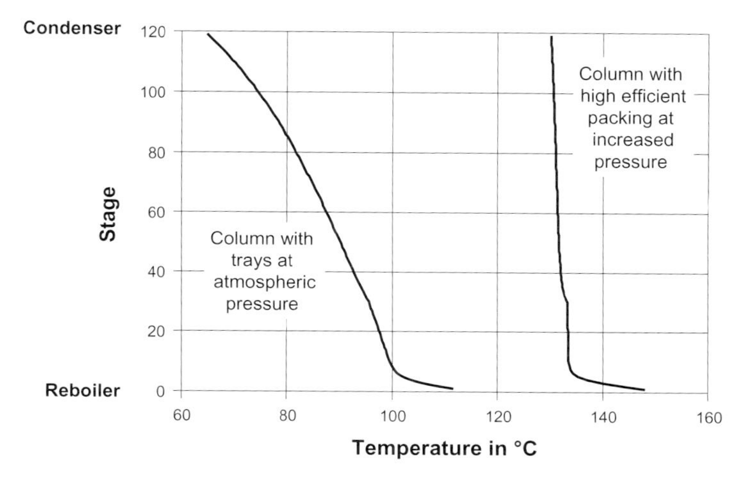

Figure 7.7 Comparison of temperature profiles in a distillation column.

usually leads to high investment costs. In addition, reducing tray spacing may be expensive. Generally, the most attractive measure is the implementation of more efficient trays or packing. An analysis of the energy savings for different trays and packing is recommended. If there is an attractive improvement, tray and packing suppliers can be contacted to determine the exact increase in efficiency.

The implementation of a packing with a higher efficiency can reduce the requirement for reflux, reduce the pressure drop and improve purities. With vacuum columns in particular, a reduction in the pressure drop can also increase the energy consumption of the vacuum pump.

Figure 7.7 shows two temperature profiles of a distillation column. The curve on the left hand side is the temperature profile of a tray column at atmospheric pressure. The condensation temperature at the top is approx. 65 °C. The vapors are condensed with cooling water. By a pressure increase and the installation of a high efficient packing reducing the pressure drop, the temperature profile gets steeper as shown in the curve on the right hand side in Figure 7.7. In addition, the condensation temperature at the condenser is approx. 130 °C. The more valuable condensation heat can now be used, for instance to generate low pressure steam or in an adsorption chiller to generate chilled water.

7.4.3.2 **Feed Preheating**

The heat requirement of a single column can also be minimized by heat exchange between the warm products and the cold feed as illustrated in Figure 7.8. In addition, other process streams or steam condensate can be used. If a process has several sources and sinks of heat, before one decides to implement feed preheating, pinch analysis should be considered to anticipate any potential pinch violations resulting from sub-optimal match of cold and hot process streams (see Chapter 6).

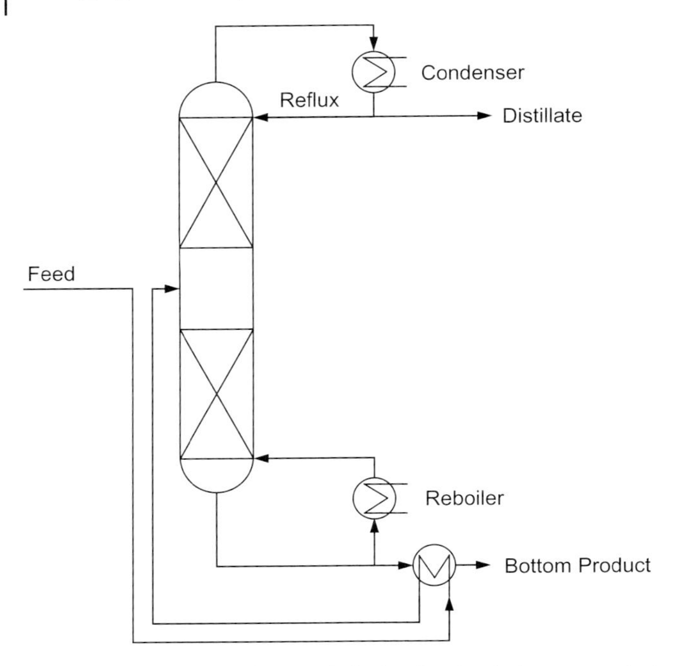

Figure 7.8 Distillation column with feed preheating by bottom product.

In the case of a vapor feed, the latent heat is used only in the rectifying section of the column. Therefore, a vapor feed can be condensed in a second reboiler before it is introduced into the column. In this way, latent heat is used in both rectifying and stripping section of the column, which can lead to a lower external heat-input requirement [2].

7.4.3.3 Vapor Recompression

The exergy describes the ability of heat to be converted into work, where the temperature difference of the heat compared with the ambient temperature is relevant. In a distillation column, a specified amount of heat is required at the bottom at a certain temperature and often an almost equivalent amount of heat is removed at the top at lower temperature level. The temperature difference of a distillation column is low for close-boiling mixtures, so exergy losses are also low [2]. However, the required heat is usually supplied in the form of steam at much higher temperature compared with the bottom temperature. In addition, the energy is removed by cooling water with a temperature much lower than the condensation temperature. In this way, larger amounts of exergy are lost during separation than necessary [2].

Often, the condensation heat of a distillation column is not used when the condensation temperature is too low. The implementation of a vapor recompression can avoid this waste of energy. After the oil crisis in 1973 vapor recompression

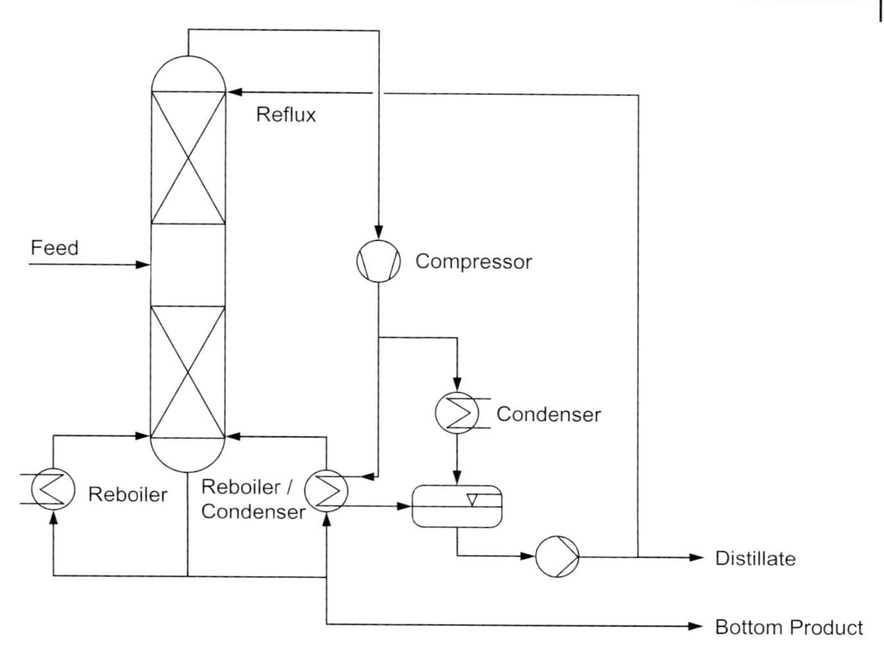

Figure 7.9 Distillation column with vapor recompression.

was introduced in chemical and petrochemical industries. As shown in Figure 7.9, the overhead vapor is compressed to the level that enables the use of the condensation heat in the reboiler. The vapor recompression separation scheme brings substantial energy savings for fractionating mixtures characterizing in close boiling points, where small temperature differences between the top and bottom of the column, small compression ratios and consequently small compressor duties are required [2, 3].

A single stage compressor with a compression factor up to three is applicable only for mixtures with a low boiling point difference. Mixtures with a high boiling point difference often require multistage compressors, which are in most cases not economical. Therefore, vapor recompression is widely used in the petrochemical industry for the separation of hydrocarbons, and in the food industry [3]. In the chemical industry, where mixtures with a high boiling point difference are often separated, vapor recompression is rarely applied. In addition, high flow rates (approx. $> 100\,000\,m^3\,h^{-1}$) or vacuum operation (approx. $< 100\,mbar$) often make vapor recompression uneconomical.

Thermal vapor recompression, beside mechanical vapor recompression, is another way of increasing energy efficiency of distillation. A common application of a thermal vapor recompression can be found at distillation or stripping columns where the heating steam is injected directly into the bottom. An example is waste water stripping. Usually, waste water strippers have a lower operating pressure

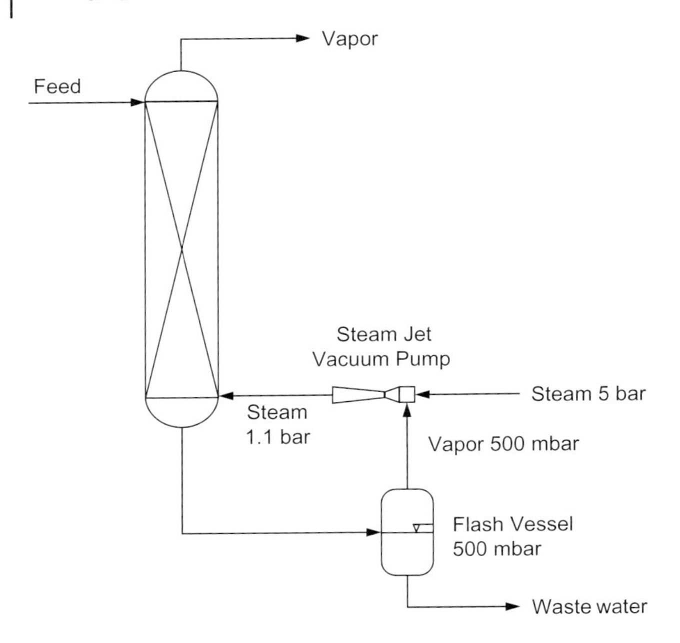

Figure 7.10 Waste water stripping column with vapor recompression.

compared with the lowest steam pressure level. By the introduction of a flash vessel with a steam jet vacuum pump on top, as illustrated in Figure 7.10, the heating steam is used to generate a certain amount of low pressure steam. By doing so, the total steam consumption can be decreased. Further benefit is that the flash vessel acts as an additional stage of the stripping column.

7.4.3.4 Intermediate Reboiler or Condenser

Exergy losses can by reduced by the use of an intermediate reboiler [2]. However, this measure does not lower the total energy requirement of a column. But a certain part of the heat is supplied at a lower temperature in reboiler 2 as illustrated in Figure 7.11. Therefore, the energy cost might be reduced. An intermediate reboiler is advantageous only if several external heat sources, for example, steam pressure levels, are available for the individual reboiler.

Analogously to an intermediate reboiler that provides heating energy at a lower temperature, an intermediate condenser can be applied to provide cooling energy at a higher temperature. This is an appropriate measure in cases where the condensation temperature at the top of the column requires a refrigerant. With an intermediate condenser a part of the required cooling energy can be provided by cooling water, which in general reduces the cooling costs.

The combination of an intermediate reboiler and vapor recompression as illustrated in Figure 7.12 can also allow the use of a single stage compressor for mixtures with a high boiling point difference.

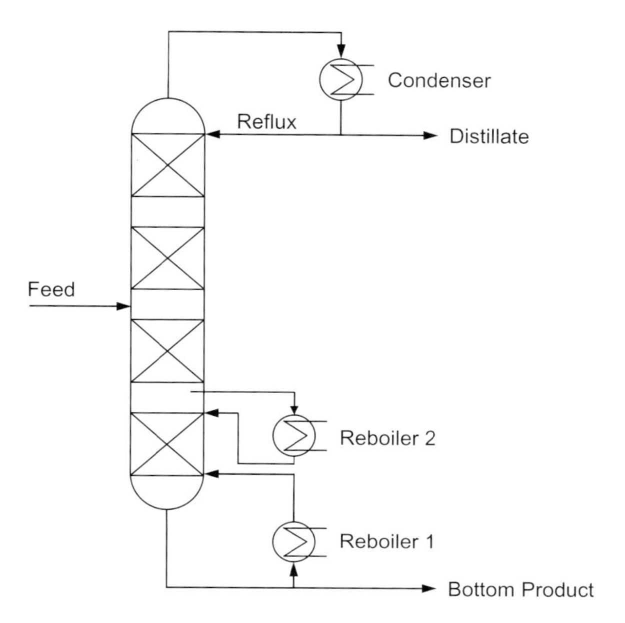

Figure 7.11 Distillation column with intermediate reboiler.

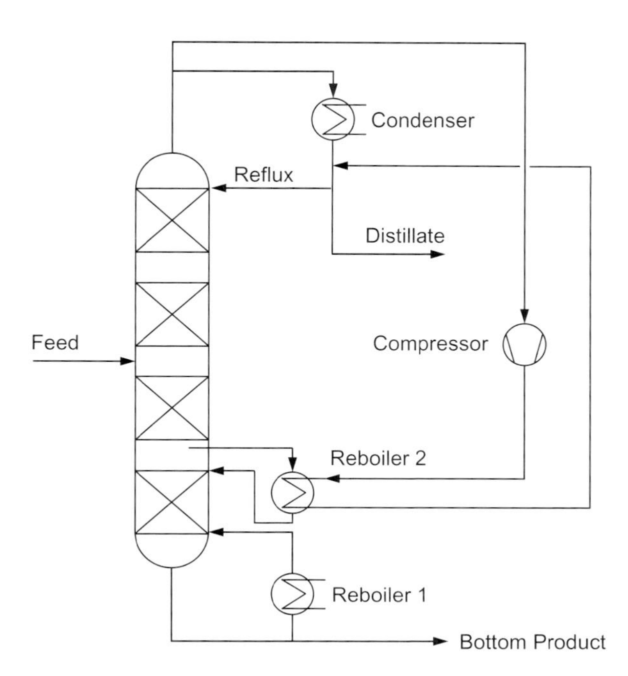

Figure 7.12 Distillation column with intermediate reboiler and vapor recompression.

7.4.3.5 Heat-Integrated Distillation Column (HIDiC)

A novel technology that significantly reduces the energy requirement – also compared with distillation columns with vapor recompression – is the heat-integrated distillation column (HIDiC) [8, 9]. Energy savings of approx. 50% can be achieved [8].

In a HIDiC system the stripping and rectifying section are divided into two different columns that are connected with each other by several internal heat exchangers as shown in Figure 7.13. The rectifying section is operated at a higher pressure and a higher temperature than the stripping section to achieve internal heat transfer from the rectifying section to the stripping section. A compressor and a throttle valve are installed between the two sections to adjust the pressures. The reflux flows in the rectifying section and the vapor flows for the stripping section are generated by a certain amount of heat which is transferred from the rectifying section to the stripping section. Therefore, in principle the condenser and the reboiler are not needed and zero external reflux operation could be realized.

Despite the fact that the HIDiC concept was introduced in 1970, it is still at the primary stage of research from a practical viewpoint and not commercialized so far for industrial application. Up till now the HIDiC technology has only been applied to a semi-industrial scale separation [8].

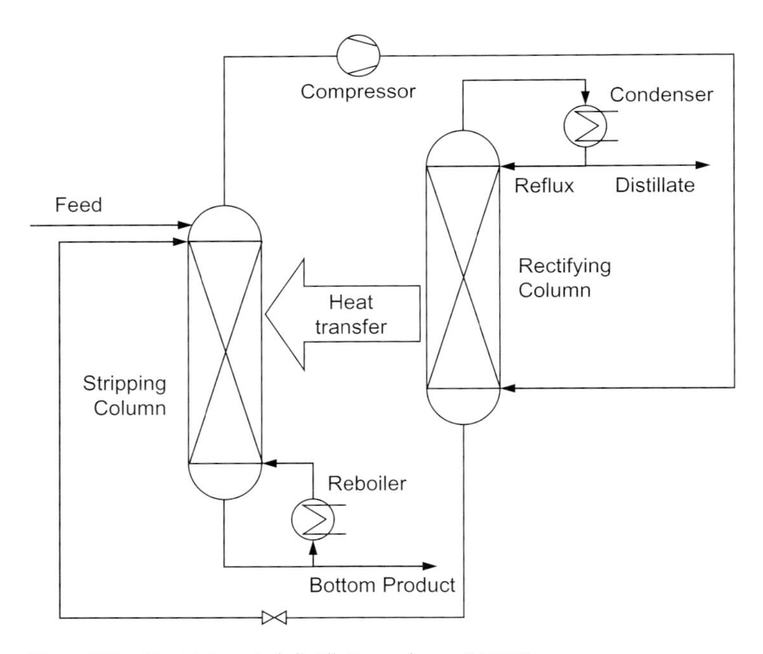

Figure 7.13 Heat-integrated distillation column (HIDiC).

7.4.4
Improved Design for Multi Columns

Several distillation steps are required for the complete separation of a multicomponent mixture into individual components. For complete fractionation of multicomponent mixtures a large number of different separating sequences are possible. The selected separating path determines the investment and operating costs. Operating costs are determined primarily by energy costs.

In processes with several distillation columns the coupling of columns is advantageous. Columns are directly coupled by means of their product streams in direct coupling. By indirect or thermal coupling the waste heat from one column is used to heat a second column [2].

In indirect coupling the main goal is to use the condensation enthalpy for other purposes in the process. In addition, a further aim is to replace the required evaporation duty at the reboiler with 'waste heat' from other parts of the process. Heat integration by thermal or indirect coupling of columns is a very effective way of reducing the heat requirement in distillation processes and is used more often than direct coupling [2]. The heat integration methodology by pinch analysis is recommended to determine the best network of indirect coupling (see Chapter 6).

A number of examples have proven that for a multicomponent separation a direct coupling of distillation columns is particularly useful and always leads to substantial reduction of the heat requirement [2]. In addition the investment costs are reduced, because significantly fewer columns and heat exchangers are required to perform separation. On the other hand, the operation and control of directly coupled distillation columns is more difficult.

7.4.4.1 Dividing Wall Column

The dividing wall column is a concept frequently used in direct column coupling. In all processes where multicomponent mixtures have to be separated into pure fractions, dividing wall columns are applicable. They are particularly suited to obtain pure medium boiling fractions.

A sequential system with at least two columns as shown in Figure 7.14 is required for the separation of a three-component mixture into its pure fractions in conventional systems.

A three-component mixture can be separated by using a dividing wall column in only one apparatus as illustrated in Figure 7.15. In the middle part of the column a vertical wall is introduced, creating a feed and draw-off section in the column. The dividing wall permits the low-energy separation of the low and high boiling fractions in the feed section. The medium boiling fraction is concentrated in the draw-off part of the dividing wall column (www.montz.de).

Dividing wall columns reduce investment and operating costs and therefore are an alternative to multicolumn systems. Investment costs can be reduced by 20 to 30%, and operating costs reduced by approx. 25% [10].

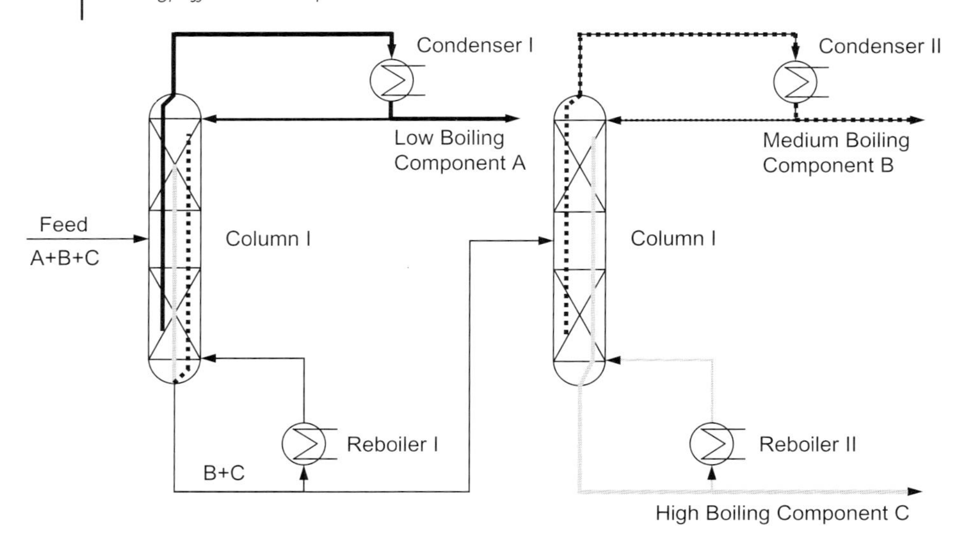

Figure 7.14 Conventional separation scheme of a ternary mixture in two columns.

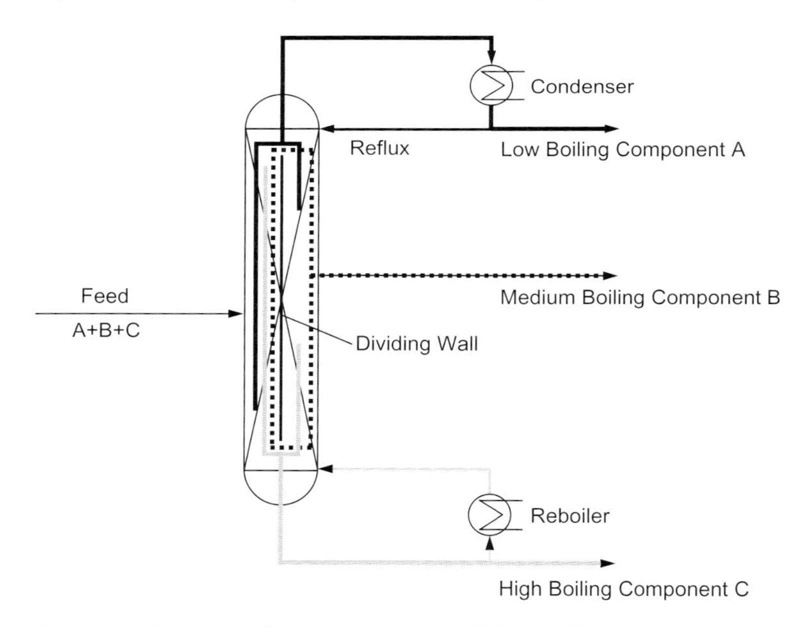

Figure 7.15 Separation of a ternary mixture in a dividing wall column.

However, the drawback of a dividing wall column is that all the heat is consumed at the highest temperature level and removed at the lowest temperature level. In addition, the different separation tasks on the right and left sides of the partition must be carried out at the same packing height. Thus, the amounts of liquid in both parts of the column must be controlled carefully.

7.4.4.2 Indirect Coupling of Columns

In recent years, multistage distillation for the separation of multicomponent mixtures has received increasing attention. The basic idea is to use the overhead vapor of one column as the heat source in the reboiler of the next column. The columns can be heat integrated in the direction of the mass flow (forward integration) or in the opposite direction (backward integration). A common practice is to operate the columns at different pressures to obtain the necessary temperature gradient. Relatively large pressure differences are required to obtain a significant temperature change because the vapor pressure of liquids increases exponentially with temperature.

The separation in two thermally coupled columns is often more economical than separation in a single column [2]. A serial and a parallel connection of two thermally coupled columns are shown in Figure 7.16. The pressure of column 2 is lower than the pressure of column 1. Therefore, the overhead temperature of column 1 is higher than the bottom temperature of column 2, enabling waste heat use from column 1 to heat column 2. As a result, the heat requirements can be almost halved.

A pinch analysis can be used to systematically establish ways of integrating energy streams into the entire process and to show concrete interconnection options. A pinch analysis requires merely a list of the hot and cold streams present in the process (inlet and outlet temperature, mass flow and heat capacity). This method is discussed in detail in Chapter 6.

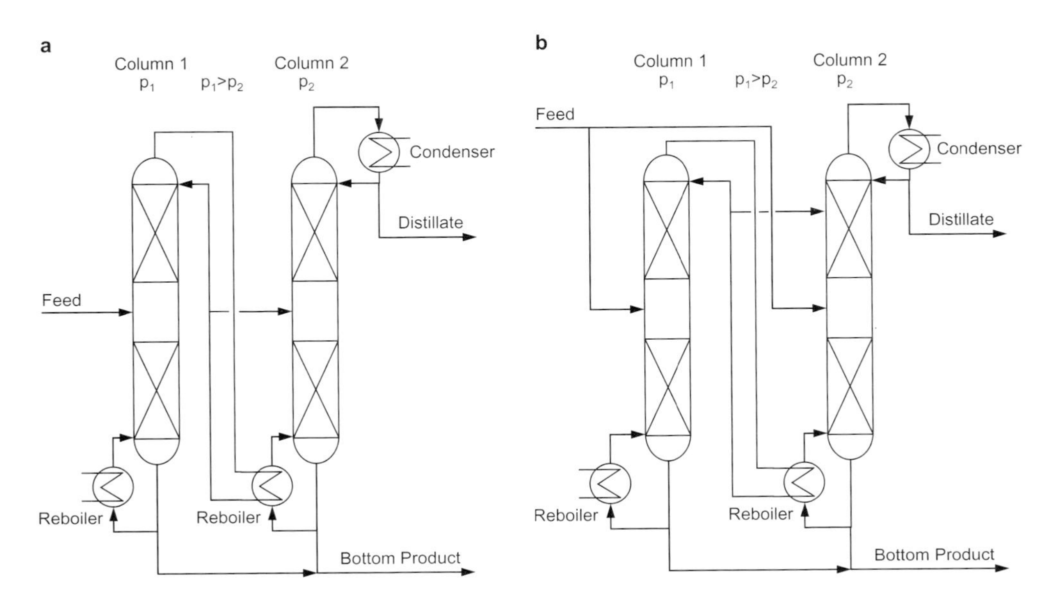

Figure 7.16 Indirect thermal column coupling; (a) serial connection, (b) parallel connection.

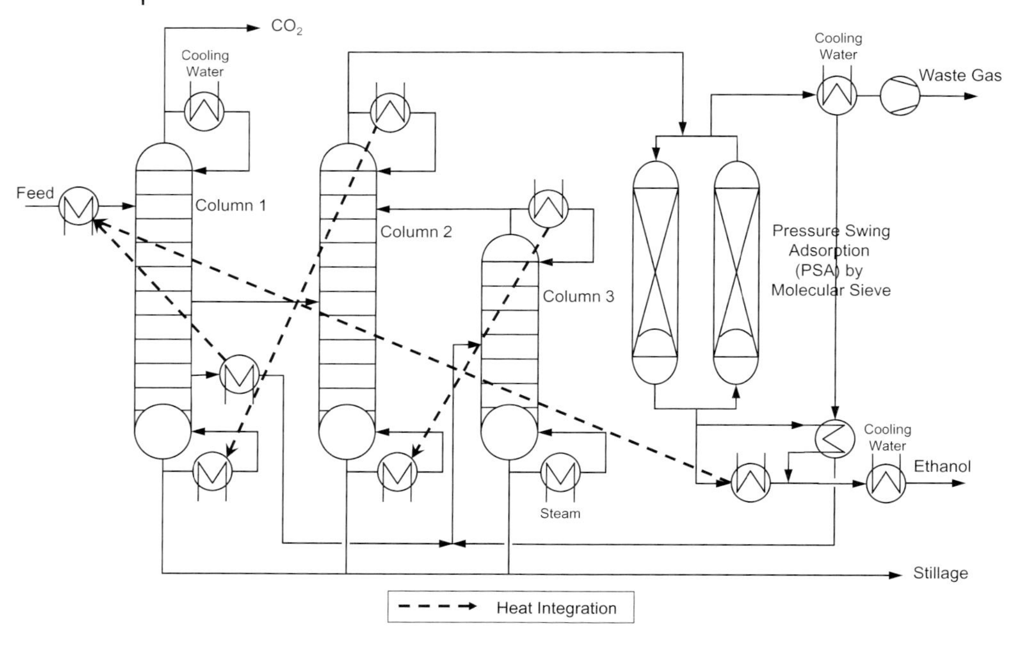

Figure 7.17 Typical distillation and purification sequence for bioethanol.

An example: Increasing oil prices and growing environmental concerns in recent years have been the major driver in the development of renewable biofuels. The use of ethanol as a fuel has been growing exponentially around the world. Until now, most of the bioethanol production concepts are based on sugar and starch crops as feedstock. Bioethanol is produced by fermentation technology leading to a product concentration of approx. 10% ethanol in water. The raw alcohol is separated by distillation and purified to fuel ethanol by dehydration, usually by means of a molecular sieve in a pressure swing adsorber. Distillation and dehydration represents the largest fraction of the energy used in the production of ethanol. The design is therefore approached with energy conservation in mind, using combinations of multipressure cascades, heat recovery and thermal coupling. Figure 7.17 shows a typical distillation and purification sequence for bioethanol. Here, the overhead vapors from columns 2 and 3 are reused to heat columns 1 and 2. The energy used to regenerate a molecular sieve dehydration unit is also recovered to preheat the feed.

The composite curves of this typical distillation and purification sequence for bioethanol are shown in Figure 7.18. Due to the heat integration, a reduction of approx. two-thirds of the energy consumption without heat integration can be achieved.

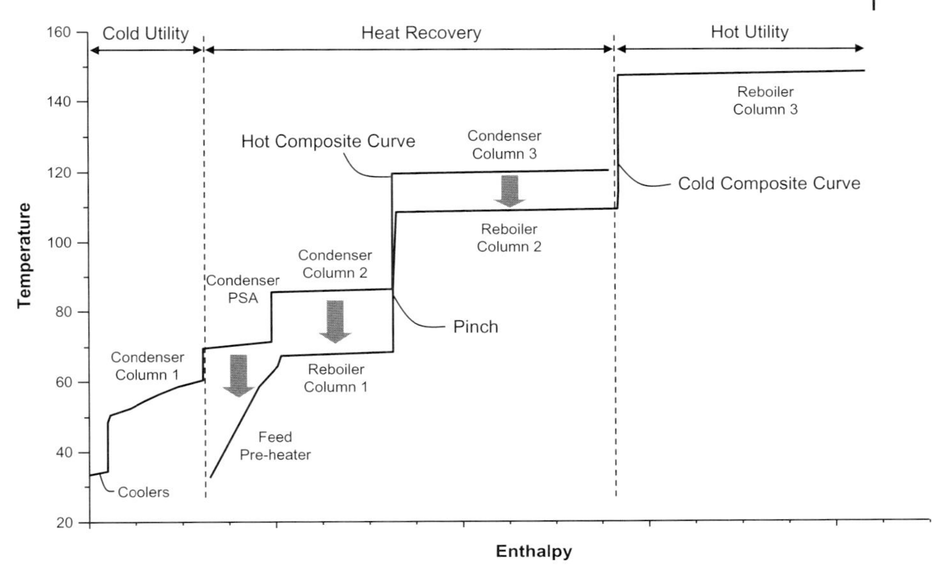

Figure 7.18 Composite curves of a typical distillation and purification sequence for bioethanol.

7.4.4.3 Design of Distillation Processes

The competent design of an optimized distillation process can bring significant financial and energetic benefits. The heat integration possibilities are defined in the design phase. Design of a distillation process in industrial practice is usually conducted by simulation studies that require detailed design specifications and detailed thermodynamic properties in an early phase.

Several shortcut evaluation methods have been developed for the fast evaluation of the cost of separation sequences. A review of shortcut methods for distillation is given by Bausa *et al.* [11]. In addition, Bausa *et al.* developed the rectification body method, which can determine the minimum energy demand algorithmically for non-ideal mixtures with an arbitrary number of components [11].

An economically optimal design of a distillation process can be obtained by rigorous process optimization, where the columns are described by detailed models. The resulting large-scale nonlinear optimization problems are discrete–continuous and are usually solved with mixed-integer nonlinear programming techniques (MINLP). An efficient optimization-based design of distillation processes for homogeneous azeotropic mixtures has been presented recently by Kraemer *et al.* [12].

7.4.5 Reactive Distillation

Reactive distillation is a process in which a catalytic chemical reaction and distillation steps take place simultaneously in one single module. Reactive distillation combines two process stages and so is classed with the process intensification technologies. In the literature this integrated reaction–separation technique is also known as catalytic distillation or reaction with distillation. An overview on the current status of reactive distillation technologies, modeling, industrial applications, etc., can be found in literature [13].

For equilibrium-controlled reactions the major advantage of reactive distillation is the possibility to achieve higher conversion in the reaction, because it is no longer limited by the equilibrium between substrates and products. As the reaction proceeds the products are removed from the reaction mixture.

Reactive distillation is also more advantageous over conventional operating methods, such as a fixed-bed reactor connected to a distillation column, in which the distillate or bottoms have to be recycled after further separation steps for a total overall conversion. Besides higher conversion, further benefits of reactive distillation are [14]:

- Possibility of using the heat of exothermic reaction to vaporize part of the liquid.

  The reaction heat is integrated with the distillation process, which leads to energy savings by reducing reboiler duty. In addition, the direct use of the reaction heat without heat transfer across walls results is a higher efficiency.

- Lower capital investment, due to the combination of two process steps carried out in the same device.

- Increased product selectivity due to the removal of reactants or products from the reaction zone.

Catalytic chemical reactions which are limited by a chemical equilibrium can be improved by reactive distillation. On the industrial scale, this technology was applied mainly for etherification (synthesis of methyl tert-butyl ether (MTBE)), esterification and alkylation (synthesis of ethylbenzene or cumene). An overview of applications can be found in literature [13, 14].

7.5 Evaporation

Evaporation is a widely used and energy-intensive unit operation in the chemical and pharmaceutical as well as in the food and beverage industry. Evaporation is used as a thermal separation technology, in which thermal energy is applied to concentrate or separate liquid solutions, suspensions and emulsions. Solutions consisting of a solvent and dissolved solid substance are vaporized to concentrate the residual solid.

Efficiency improvement measures can reduce the energy consumption considerably. There are three basic categories, in which energy savings can be realized:

- feed preheating;
- multistage evaporation;
- vapor recompression.

Combination of two of these measures is often feasible and is a common practice to minimize energy costs. All three techniques can be found in highly sophisticated evaporation plants [15]. Detailed energy efficiency studies can help to decide, whether a multistage evaporation or a vapor recompression system is the more attractive option.

However, boiling-point elevation, heat of mixing and other thermodynamic non-idealities need to be considered while improving the energy efficiency of evaporation plants.

7.5.1 Feed Preheating

In the evaporation of aqueous solutions, the ratio of heating steam to generated vapors is in the region of 1.2. By preheating the fresh feed solution by condensing a small portion of the vaporized solvent, it is possible to achieve efficiency improvement, of the order of 10 to 15% [15]. A diagram representing the principle of introducing this technique is presented on the Figure 7.19. A small portion of the vapor preheats the feed by condensing in the preheater. A cooling medium is used to condense the residual vapors in the condenser.

Single-stage evaporators are not often used because of their low thermal efficiency. Their application is limited to systems of small capacity, or those in which the product requires short residence time due to temperature sensitivity [15].

7.5.2 Multistage Evaporation

Multistage evaporation exploits the latent evaporation heat of the solvent vapors from the first stage as a heating medium for a succeeding stage, where it is condensed. Such systems have a reduced energy consumption compared with the traditional single-stage systems, as only the first stage needs to be heated with steam. Vapors generated in the last stage must be condensed by a cooling medium, thus reducing cooling water usage. Figure 7.20 shows an evaporation unit with four stages and heat integration. As a consequence, the energy consumption decreases with the number of stages as shown in Figure 7.21. In a multistage plant, with n numbers of stages, the steam consumption is approx. only $1/n$ of the consumption of a single stage evaporation unit neglecting thermodynamic non-idealities.

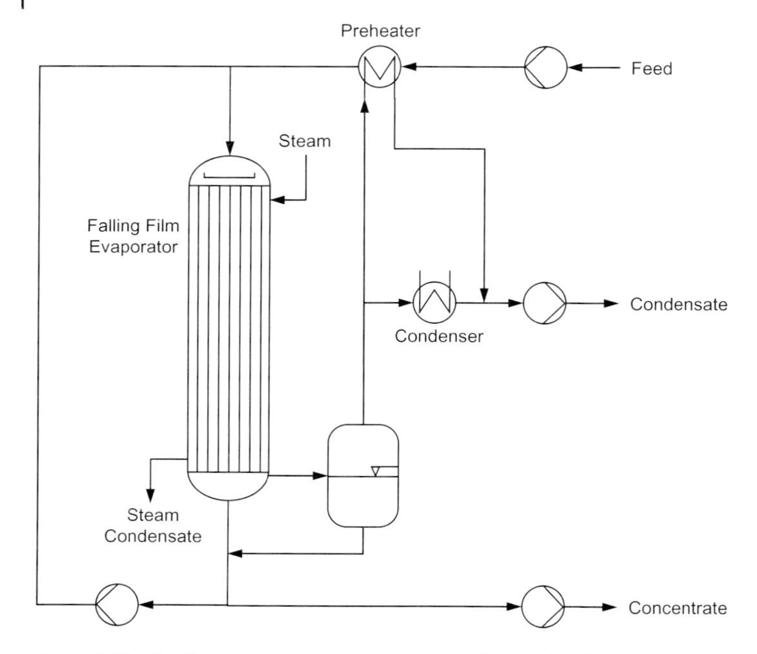

Figure 7.19 Single-stage evaporator system with feed preheating.

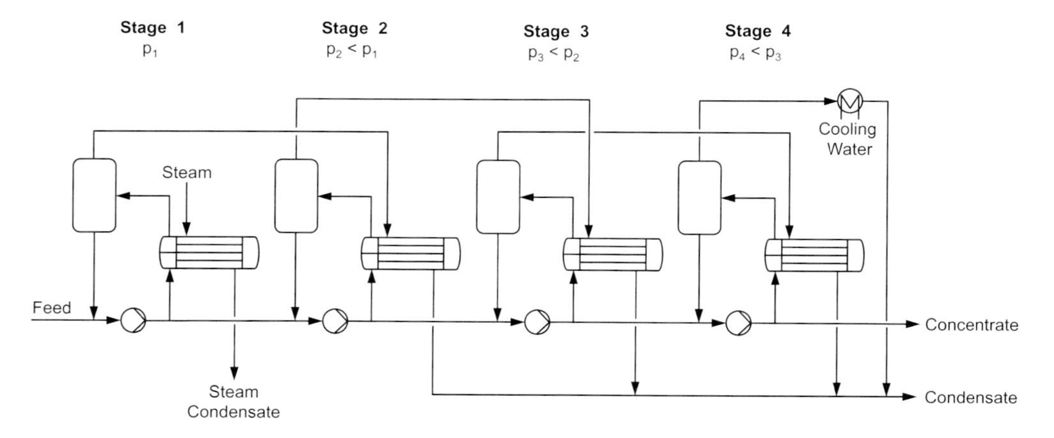

Figure 7.20 Evaporation unit with four stages and heat integration; co-current flow of liquid and vapor.

The overall temperature difference, which can be divided among the individual stages, is determined by the highest allowable product temperature of the first stage and the lowest boiling temperature of the final stage [15]. As a result, the temperature difference per stage decreases with an increasing number of stages and the heating surfaces of the individual stages must be made larger accordingly

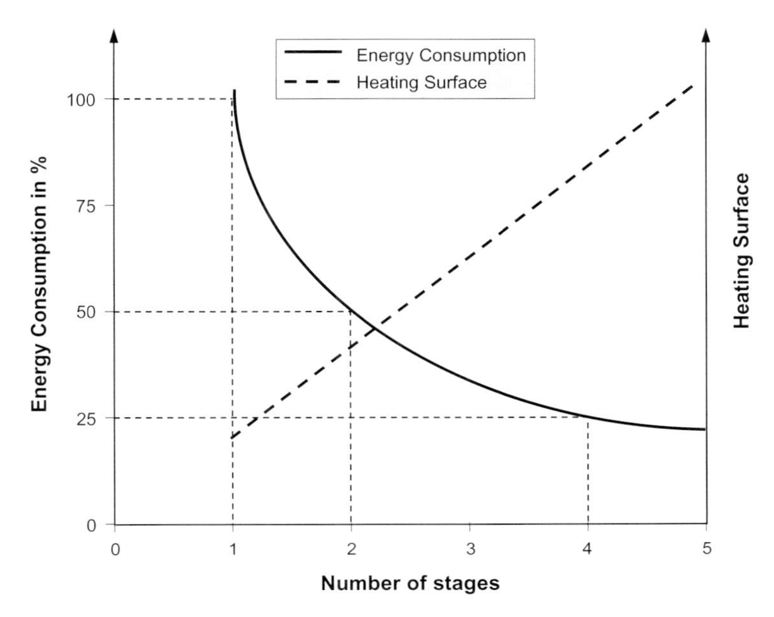

Figure 7.21 Relationship of energy consumption and heating surface as a function of the number of stages.

to achieve the required evaporation rate, but with a lower temperature difference. Approximately, the total heating surface of all stages increases proportionally with the number of stages. As a result, the investment costs rise whereas the energy consumption decreases with the number of stages (as shown in Figure 7.21).

For a specified temperature difference between the heating steam in the first stage and the boiling point in the last stage, an optimum number of evaporator stages can be found, subject to given economic and operating constraints.

Pumps can be installed to force liquid to circulate in the evaporator, in cases where the temperature difference in an individual stage is insufficient to provide a natural thermosiphon circulation. Forced circulation leads to an improved heat transfer through increased liquid velocity, and therefore reduced heat transfer area.

7.5.3 Vapor Recompression

In a conventional evaporator, the vapor stream is condensed, so that its valuable energy content in form of condensation heat is lost. Vapor recompression evaporation is based on the heat-pumping technique. A portion or the total vapor flow is compressed to a higher pressure and subsequently condensed in the same evaporator. In many cases, only multistage evaporator systems with three to five stages achieve the same specific efficiencies [15].

For the recompression of solvent vapors, steam jet ejectors operated by high pressure steam can be applied. Alternatively, this can be done mechanically with centrifugal compressors or with positive displacement machines. However, the energy consumption of the compressor increases with increasing temperature difference between the condensing compressed vapors and the boiling liquid. Therefore, vapor recompression is not economical for solutions with high boiling point elevation [15].

The operating principle of a steam jet ejector is the same as that of a steam jet pump. Since it is simple and has no moving parts, its operation is free of breakdowns and it is relatively cheap. However, the efficiency is low and usually decreases in operation due to nozzle erosion, which is caused by water droplets in the drive steam. Steam jet ejectors are more favorable for vapor recompression in the area of lower working pressures from the perspective of energy efficiency. Higher flow rates, above 5000 $Nm^3 h^{-1}$, are the field of application for mechanical compressors [15].

Mechanical vapor compressors use all the vapor compared with steam jet compressors, which only compress a part of the vapor. Therefore, compared with multistage and thermal vapor recompression the wasted condensation heat is significantly lower in mechanical vapor recompression. As a result, evaporation plants with mechanical vapor recompression require particularly low amounts of energy [15].

Single stage centrifugal fans are used in evaporation plants because they are simple and maintenance friendly [15]. For high pressure increases, multiple-stage compressors can be used. Frequently, the energy of the vapor condensate is used to preheat the feed.

Economically, mechanical vapor recompression is worth considering only when the temperature difference between the condensing compressed vapor and the boiling solution is small, and a single-stage compressor is sufficient. In practice, the temperature difference must be less than 20 °C, which is the case only for solutions having low boiling point elevation [15].

7.6
Drying

In the drying process volatile liquids are separated from solid materials by vaporizing the liquid and removing the vapor. During the process, heat is supplied to vaporize the liquids. Drying accounts for a large part of the energy usage in the chemical industry. Dryers can be separated into five main categories, depending on the method of heat transfer [3, 16]:

- convective dryers;
- contact dryers;

- radiant heat dryers;
- high-frequency dryers;
- freeze dryers.

In a convective dryer the vaporization of liquid is carried out by direct contact of the wet solid and a hot gas stream. In contact dryers the heat is conducted from a heated surface to the wet solid. The principle of radiation is exploited in radiant heat dryers, in which heat is supplied from a radiation source. High-frequency drying is characterized by dielectric heating of the wet solid for example, microwave drying. Finally, in freeze dryers the removal of water is carried out by sublimation of ice from the solid at temperatures below 0 °C and reduced pressure.

The most popular type of large scale dryer in chemical industry is the convective dryer, used to perform evaporation of water. Additionally, dryers that evaporate solvents can be found in polymer, fine chemical and pharmaceutical processes [16].

The ratio of the heat required to vaporize the solvent in relation to the heat actually consumed can be defined as the thermal efficiency of dryers. Contact dryers have the highest efficiency. However, the thermal efficiency of many convective dryers is often low. Thermal efficiency of convective dryers can be enhanced by improved operation and control and by heat recovery from exhaust gases.

7.6.1
Operational Improvements for Convective Dryers

Often convective dryers are operated with a hot gas stream which is directly or indirectly heated. Usually the heating medium is air. But inert gases, exhaust gases or superheated steam are also applied. Superheated steam as the heating medium has energetic advantages. By condensation of the used steam and the evaporated liquids from the wet solid a high rate of energy recovery of approx. 90% is possible [3]. On the other hand higher temperature ranges have to be applied resulting in higher thermal stress for the product.

The main operating parameters in convective dryers are the flowrate and inlet temperature of the hot gas stream. To improve the energy efficiency it is recommended that the air flowrate be reduced to a minimum and to increase the inlet temperature to the allowed maximum.

At partial load, many convective dryers are operated at excessive air flow rates. Therefore, optimization of operating parameters for the entire throughput range is recommended with regard to sufficient gas velocities in dryer and gas/solid separators for example, cyclones.

To enable operation at the energy optimum, convective dryers should be equipped with temperature measurement of inlet and outlet air, as well as with flow control. In addition, an online measurement of exhaust air humidity enables closer approach to the optimum operation point [1].

Often, spray dryers and fluidized-bed dryers are characterized by only little flexibility in changing the air flow rate without affecting particle characteristics within the dryer. Consequently, the air inlet temperature may be the only variable that can be controlled [1].

7.6.2
Heat Recovery from Convective Dryers

A measure to achieve significant energy savings is the recirculation of a fraction of the exhaust gas stream back to the inlet air to the heater. However, due to the increased moisture content of the hot air entering the dryer, higher temperatures are required to achieve the same drying rate. In addition, the investment costs for the ductwork and the control of the air recirculation is often high.

An additional measure to improve the energy efficiency of a convective dryer is to recover heat from the waste gas by means of heat exchangers. Usually the heat is used to preheat the inlet air (as illustrated in Figure 7.22) or the wet product feed, although it can also be used for other process streams.

In general, a major limitation for heat recovery from convective dryers is fouling of heat exchanger by particles. Therefore, cyclones or filters must be used for particle removal from the gas stream before heat recovery. However, these additional installations reduce the overall profitability of this measure. Furthermore, cleaning in place (CIP) strategies can be applied to frequently clean the heat exchanger surfaces to ensure best possible heat transfer. For avoiding fouling and blocking in the off-gas system scrubbers are often used so that a heat recovery is not possible.

Usually for heat recovery, the temperature of the hot gas stream is reduced without condensing any vapor, as convective dryers are often run with off-gas temperatures far above the dew point. An additional potential for heat recovery can be achieved by cooling the exhaust gas below its dew point.

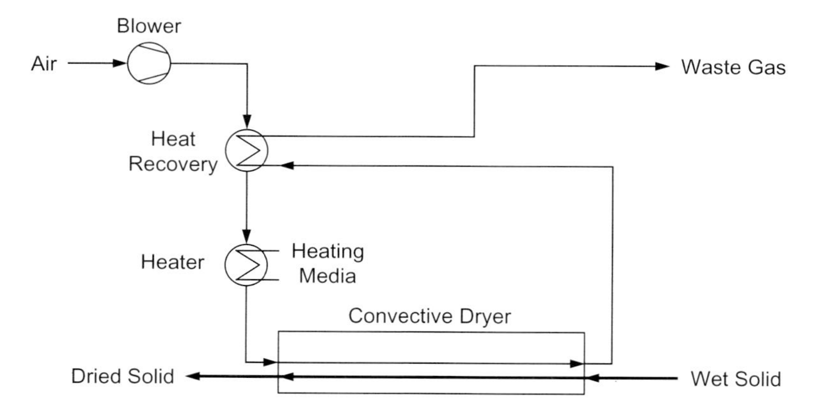

Figure 7.22 Convective dryer with waste gas heat recovery.

7.6.3
Additional Measures for Improving the Energy Efficiency of Dryers

Additional measures for improving the energy efficiency of dryers can be summarized:

Avoidance of over-drying

Avoidance of over-drying is another aspect of further improvement of dryers' operation. This may be achieved by online measurement of product moisture content. In addition, the product moisture often lies below the required specification for further processing or sale, which also contributes to higher energy consumption as not only the additional solvent is to be evaporated but also higher off-gas temperatures are necessary increasing the heat losses over the gas outlet. For this reason, the dryer should be operated at the maximum allowed product moisture.

Mechanical dehumidification

An aspect that one should consider is the reduced initial moisture of the wet material that is dried mechanically before entering the dryer. Mechanical dehumidification can be performed for example in press filters and centrifuges. The economical optimum of the moisture of the dry material in pre-humidification stage affects the total costs of the process.

Heat recovery from dried product

The dried product can also be used for heat recovery, for instance to preheat the cold inlet gas stream. However, only low recovery rates are possible.

Spray dryers

In spray dryers large amounts of pressurized air are often consumed in the pneumatic nozzles that are used for the atomization of the slurry. Pressurized air can be saved by using more efficient high-pressure jets, where the energy input is realized by the pressure of a pump. For media with a higher viscosity it is worth considering alternative atomization techniques.

7.7
Crystallization

Crystallization is based on the formation of solid crystals precipitating from a solution or melt or more rarely deposited directly from a gas. The physical principle of crystallization is a mass transfer of a solute from the liquid solution to a pure solid crystalline phase. Supersaturation must be achieved before crystals can grow.

The supersaturated condition can be achieved by two main alternatives. One is to decrease the solution temperature carried out in cooling crystallization. However, if the dependence of the saturation concentration on the temperature is not large enough, this technique cannot be applied for crystallization. In such a case, evaporative crystallization can be used. In addition, supersaturation can be achieved by solvent shift, by pH shift or by a chemical reaction between two or more substances with precipitation of one of the reaction products [17].

Crystallizers are generally classified based on the technique by which supersaturation is achieved, for example, cooling, evaporation, vacuum (adiabatic cooling), reaction, salting out.

While many chemical processes can be optimized purely on the basis of energy considerations, in crystallization plants a number of process parameters must be taken into account. Thus the operating temperature, temperature gradients, the volume of the apparatus etc. are closely related to the kinetic data of the crystallization system and in certain cases can be varied only within very narrow limits.

The crystallization process (evaporation or cooling) is usually determined by the solubility properties of the product that is being crystallized. The first selection criterion is: If the saturation concentration has low temperature dependency then evaporation–crystallization should be carried out. If the saturation concentration varies strongly with temperature, then cooling crystallization should be used.

7.7.1 Melt Crystallization by Cooling

Distillation is energy-intensive for solutions that form isomers and azeotropes. In these cases, crystallization can provide better separation and a more energy efficient process. Melt crystallization is the process of separating the components of a liquid mixture by cooling until a quantity of crystallized solid is deposited from the liquid phase. Therefore, melt crystallization is often considered to be commercially attractive, compared with distillation, for the separation of close-boiling organic substances. The relatively low energy demand of the freezing process and high selectivity of crystallization are regarded as the main advantages of melt crystallization [17]. The operation at much lower temperatures than distillation is a further benefit, which makes it a very valuable operation for processing thermally unstable substances [18]. The heats of fusion for the majority of compounds with industrial importance are by the factor 0.2 to 0.5 lower than the heats of vaporization. The example of water can demonstrate the low energy demand: the heat of vaporization is $2260\,kJ\,kg^{-1}$, whereas the heat of fusion is only $334\,kJ\,kg^{-1}$.

However, in batch crystallization the energy requirements of heating and cooling of the whole crystallizer equipment has to be considered, which can reduce the energy benefits of melt crystallization compared with distillation [19].

A hybrid purification process consisting of a combination of crystallization with another separation process can provide an effective measure for the production of high purity products with much greater efficiency than the single process alone. An example is a distillation process as an initial separation step followed by the crystallization-based process as a second step to achieve the pure product. A typical hybrid purification process with a distillation column and crystallization

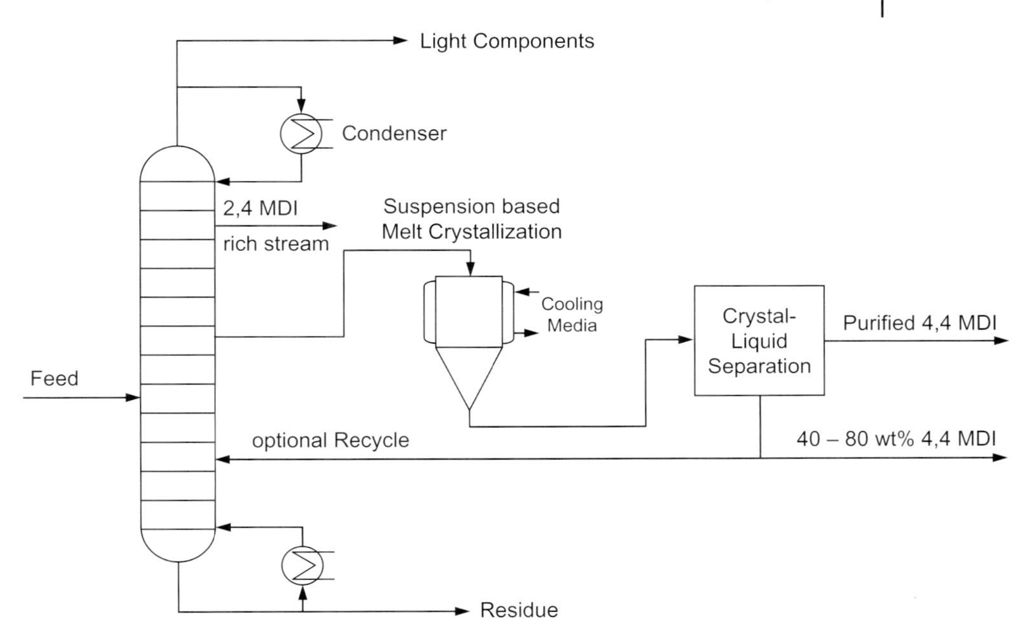

Figure 7.23 Typical hybrid purification process with distillation column and crystallization for the separation of diphenylmethane diisocyanate (MDI) [18].

for the separation of diphenylmethane diisocyanate (MDI) is shown in Figure 7.23 [18].

7.7.2
Evaporative Crystallization from Solutions

In evaporative crystallization, the solution is heated and evaporated to remove the solvent from the solution. One of the disadvantages of evaporative crystallization is therefore its large energy requirement. Comparable to evaporation (see Section 7.5.2) there are three basic possibilities to save energy:

- feed preheating;
- multistage evaporation;
- vapor recompression.

The energy efficiency of the evaporative crystallization can be improved by feed preheating and reusing the hot vapor discharged from a crystallizer. In a typical process for crystallization of salt for example, usually three or four evaporative crystallizers are configured in a multistage concept, comparable to multistage evaporation. The energy efficiency of the multistage process can be improved further by vapor recompression and recovering the enthalpy from the condensed liquid (see Section 7.5.2).

7.7.3 Freeze Crystallization

In crystallization by freezing, so-called freeze crystallization, the heat is removed from a solution to form crystals of the solvent. Similarly to melt crystallization, the advantage of freezing compared with evaporation is the potential for saving energy, since the enthalpy of crystallization is only one-seventh of the enthalpy of vaporization of water [17]. However, often the costs of refrigerants are higher compared with heating media like steam.

7.7.4 Additional Measures for Improving the Energy Efficiency of Crystallization

Additional measures for improving the energy efficiency of crystallization can be summarized:

Increasing plant availability

The up-time of crystallization plants is often limited by crust formation. The necessary cleaning mean a loss of production due to interruption of the operation, and the added washing solution has to be evaporated again. This leads to the necessity of designing a crystallization plants in such a way that scaling and deposits in flow sections are avoided as far as possible [20].

Avoidance of excessive quality requirements of the product

The product specifications naturally have a high priority in determining the choice of the most suitable process. However, when setting these specifications one must keep in mind that excessive quality requirement can possibly lead to disproportionately higher energy consumptions. A critical examination of the requirement for crystal purity can lead to saving of large amounts of steam and cooling water.

Avoidance of excessive water addition to the plant

A common practice is to carry out preventive washing in crystallization, which increases the production safety. However, excessive care can sometimes be carried to such extremes that the avoidance of product losses that might occur may lead to highly increased energy consumption [20].

7.8 Membrane Separation

Membranes can be successfully applied in a variety of fields ranging from production of oxygen-enriched air for combustion, volatile organic carbon recovery, and hydrogen purification. In addition, membranes are used for desalination of sea water and for recovery and recycling of wastewater. They are efficient tools for the concentration and purification of food and pharmaceutical products and the production of base chemicals [21]. Membranes have also been combined with conventional techniques such as distillation in so-called hybrid separation processes to provide more energy efficient production routes [22].

7.8.1 Basic Principles

A key property of a membrane is its ability to allow selected components to pass through the membrane, called permselectivity. Differences in the transport rate of various components through the membrane determine the permselectivity. The processes in which membranes are used, can be classified according to the driving force used in the process. The commercially and technically most relevant processes are pressure-driven processes, such as reverse osmosis, ultra- and microfiltration or gas separation; concentration-gradient driven processes, such as dialysis; partial-pressure-driven processes, such as pervaporation; and electrical-potential-driven processes, such as electrolysis and electrodialysis [21].

In many areas of the chemical, petrochemical, oil and gas industries today, membrane separations have become standard processes used either as alternatives to or in combination with conventional process steps [23]. Membrane separations can lead to reduced energy consumption, operating costs and investment requirements. However, the overall success of the membrane technology is still lagging behind these expectations. Technical barriers for membranes that prevent them to enter other industries are fouling, instability, low flux, low separation factors, and poor durability.

7.8.2 Applications of Membrane Separation to Increase Energy Efficiency

A number of membrane processes have been developed in recent years to increase energy efficiency. Pervaporation, hybrid membrane processes and membrane reactors have gained increasing attention as efficient techniques in the chemical and petrochemical industry [21]. In the following, examples of membrane separation processes for increasing energy efficiency are described.

7.8.2.1 Separation of Organic Vapors

With membranes, organic vapors can be separated continuously from large scale production processes. Application areas in polyolefin and polymer production are raw material purification, product purification and finishing. Propylene, ethylene, and hexane, for example are separated from process streams and recycled for further use [24].

Another example of industrial application is the recovery of vinyl chloride monomer from off-gas from PVC. In the conventional recovery process, the off-gas is compressed and subsequently condensed in multistage condensers. The first condenser is operated with a cooling tower water. In the subsequent condenser refrigerants are used. The advent of new membrane technology enables the second and third condensation stages to be replaced by a membrane stage. Depending on the membrane stage layout an increased recovery of vinyl chloride monomer and reduced energy consumption can be achieved [24].

7.8.2.2 Pervaporation

Distillation is energy-intensive for solutions that form azeotropes. In order to separate such components through distillation, it is necessary to eliminate the azeotropic point by adding an entrainer and perform an extractive or azeotropic distillation, which increases the complexity of the separation [2].

The application of vapor permeation and pervaporation membranes has become an increasingly popular solution for the process industry during the last couple of years, because the separation mechanism at membranes is not governed by the thermodynamic equilibrium between vapor and liquid.

In principle, pervaporation is based on the solution–diffusion mechanism. Its driving force is the gradient of the chemical potential between the feed and the permeate sides of the membrane. Membranes can be either hydrophilic or hydrophobic. In general, most membranes are hydrophilic or water permselective due to the smaller molecular size of water. Few membranes are hydrophobic or

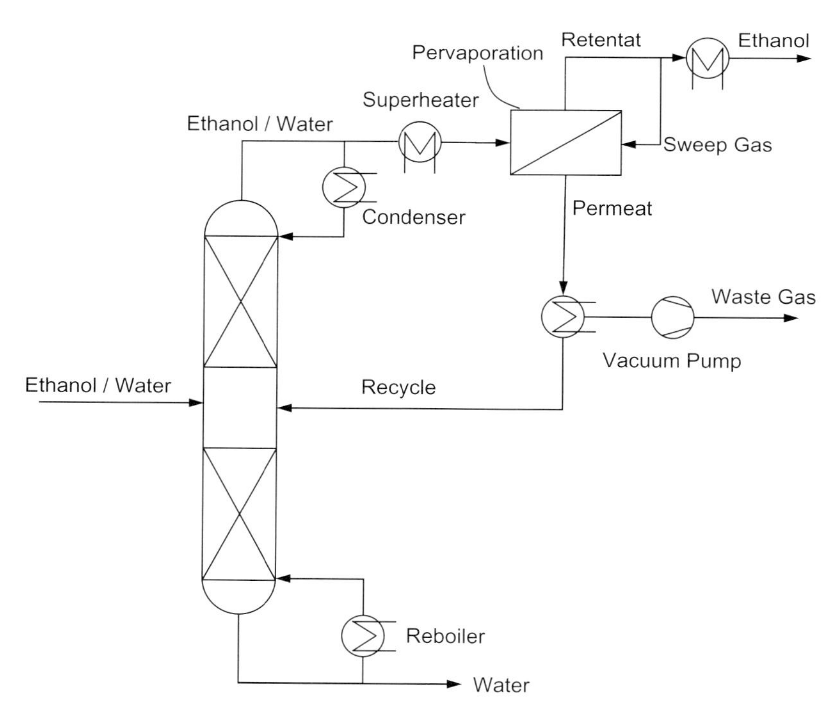

Figure 7.24 Typical hybrid purification process with distillation column and pervaporation for the separation of water and ethanol.

permselective for organic components. There are three categories of membrane materials: inorganic, polymeric and composite membrane [25].

Currently the dehydration of solvent mixtures is the biggest field of application of membrane-assisted distillation. Ethanol–water for example forms an azeotrope at approx. 96% ethanol. A typical distillation column with pervaporation for the separation of water and ethanol is shown in Figure 7.24. So far, over 100 plants worldwide use the pervaporation technique for alcohol dehydration [26]. Additional applications are given by Ohlrogge *et al.* [24].

7.9 Reaction and Entire Processes

For many chemical processes, improvements in the reaction and modification in the overall process design mostly offer the biggest potential for reducing energy consumption. The reaction and the entire process are closely related, since measures in the reaction usually have a large influence on other parts of the process. The separation and purification of the product often dominates the total energy costs of the entire process. A yield increase in the reaction often reduces the process energy consumption significantly. Therefore, the reaction and the entire process are discussed together in the following.

Substantial energy efficiency improvements are still possible even in well-established processes, which have undergone extensive optimization. In addition, many chemical processes still have energy requirements that are much higher compared with the theoretical minimum, offering potential for improvement. Processes can be improved in many ways. In the following improvement examples for the reaction part as well as for the entire process are discussed to illustrate the wide field of opportunities.

7.9.1 Recovery of Reaction Heat

Often an appropriate option to reduce the energy demand of an entire process is use of the heat of an exothermal reaction. This can be done directly in the process or in other processes on the same site (see Section 7.10). In the latter case this is realized in general by means of heating utilities, for example steam.

There are manifold options for the recovery of reaction heat. The energy can be used to preheat feed streams, to generate steam or to support the subsequent separation.

An example for the recovery of reaction heat is the nitration process as shown in Figure 7.25. In the presence of nitric acid and sulfuric acid, large quantities of heat are generated in the reactor as the exothermal reaction proceeds. In a conventional isothermal process, the heat of reaction is removed from the system by cooling water. After product separation, water (also water coming from nitric acid and the reaction) is separated from diluted sulfuric acid in an energy intense

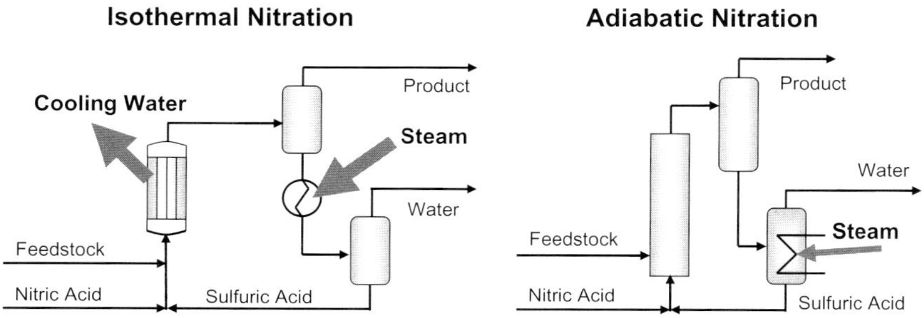

Figure 7.25 Comparison of isothermal and adiabatic nitration.

manner. Such concentrated sulfuric acid is sent back to the reactor. The heat generated in the reaction is essentially wasted in this process. For this reason efforts have been made to modify the process in a way that would enable the valuable heat of exothermal reaction to be recovered. The result of this work is a new adiabatic nitration, in which the heat of reaction instead of being wasted by cooling water is used in concentration of sulfuric acid, consequently lowering the energy demand of the entire system. For a Bayer plant in Leverkusen, it was possible to reduce the steam demand by 240 000 t/a, which translates to 40 000 t/a of CO_2 emissions [27].

7.9.2 Heat Integration

If a number of hot and cold streams exists within the plant, between which heat exchange is possible, one of the classical methods of use of available energy in the plant is optimal design of heat exchanger network. Most chemical plants have numerous heat sources and heat sinks at different temperature levels, that are satisfied by different steam and cooling medium networks. The design of the heat exchanger network focuses on identification of profitable and feasible combinations of heat exchanger network that would minimize the amount of energy imported from outside, through heat transfer between two process streams. Pinch analysis is the most established tool used to identify the optimal combination of hot and cold streams. The development of the pinch principle by Linnhoff *et al.* has provided engineers with a scientific design methodology which has achieved outstanding results across the range of process industries [28]. This methodology is discussed in detail in Chapter 6.

The resulting heat exchanger network is characterized by a minimum energy requirement that needs to be imported. However, the balance between energy savings and capital costs, as well as operability and safety considerations for the network needs to be taken into account.

Pinch analysis is an established tool in the design phase for new plants. In addition, retrofitting of the heat exchanger network of an existing plant is highly recommended. Since energy prices are rising in the long term and many older plants were not designed considering energy minimization, and often have been de-bottlenecked, there is usually a large optimization potential to be identified.

A possibility to improve heat integration is intensification of heat transfer. A reduced temperature difference between the hot and the cold process streams of the heat exchanger network allows increased energy recovery. The possibilities to improve heat transfer are increased heat transfer coefficients or increased heat transfer surfaces. Heat transfer coefficients can be increased for instance by turbulence inducing elements in pipes or by structured heat exchanger surfaces. Plate heat exchangers or microstructured equipment (see Section 7.9.6) can help to increase heat transfer surfaces.

Besides thermal coupling by heat exchange within the process it is also worth considering the use of waste heat to generate other types of energy carrier. An example is the increasing usage of waste heat at a temperature of approx. 120 °C to 150 °C to generate chilled water of approx. 5 °C by an adsorption chiller. An additional example is the usage of low pressure steam in a steam turbine to generate electricity.

7.9.3 Increased Conversion and Selectivity

In many processes, downstream processing might constitute the major energy requirement of the plant. Separation also frequently dominates the capital costs of the process plant. Increased conversions in the reactor, for example, by better and more selective catalysts, may substantially reduce energy requirements for downstream separations.

The chemical industry has been very successful in identification and implementation of new catalysts as a way of improving its processes. As a result, catalyzed chemical reactions now constitute about 80% of all reactions in the chemical industry [10]. Nevertheless, progress in the development of new catalysts represents a substantial lever in efficient use of materials and energy. Not only catalysts do reduce the energy barrier required for the reaction to take place, but they also reduce the usage of raw materials by increased selectivity.

To demonstrate how important the appropriate usage of catalysts can be, an example of acrylic acid production will be given. Through the continuous improvement of the catalyst over a period of 20 years, the yield of reaction has increased by more then 11%. Through targeted selectivity of the reaction, that favors intermediate reaction of acrolein, it was possible to stifle formation of side reaction and consequently shift the equilibrium towards the desired product. Given the total annual capacity of 800 000 tones of acrylic acid, CO_2 savings of 230 000 tones per year were achieved [10].

An additional example is the production of styrene monomer from ethylene and benzene. The stages of the reaction are alkylation to produce ethylbenzene followed by dehydrogenation to produce styrene. The dehydrogenation reaction is not complete. The products are condensed and give a mixture of ethylbenzene, styrene, and other hydrocarbons. Most of the energy used for the entire process is required to separate this mixture by distillation. Improvements in operating conditions and catalysts have increased the conversion to styrene from approx. 40% to over 75% [1].

Considering the operating conditions of reactors, it can be observed that they are frequently operated with an excess of reactant or solvent compared with the required process conditions. This leads to an increased load of the subsequent separation of the product form raw materials, by-products and solvents. Only by operating reactors at their design conditions a large amount of energy can be saved.

7.9.4 Solvent Selection

In the chemical industry solvents are used in various process steps such as reaction and separation. To minimize the energy demand of the entire process the goal is to select a solvent that promotes the chemical reactivity as well as the subsequent downstream processing. Solvents that improve the energy efficiency of thermal separation are, for instance, components with a low evaporation enthalpy or a high separation factor. In recent years methods for selecting appropriate solvents have been developed which are often based on computer-aided property estimation tools [29].

Ionic liquids have frequently been proposed to improve the efficiency of chemical processes. Ionic liquids are salts with boiling point below 100 °C and negligible vapor pressure. Therefore, they are not volatile at process conditions. There is a large field of potentially interesting applications although only a few industrial applications have been published so far. Mostly, ionic liquids are used as solvents or catalysts to improve chemical reactions. An overview is given by Meindersma *et al.* [30]. One commercial application of ionic liquids is acid scavenging [10, 30].

In addition, ionic liquids can be used as additives to improve distillation processes, especially azeotropes and close boiling mixtures. Usually ionic liquids are hygroscopic with a strong affinity to water. Therefore, if water is part of the azeotrope high separation factors can be possible leading to high potential energy savings [30].

7.9.5 Optimized Process Conditions

The energy requirements may also be substantially reduced if processes can be operated at less extreme conditions of temperature or pressure. In addition the change from a liquid phase reaction to gas phase reaction can help to reduce the energy requirements.

Isocyanates are pre-products of many substances such as glues, paints and polyurethane foams. In the conventional process, as illustrated in Figure 7.26, they are manufactured by contacting amines with phosgene in a solution at the presence of a solvent. Due to its energy-intensive nature of downstream processing, where the solvent is separated from the product, industry was motivated to develop an alternative, less energy-intensive process. In such a way a gas phase synthesis was developed. With the new process design it is possible to reduce the amount of solvent by 90% and the excess of the reactant phosgene by 70%. Due to the reduced amounts of solvent and reactant the load of the energy-intensive multistage distillation route is also lower, and large recycle streams of phosgene in the system are eliminated. Therefore, the new process requires only half of the energy demand, compared with the conventional process and characterize in increased safety due to reduced hold-up [27, 31, 32].

The example from the isocyanate production mentioned above, illustrated how steam savings have been achieved. One of the biggest electricity consumers in the chemical industry is chlorine manufacturing. The main process is electrolysis of NaCl, during which chlorine, hydrogen and caustic soda are produced. Alternatively, chlorine is produced by electrolysis of HCl solution, in which hydrogen is a by-product. A second process is applied on all sites where chlorine is used as a raw material, and HCl as a side product of the reaction cannot be used for other purposes. On such plants, HCl is sent back to the electrolysis, where it is used for chlorine production.

The oldest process is the amalgam process, where the specific energy consumption is in the region of 3.6 MWh t^{-1} chlorine. The amalgam process has been consequently replaced by a more efficient membrane process, characterized by about 30% lower electricity consumption, of c.a. 2.5 MWh t^{-1} chlorine as illustrated in Figure 7.27. Over recent years, membrane technology has been optimized to such an extent that no substantial reduction of the energy demand can be expected from further process modifications [33].

The latest technology is so-called oxygen depolarized cathode (ODC), which combines principles of chlorine-alkali electrolysis with fuel cells, in which electric-

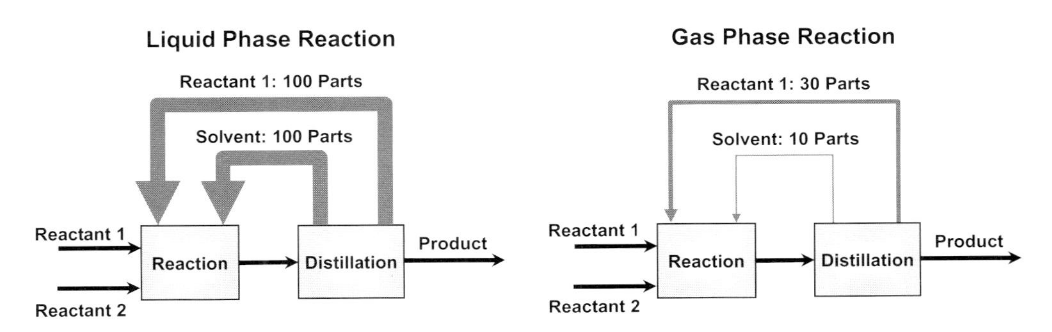

Figure 7.26 Comparison of liquid and gas phase phosgenation of isocyanates.

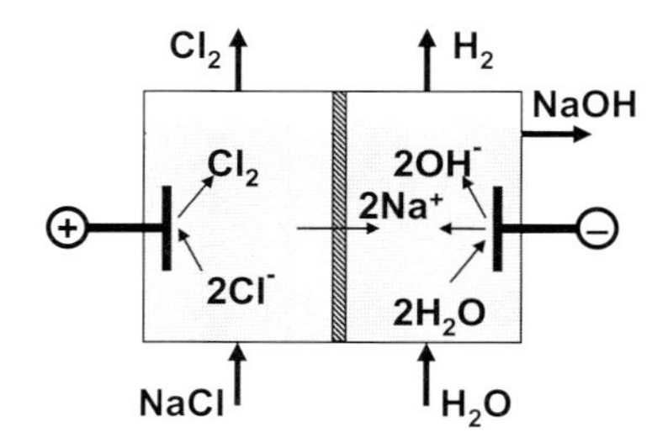

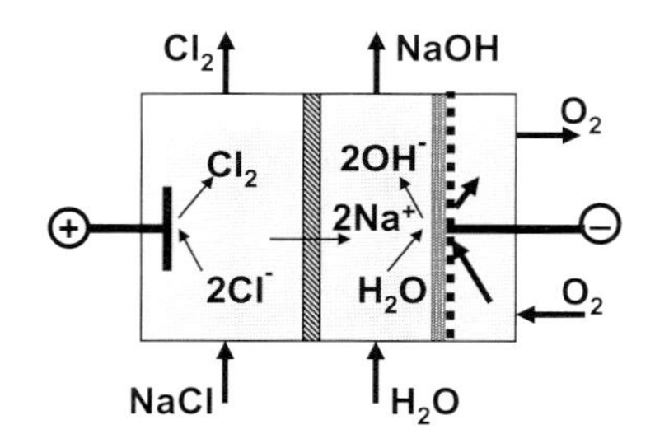

Figure 7.27 Comparison of conventional membrane and oxygen depletion cathode process for chlorine production.

ity is produced during synthesis of hydrogen with oxygen. Oxygen is supplied to the cell and is reduced on the cathode to hydroxide (OH^-). Thanks to this principle, the co-product hydrogen is used in the electrolysis to recover electricity, which consequently reduces the specific energy consumption by another 30% to 1.8 kWh t^{-1} chlorine. Of course, hydrogen is no longer a product of such a cell. The technology can be applied both for HCl electrolysis as well as conventional chlorine-alkali electrolysis [27, 31].

7.9.6
Microstructured Equipment

Microstructured process equipment, such as reactors, mixers or heat exchangers, have been introduced in recent years as a novel approach to improve the efficiency of chemical processes. The dramatic increase of specific surface areas in microstructured equipment enables a more precise control of process conditions, which is often the key to significant increases in yield and selectivity of chemical reactions, reduced solvent and reactant excesses or improved heat transfer rates and recovery of waste heat streams. For example, an improved temperature control in a microstructured reactor can result in significantly reduced side reactions and less pronounced hot-spots for rapid, exothermic reactions. As the separation of unwanted side products often is the major cause for energy-intensive downstream purification steps, this can offer a tremendous potential for improved energy efficiency [34].

In addition, intensified heat exchanger/reactor concepts (HEX reactor) are the subject of process development [35]. The primary aim is to improve the temperature control leading to higher yield and selectivity or to directly use the heat of an exothermal reaction in an endothermal reaction to improve the energy efficiency.

7.10
Total Site Network

In comparison to other branches in industry, the production chains in chemical industry are characterized by high complexity. The manufacturing processes of fine chemicals, basic chemicals, polymers or even pharmaceuticals are integrated in complex systems. This applies not only to physical integration of the production line but also integration resulting from using the same platform for energy and utilities supply. For this reason, chemical industries tend to cluster together, to benefit from the possibility of production integration, which consequently contributes to reduced energy demand, use of by-products and increases competitiveness.

Total site integration offers huge potentials for energy savings, for instance through the energetic integration between processes. The heat of exothermal processes can be used in endothermal reaction or separation processes. In doing so, a proper design of steam pressure levels of the steam network on the site is essential. In a large manufacturing site approx. 50% of thermal energy can be satisfied by heat integration between exothermal and endothermal processes [10]. In addition, a well-balanced material usage is recommended so that most of the by-products can be used. Heat integration by pinch analysis for a total site is discussed in Chapter 6.

Additionally, a pivotal point of total site network is highly efficient energy supply system, based on power plants using cogeneration (combined heat and power generation). With cogeneration it is possible to achieve a total efficiency in the region of 90% compared with modern power generation plants with efficiencies in the range of 45% to 55%. Another dimension of integration is the thermal use of by-products for instance from the refinery as alternative fuels.

The physical and energetic integration within a chemical manufacturing site offers the advantage of use of by- and co-products of different plants. This enables reduction of waste of raw materials and increase in the overall profitability of the plant. Energy consumption expressed in primary energy can be reduced by integrating energy consumers and energy suppliers by utility networks (e.g., steam network, but also cooling media and ammonia network). The impact on integration (total site network) must be thoroughly considered in cases where, for example, projects leading to reduced energy consumption are carried out. Due to the integrated nature of the system, energy savings and energy integration must always be considered simultaneously. Leimkühler *et al.* discuss this aspect at the example of a sulfuric acid network [27].

7.11
Advanced Process Control and Performance Monitoring

The aim of advanced process control (APC) is to move the operating parameters to the best process conditions and to keep them regardless of disturbances, while satisfying process constraints and production goals. Therefore, APC offers considerable potential for reducing energy requirements and operating costs.

Advanced process control consists of control structures that go beyond ordinary control strategies like PID controllers. It incorporates for instance cascade controllers, feed forward control, as well as model predictive control and online optimization-based on process models. Model-predictive multivariable control technology was introduced into industry about 25 years ago and was primarily used in refineries. Nowadays APC is a well-established technology in the chemical industry. It can be applied to a single unit operation or to a whole plant. In addition, APC reduces process variability. Therefore, the processes can be operated closer to their optimum operating conditions. An example of APC for distillation is give in Section 7.4.2.6.

Increasing number of companies in the chemical industry recognize the significance of information as an important factor for optimizing their production. Beside classical automation technology, modern operating systems delivering real-time information about the actual status of the process attract more and more attention. Performance monitoring, based on the monitoring of selected variables in the process, so-called KPI (key performance indicators), plays a dominant role in increasing productivity and reduction of energy costs [36].

Energy efficiency improvements can be achieved by optimized design of the equipment or the entire process as well as by optimized operation. Optimized operation is achieved by improved automation and process control or directly by better manual operation.

What can often be observed is that processes are operated at a comfort zone. The energy-influencing operating parameters are not challenged and not load-dependent. After a failure the process often remains in a 'safe' operating point. Examples are distillation columns with too high reflux flows (see Section 7.4.2) or reactors, operated with an excess of solvent or reactant leading to large circulation flows in the purifying section. Furthermore, the product specification of a separation is often slightly higher than the required specification for further processing. In addition, the specific energy consumption is frequently load dependent and shows fluctuations.

The basic principle of performance monitoring is continuous data logging of physical data (process monitoring) and technical condition of the plant (condition monitoring). An example of the first is the definition of reaction-relevant data, such as activity of catalyst, selectivity and yield. The example of the latter is the definition of equipment-relevant data, for example fouling of heat exchangers, control and measuring equipment [36]. Performance monitoring answers the questions [36]: At what state is the process currently? How far is the process from optimum or from a pre-defined benchmark? What are the reasons for deviations from the set points? How fast can the critical point of the plant be achieved and what are the measures to prevent this from happening?

The goal of compressing the information is usually to extract from the vast amount of process data only data with highest importance to the system (KPI).

The KPIs, diagnosis information and improvement proposals are visualized in an appropriate form, for example, trends or tables [36].

A pivotal condition for performance monitoring is the real-time processing and interpretation of process data with respective process know-how and process control knowledge. The technical conditions are therefore created through implementation of modern process database (PIMS, process information management system) and process operating system on the MES level (manufacturing execution systems).

By a potential analysis with respect to the energy consumption for the typical chemical plant, the situation pictured on the Figure 7.28 can often be observed. The graph represents specific energy consumption plotted over the range of plant load.

Usually, the plant operates at higher energy efficiency at higher plant loads. Therefore, operation of the plant at high output levels represents an important measure for improving energy efficiency. If required production levels are significantly lower than capacity, then a variety of operating strategies, including campaign operation, should be considered.

Often fluctuations in energy consumption can be observed that contribute to much higher energy consumption compared with the best practice level as shown in Figure 7.28. The reasons for that happening can range from for example, non-optimal process operation by manual operation, fluctuating purities of raw materials, changing ambient conditions or fouling of the equipment. Therefore, identification of the correct energy-influencing operating parameters by process analysis is important for proper performance monitoring. Once the calculated relationship is presented in a graphical form, the fluctuations can be clearly

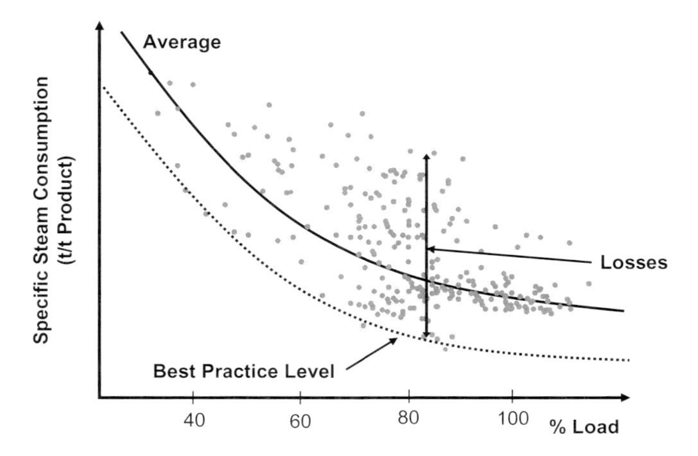

Figure 7.28 Specific steam consumption versus the load of a production process.

observed and the assessment of energy efficiency can be carried out very quickly. In addition, it is possible, for instance, to estimate the intensity of fouling and its effect on specific energy consumption by on-line simulation and to visualize them in a clear way [36, 37]. Based on this, it is possible to achieve improved operating conditions of the plant, and consequently reduce the energy consumption. Performance monitoring delivers information about wanted and unwanted variances at an early stage and provides reasons – not only 'gut feeling' – for necessary decisions concerning how to operate the process. The system can also be used as a supervisory control system, for example, by changing set points on programmable controllers.

By creating transparency with performance monitoring the operator only gets the ability to improve. The crucial point to achieve energy efficiency is the regular evaluation of the monitoring, diagnosis and guidance information that should become a daily routine in the production workflow. Differences between actual and expected status should act as a motivation for the operators to undertake appropriate action. In order to achieve sustainable improvements, it is necessary that experts accept and implement the proposed improvements. This can by facilitated by a change program.

References

1. Grant, C.D. (2005) Energy management in chemical industry, in *Ullmann's Encyclopedia of Industrial Chemistry*, Wiley-VCH Verlag GmbH, Weinheim.
2. Stichlmair, J. (2005) Distillation and rectification, in *Ullmann's Encyclopedia of Industrial Chemistry*, Wiley-VCH Verlag GmbH, Weinheim.
3. Sattler, K. (2001) *Thermische Trennverfahren: Grundlagen, Auslegung, Apparate*, Wiley-VCH Verlag GmbH, Weinheim.
4. Fenske, M.R. (1932) *Industrial and Engineering Chemistry*, **24**, 482.
5. Underwood, A.J.V. (1948) *Chemical Engineering Progress*, **44** (8), 603–614.
6. Dünnebier, G., Friedrich, M., Gerlach, M., Hotop, R., and Schmidt, R. (2003) Online Analytik zur Prozessführung (Teil 1). *atp – Automatisierungstechnische Praxis*, **45**, 24–32.
7. Dünnebier, G., Friedrich, M., Gerlach, M., Hotop, R., and Schmidt, R. (2004) Online Analytik zur Prozessführung (Teil 2). *atp – Automatisierungstechnische Praxis*, **46**, 8–22.
8. Nakaiwa, M., Huang, K., Endo, A., Ohmori, T., Akiya, T., and Takamatsu, T. (2003) Internally heat-integrated distillation columns: a review. *Chemical Engineering Research and Design*, **81** (1), 162–177.
9. Jana, A.K. (2010) Heat integrated distillation operation. *Applied Energy*, **87**, 1477–1494.
10. Moutz dividing wall columns, www.moutz.de
11. Rudolph, J. (2006) Die chemische Industrie als Innovationsmotor für den Klimaschutz. *Chemie Ingenieur Technik*, **78** (4), 381–388.
12. Bausa, J., von Watzdorf, R., and Marquardt, W. (1998) Shortcut methods for nonideal multicomponent distillation: 1. Simple columns. *AIChE Journal*, **44**, 2181.
13. Kraemer, K., Kossack, S., and Marquardt, W. (2009) Efficient optimization-based design of distillation processes for homogeneous azeotropic mixtures. *Industrial and Engineering Chemistry*, **48** (14), 6749–6764.
14. Sundmacher, K., and Kienle, A. (2003) *Reactive Distillation*, Wiley-VCH Verlag GmbH, Weinheim.

15 Sakuth, M., Reusch, D., and Janowsky, R. (2008) Reactive distillation, in *Ullmann's Encyclopedia of Industrial Chemistry*, Wiley-VCH Verlag GmbH, Weinheim.

16 Billet, B. (2005) Evaporation, in *Ullmann's Encyclopedia of Industrial Chemistry*, Wiley-VCH Verlag GmbH, Weinheim.

17 GEA Evaporation Technology, www.gea-wiegand.com

18 Tsotsas, E., Gnielinski, V., and Schlünder, E.-U. (2005) Drying of solid materials, in *Ullmann's Encyclopedia of Industrial Chemistry*, Wiley-VCH Verlag GmbH, Weinheim.

19 Mullin, J.W. (2005) Crystallization and precipitation, in *Ullmann's Encyclopedia of Industrial Chemistry*, Wiley-VCH Verlag GmbH, Weinheim.

20 Ruemekorf, R., and Scholz, R. (2004) Crystal clear separation. *Hydrocarbon Engineering*, **9** (6), 75–78.

21 Wintermantel, K., and Wellingdorf, G. (1990) *Industrial Crystallization 90* (ed. A. Mersmann), GVC-VDI, Düsseldorf, pp. 703–708.

22 Wöhlk, W. (1982) Industrial Crystallization under the aspect of energy economic operation, www.gea-crystallization.com (accessed 29 April 2010).

23 Strathmann, H. (2005) Membranes and membrane separation processes, in *Ullmann's Encyclopedia of Industrial Chemistry*, Wiley-VCH Verlag GmbH, Weinheim.

24 Gorak, A., Hoffmann, A., and Kreis, P. (2007) Prozessintensivierung: reaktive und membranunterstützte Rektifikation. *Chemie Ingenieur Technik*, **79** (10), 1581–1600.

25 Strathmann, H. (2001) Membrane separation processes: current relevance and future opportunities. *AIChE Journal*, **47** (5), 1077–1087.

26 Ohlrogge, K., and Stürken, K. (2005) Membranes: separation of organic vapors from gas streams, in *Ullmann's Encyclopedia of Industrial Chemistry*, Wiley-VCH Verlag GmbH, Weinheim.

27 Huang, H.-J., Ramaswamy, S., Tschirner, U.W., and Ramarao, B.V. (2008) A review of separation technologies in current and future biorefineries. *Separation and Purification Technology*, **62**, 1–21.

28 Dong, Y.Q., Zhang, L., Shen, J.N., Song, M.Y., and Chen, H.L. (2006) Preparation of poly(vinyl alcohol)-sodium alginate hollow-fiber composite membranes and pervaporation dehydration characterization of aqueous alcohol mixtures. *Desalination*, **193**, 202–210.

29 Leimkühler, H.-J., and Helbig, J. (2006) CO2-Minderung in der chemischen Industrie. *Chemie Ingenieur Technik*, **78** (4), 367–380.

30 Linnhoff, B., Townsend, D.W., Boland, D., Hewitt, G.F., Thomas, B.E.A., Guy, A.R., and Marsland, R.H. (1982) User guide on process integration for the efficient use of energy, IChemE, Rugby, UK.

31 Gani, R., Jimenez-Gonzalez, C., and Constable, D.J.C. (2005) Method for selection of solvents for promotion of organic reactions. *Computers and Chemical Engineering*, **29**, 1661–1676.

32 Meindersma, G.W., Maase, M., and De Haan, A.B. (2007) Ionic liquids, in *Ullmann's Encyclopedia of Industrial Chemistry*, Wiley-VCH Verlag GmbH, Weinheim.

33 Pabst (2007) Sparsame verfahren, Chemie Technik, 44, www.chemietechnik.de (accessed 01 February 2010).

34 Wolf, M. (2007) Energieoptimierung in der chemischen Industrie, in Potenziale der Chemie für mehr Energieeffizienz, Nachrichten aus der GDCh-Energieinitiative, www.gdch.de (accessed 01 February 2010).

35 Moussallem, I., Jörissen, J., Kunz, U., Pinnow, S., and Turek, T. (2008) Chlor-alkali electrolysis with oxygen depolarized cathodes: history, present status and future prospects. *Journal of Applied Electrochemistry*, **38** (9), 1177–1194.

36 Kiwi-Minsker, L., and Renken, A. (2005) Microstructured reactors for catalytic reactions. *Catalysis Today*, **110**, 2–14.

37 Anxionnaza, Z., Cabassud, M., Gourdon, C., and Tochon, P. (2008) Heat exchanger/reactors (HEX reactors): concepts, technologies: state-of-the-art.

Chemical Engineering and Processing, **47**, 2029–2050.

38 Dünnebier, G., and vom Felde, M. (2003) Performance Monitoring – Ein entscheidender Beitrag zur Optimierung der Betriebsführung. *Chemie Ingenieur Technik*, **75** (5), 528–533.

39 Joshi, S., Su, M., Potaraju, S., Daszkowski, T., and vom Felde, M. (2003) eProcess – a new was to Benchmark and improve your production process. FOCAPO, Florida.

8
Energy Efficient Equipment

Roger Grundy

8.1
Introduction

Most existing installations were built before the latest energy efficient technology was available, and also at a time when energy costs were lower. As a result, the main focus for 'new build' projects was (and often still is) to minimize investment cost.

This chapter therefore examines how existing equipment can be improved as well as the choices available for energy efficient new equipment.

There are many references in this chapter to new product types. Most, if not all of these can be researched easily using the Internet.

8.2
Rotating Equipment

8.2.1
Compressors

8.2.1.1 Reciprocating Compressors

Large, low speed machines are more efficient than smaller (cheaper) high speed units. Also, valve and bearing life is longer giving reduced maintenance cost.

Traditional compressor control is step-wise through the use of suction valve unloaders. Capacity is therefore typically 0% (at start-up) ~25%, ~50%, ~75% and 100%. Where the process requires some intermediate flow, it is normally achieved by spilling flow back from discharge to suction. *This wastes energy*. Further, operators often do not make best use of the step controls, needlessly increasing spillback and energy wastage.

There are two solutions:

- Variable volume clearance pockets at the outboard end of the cylinders – manual or automatic.

Managing CO_2 Emissions in the Chemical Industry. Edited by Leimkühler

ISBN: 978-3-527-32659-4

- Rapid acting hydraulic suction valve unloaders which operate at a variable percentage of each piston stroke.

It is often found that the energy saved when using valve unloaders is far less than expected. This is usually due to excessive pressure drop across the valves *even when being held open*. The solution is to select valves properly engineered for the service.

For existing machines, check the effect on motor amps at the various unloader settings. If the reduction falls short of expectations, investigate the valve design.

8.2.1.2 **Centrifugal Compressors**

Apart from the minimum theoretical energy for compression, energy consumption also results from:

- gas friction in impellers and diffusers;
- bearing and seal friction;
- labyrinth leakage both between wheels and into the seal system;
- balance piston leakage from discharge to suction;
- fouling;
- leaking anti-surge valves and/or poorly calibrated anti-surge protection.

Minimizing these losses can be achieved as follows:

It is desirable that the compressor's interior surfaces with a positive pressure gradient are highly polished to reduce friction. Note that polishing areas with a negative pressure gradient may risk premature onset of surge (discussed below) – consider the top surface of an aircraft's wing which has a matt finish (negative pressure above the wing) compared to the leading edge which is polished (positive pressure). An alternative to polishing is to coat the surfaces of the gas flow path (rotor and stator) and a number of proprietary products are available.

Energy loss due to oil seals can be considerable. The oil supply must be pumped to above the compressor's suction pressure and the return sweet oil must be cooled since the tight clearances in an oil seal result in heating due to the high shear rate. The use of dry gas seals reduces these losses to near zero. Dry seals are not only far cheaper than seal oil systems, but many users have found that retrofitting is justified.

The vast majority of compressors use oil-lubricated bearings. The associated energy loss can be seen in the power required to pump the oil and the amount of heat removed in the oil cooler. Slowly, magnetic bearings are being used, initially for pipeline compressors in cold remote locations and are now available for most process applications. Unfortunately, these generally cannot be retrofitted to existing compressors since the bearings are physically larger – especially the thrust bearings. For new installations, magnetic bearings plus dry gas seals result in an oil-free compressor. Surprisingly, magnetic bearings control shaft position to a tighter tolerance than oil-lubricated sleeve bearings and can even give active vibration control. The tighter tolerance between the rotor and labyrinths also improves efficiency through reduced internal leakage.

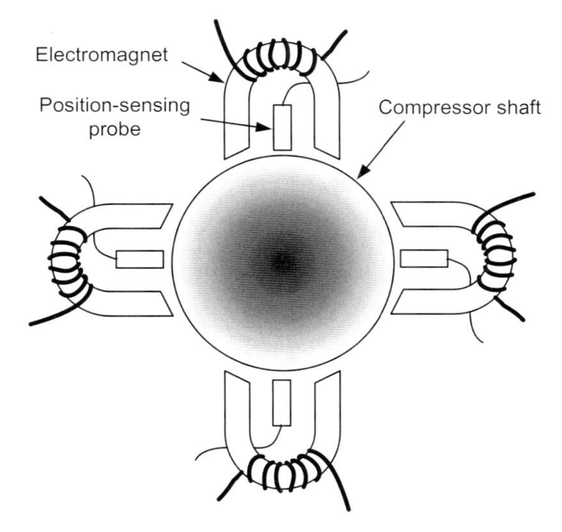

Figure 8.1 Magnetic journal bearing.

Magnetic bearings operate by sensing the distance between the probe(s) and shaft, and adjusting current to the electro magnets to maintain the shaft position. Figure 8.1 illustrates the general principle of operation. If current is increased to all the magnets at the same time, the bearing stiffness is increased. Because there is close to zero friction in magnetic bearings and dry gas seals, the pressure-velocity limit (PV Limit) inherent in oil-lubricated sleeve and thrust bearings disappears, allowing increased speed. Finally, because of the rapid response of the magnets, these bearings can be used for active vibration control.

A traditional labyrinth is simply a series of knife edges set into a ring mounted inside each diaphragm, with the edges facing inwards towards the rotating shaft. Typical clearance is 1.3 to 1.5% of the compressor shaft diameter and so there is always *some* leakage. Leakage increases if the labyrinths are fouled or worn (such as can happen due to surge or not accelerating quickly through first lateral critical speed). This leakage can be reduced to near zero through retrofitting improved design such as abradable labyrinths where the knife edges are mounted on the shaft and allowed to cut into an abradable ring mounted in the diaphragm.

Some compressor services carry the risk of internal fouling. Rotor and stator coatings can reduce this tendency. It is common to inject wash oil into the suction of compressors that foul (for example, cracked gas compressors), but this wash oil increases power requirements and so should be minimized consistent with achieving the desired result. Since polymer fouling increases with gas temperature, suction cooling and inter-cooling should be maximized.

All dynamic compressors (centrifugal and axial) exhibit instability if the suction volume flow is too low. This is known as surge, resulting in periodic flow reversal. It can be violent for axial compressors, and for both centrifugal and axial there is a risk of thrust bearing damage. The protection consists of ensuring that the suction flow is safely above the surge point over the full operating speed range. Other than air compressors, which avoid surge by venting from the discharge, process compressors take some of the discharge flow and spill it back into the compressor suction, usually via a cooler. Anti-surge protection should ideally not normally require continuous spillback except when the plant is running at low throughput. The surge protection margin should be no greater than 5% above flow at surge. Advanced anti-surge controls are self-calibrating to ensure that spillback flow (which wastes compression energy) is kept to the safe minimum. If the process demand is always small enough to result in anti-surge spillback, then in the short term investigate suction throttling and long term the compressor should be modified.

Low flow, high pressure ratio is more efficiently handled by reciprocating compressors. However, mainly on cost grounds a number of these applications have used centrifugal machines, especially offshore where the weight and vibration of reciprocating compressors is more difficult to accommodate.

8.2.1.3 Axial Compressors

These machines have around 6 to 8% better polytropic efficiency than centrifugal compressors but are highly susceptible to loss of performance due to fouling.

The selection of a more efficient axial type over centrifugal is partly driven by the duty in terms of volume flow rate and pressure ratio. Nevertheless, a number of users have found it economically attractive to retrofit axial machines to replace centrifugal machines, for instance FCC air blowers.

Axial compressors are often fitted with variable angle stator vanes, especially when operating with a motor drive at fixed speed. Though increasing initial cost, the added flexibility over the operating range of pressure ratio and flow is attractive.

As above, coating the rotor and stator can prevent or reduce fouling and maintain efficiency. Many applications use axial compressors in air service. In these cases, it is possible to monitor efficiency easily and clean when required. The polytropic efficiency of a compressor is determined simply by the inlet and discharge pressures and temperatures.

$$\eta_{pol} = \frac{k-1}{k}\left[\frac{\log\left(\frac{P_2}{P_1}\right)}{\log\left(\frac{T_2}{T_1}\right)}\right]$$

η_{pol} = polytropic efficiency (multiply by 100 to get %)

k = C_p/C_v: For air use 1.4 which is good enough for the P and T of an air compressor since it is monitoring *change* rather than absolute value which is important.

P = absolute pressure

T = absolute temperature

8.2.1.4 General Considerations

Gas temperature greatly affects compressor power requirement, regardless of compressor type. As a general rule, suction temperature should be as low as possible for each stage. Keeping coolers internally clean and optimizing cooling water flow to these critical exchangers therefore saves energy.

Similarly, incorrect use of suction strainers – often left in place after pre-commissioning – results in higher than necessary pressure ratio and increased energy consumption. Generally, suction strainers are not required since suction catchpots with demisters (and properly designed holding-down grids) prevent foreign object damage.

One area often overlooked is to minimize the required compressor pressure ratio. A single gage survey through the process equipment upstream and downstream of the compressor can highlight excessive pressure drop. Calculating the energy cost may justify de-bottlenecking – for example, lower pressure drop heat exchangers, increased line size etc.

8.2.2 Pumps

8.2.2.1 Centrifugal Pumps

This work horse of the process industries offers one of the largest opportunities for energy saving.

The design process is usually conservative when specifying pumps, especially when calculating the required discharge pressure. As a result, the pressure drop across the discharge control valve is commonly higher (sometimes much higher) than needed for process control.

Many, if not the majority, of centrifugal pumps in the oil, gas and petrochemical industries are specified to meet the Code API-610 or its equivalent. The prime function of the Code is to ensure safe and reliable operation. Whether the pump operates efficiently is mainly the responsibility of the engineer making the selection.

Using a typical single stage centrifugal pump as an example, a selection chart looks like Figure 8.2.

Obviously, the ideal situation would be for the pump to have the maximum impeller size and operate at a flow corresponding to the best efficiency point (BEP). However, it is usual to select a smaller impeller (typically between 1/3 and 2/3 of the size range). Also, the design operating point is chosen to the *left* of BEP. Both of these decisions are taken to provide for possible future upgrading in both head and flow, associated with plant de-bottlenecking – without having to replace the pump. Note also that the pump head includes an allowance for pressure drop across the discharge control valve.

Taken altogether, the energy consumed can easily be 20% more than the achievable minimum.

The best way to minimize pump energy consumption of an existing unit is to fit the maximum impeller size for the casing and convert to variable frequency

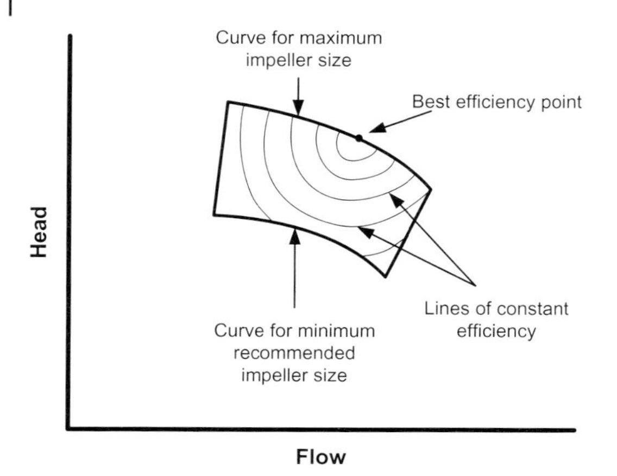

Figure 8.2 Pump selection chart.

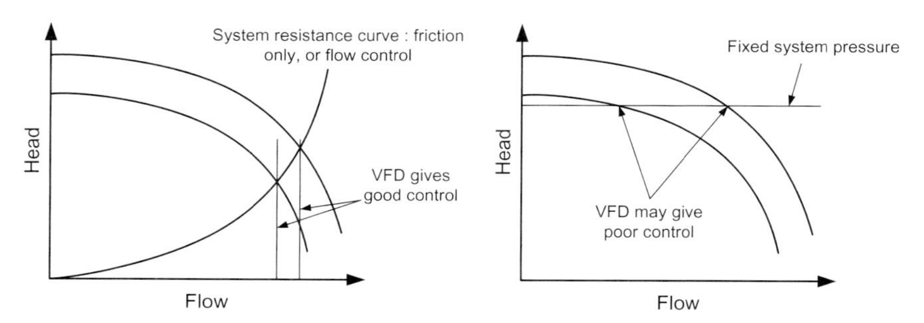

Figure 8.3 VFD pump curves.

drive (VFD). Pump speed is then determined by the required process control parameter – flow, drum level, pressure etc. This eliminates the control valve pressure loss. Speed is also lower without a control valve and so bearing and seal life is improved. Some operators leave a control valve in place – full open but set to operate if the VFD system defaults to full speed if a fault is detected.

Application of VFD must be done with care. If pumping into a system where pressure drop is defined by friction (for example a transmission pipeline), VFD is ideal. However, if pumping into a fixed pressure system, then depending on the shape of the pump head-flow curve, small changes in speed can have a large impact on flow and care must be taken in designing the control system. In this case, VFD can be linked to the pump absorbed power to avoid instability (Figure 8.3).

There is a further problem to avoid. If the speed control is set within too tight a range, then sudden and continuing changes in transmitted torque can result in coupling failure or even shaft failure. Though rare, this has occurred leading to a reluctance among engineers who have experienced the problem to apply VFD – thereby losing the energy saving opportunities.

Where two pumps are run in parallel, then VFD can only be used in conjunction with flow metering of the individual pumps to ensure load sharing. This becomes complicated if flow is not the control parameter and so VFD of paired running pumps is rarely used.

There is the potential for considerable energy saving when the system is applied to boiler feed pumps linked to steam drum level. For a long time, the 'rule' was that the pump discharge pressure should be high enough to get water into the drum *with the drum relief valves blowing*. Pump discharge pressure in this case is therefore often 50% higher than necessary with VFD.

Centrifugal pumps are generally specified to have a continuously rising head-flow curve from operating point to cut-off (zero flow). This is done for stability since a rising-falling curve can have two operating points at the same pressure (Figure 8.4).

For high head, low capacity pumps, the most efficient impeller design results in a rising-falling curve. To get a continuously rising curve a lower exit angle from the impeller vanes is needed. This reduces efficiency and head due to higher velocities in the impeller passages, and so a larger diameter impeller is required. (N.B. This refers to conventional impeller plus volute design, not the Barske type which uses a different principle.)

It is possible to adopt the more efficient design by ensuring the control system maintains flow to the right of the peak head of the curve. Since this type of pump needs a smaller impeller there is the added benefit of cost saving since the casing size is also smaller.

For new installations, there is a case for increasing impeller diameter and running at lower speed. The reduced fluid velocity in the impeller passages increases efficiency.

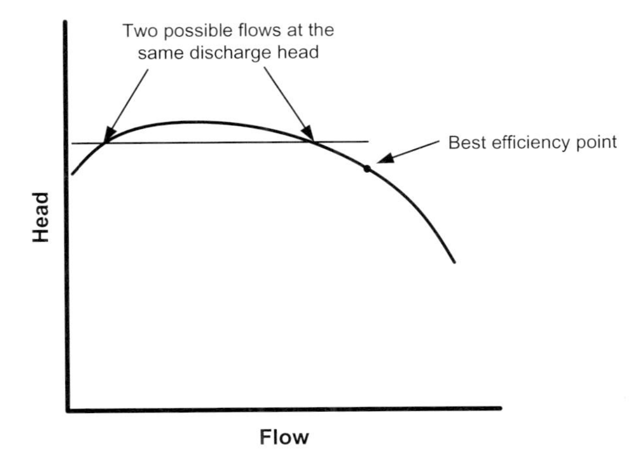

Figure 8.4 Rising-falling pump curve.

Pump efficiency can be greatly improved by applying a suitable smooth coating to the impeller and casing flow passages. Compare the surface roughness of typical castings at ~18 μm, mildly rusty sheet steel at ~5 μm and a modern coating at ~0.1 μm. Power savings of between 4% and 16% are reported depending on the size and type of pump.

8.2.2.2 Reciprocating Pumps

Larger, slower pumps are more efficient but in all cases (except for small chemical dosing pumps), the important area is pulsation damping. The dampeners should be properly sized and maintained to minimize energy-wasting pressure spikes plus unsteady flow in the discharge piping. It is relatively easy to smooth flow to the suction side but the discharge often requires dampeners to be tuned to match the pipework design – avoiding natural frequencies that create vibration.

8.2.2.3 Other Types of Pump

There are many types of pump including screw, rotary vane, axial-elbow and diaphragm. Energy consumption and efficiency vary considerably according to type and so it is important to ensure that the type selected has the best cost/efficiency combination yet is still well matched to the required duty.

8.2.3 Fans

8.2.3.1 Centrifugal Fans

In process plants, centrifugal fans are mostly found providing air to boilers and furnace air preheaters. Where inlet air filters are required, these should be elevated to reduce the risk of sucking in airborne dust and generously sized. Since fans operate with relatively low pressure ratio, even apparently modest pressure drop across a filter can result in a large increase in fan power required.

Fans have either straight radial vanes, forward curved vanes or backward curved vanes. Straight vanes are generally used on smaller fans with higher pressure ratio and speed than large industrial fans.

Impeller vanes may be either curved forward or backward as shown in Figure 8.5.

Forward curved vanes are used in low pressure, high flow applications.

Backward curved vane impellers can be designed for a wide range of specific speeds but are mostly selected at medium specific speed for medium flow high pressure service.

Figure 8.6 compares the characteristics of the three types – all with the same impeller diameter and speed.

It can be seen that the forward curved unit is capable of higher head and flow than the others and so needs a physically smaller impeller and casing for the same duty provided by a radial or backward curved vane design. This is therefore a low cost option.

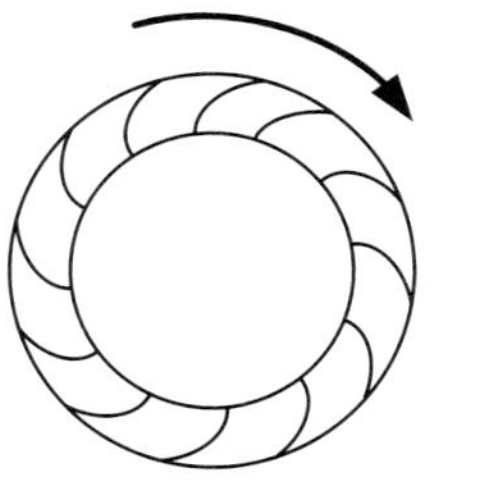

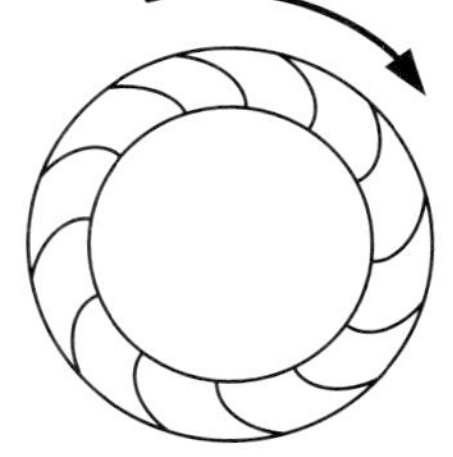

Figure 8.5 Centrifugal fan types.

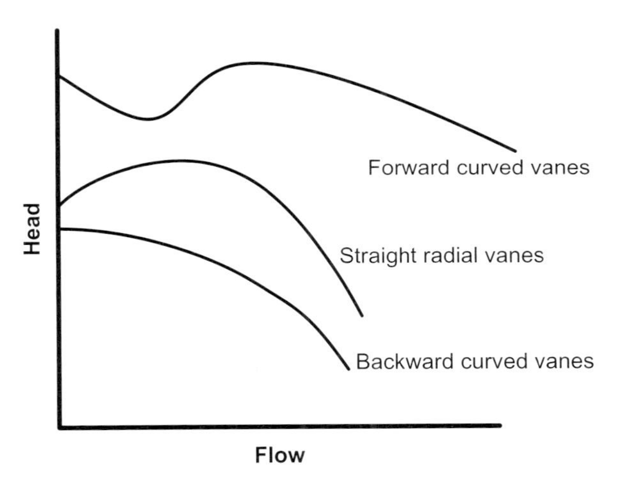

Figure 8.6 Comparative fan performance curves.

Figure 8.7 shows the typical performance of an impeller with forward curved vanes. Efficiency peaks around 65% *but at less than half maximum flow and about 30%of maximum power*. If the process duty is selected for best efficiency operation, it is important that flow is controlled to avoid overloading the driver.

Figure 8.8 shows that the backward curved vane design peaks at over 80% efficiency, and has a non-overloading characteristic. It is therefore better, though initially more expensive due to increased size, to select the backward curve option.

Fan control has traditionally used dampers in the ductwork. There are three options – a multi-flap damper in the suction ducting, the same in the discharge, or variable angle inlet guide vanes (VIGV). From an energy standpoint, a discharge damper is the worst, suction damper better and inlet guide vanes the best. Figure 8.9 shows a typical inlet guide vane assembly.

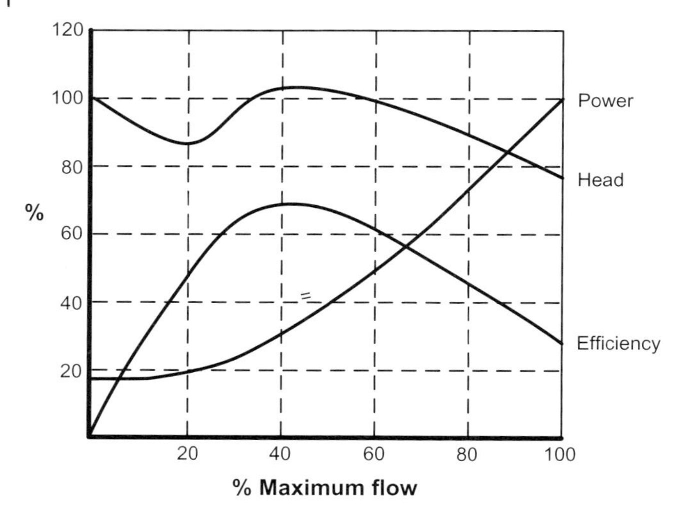

Figure 8.7 Forward curve vane characteristics.

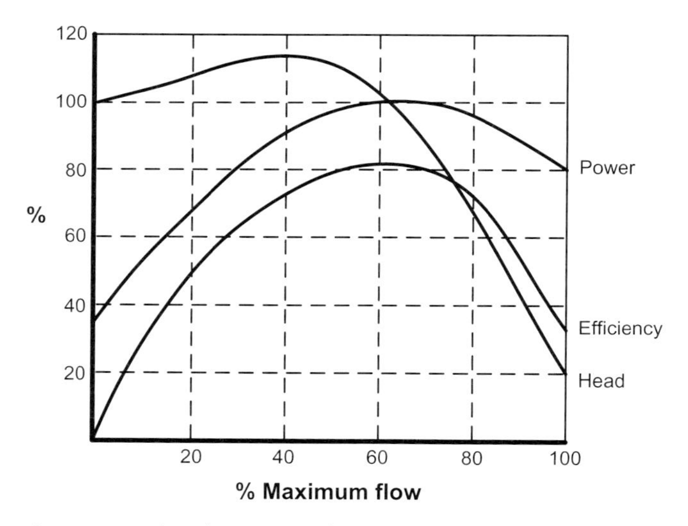

Figure 8.8 Backward curve vane characteristics.

These vanes have adjustable angle and give varying degrees of pre-rotation to the air entering the fan. This in turn adjusts the energy imparted by the impeller and so reduces the power required.

The most efficient option is to vary the fan speed to match process demand. For instance, a good example is to use a VFD electric motor driving a forced draft fan for a boiler or furnace where the speed is linked to the stack oxygen reading. In this case, any dampers or guide vanes are set to 100% open giving little or no restriction to flow.

Figure 8.9 Typical inlet guide vanes.

8.2.3.2 Axial Fans

These are generally used in high volume, low head applications. Depending on the service, VFD may be applicable. Variable angle inlet guide vanes also offer good efficiency at part load. For large fans, efficiency can be improved by selecting aerofoil section blades rather then pressed sheet. The possible efficiency improvements detailed in Section 8.2.8 should be checked.

8.2.4 Power Recovery Equipment

8.2.4.1 Turbo-Expanders

These units are used instead of letting gas pressure down through control valves or multiple orifice chambers. There are three main applications: fluid catalytic cracker (FCC) regenerator gas at around 750 °C and ~3 bar(g) letting down to near atmospheric pressure; natural gas expanders supplied to a plant at anywhere from 10 to 100 bar(g) and letting down to ~3 bar(g) for fuel gas; and cryogenic expanders linked to compressors in cold box service.

Each service has its own design requirements and constraints. There is one main point to make – if the upstream or downstream pressure has to be controlled, then it is better to make the expander slightly *undersized*. Turbo expander performance falls off rapidly if the flow and pressure ratio does not match its design. It is therefore better to have some modest by-pass flow through a control valve so that as the process requirement changes, it is this by-pass flow that alters, leaving the expander at peak performance. In the case of an FCC expander, it is particularly important since catalyst fines will either build up or erode the expander blading.

8.2.4.2 Liquid Turbines

Liquid turbines are familiar as water-powered electricity generators. There are three main types of rotor:

- Pelton wheel – used with limited supply of high pressure water.
- Francis type – used with large supply of high pressure water.
- Kaplan type – used with large supply of low pressure water.

It is rare to find any of these designs used in the process industries, though there are certainly opportunities in some high pressure plants for a Pelton wheel application. Most liquid turbines in the process industries are essentially pumps running backwards to recover pressure energy. A typical arrangement is shown in Figure 8.10.

The power recovered is a roughly a function of (flow rate)^3 and so even a small reduction in flow drops power recovered dramatically. Since it is necessary to control the upstream pressure, it is better to undersize the turbine and control flow or upstream pressure using a by-pass control valve. Figure 8.10 shows a split range control – the inlet to the hydraulic turbine being normally full open except during start-up or when running at low plant rate. The over-run clutch allows the hydraulic turbine to idle when not in use, rather than absorb power from the motor. The motor should be rated to drive the charge pump with the turbine out of commission.

Liquid turbines can be used in refrigeration systems in place of the more conventional Joule–Thompson (JT) valves. In most cases these are used to generate electricity and thus remove extra energy from the refrigerant. Depending on the size of the installation and the refrigerant used, there can be net system energy savings of up to 15%.

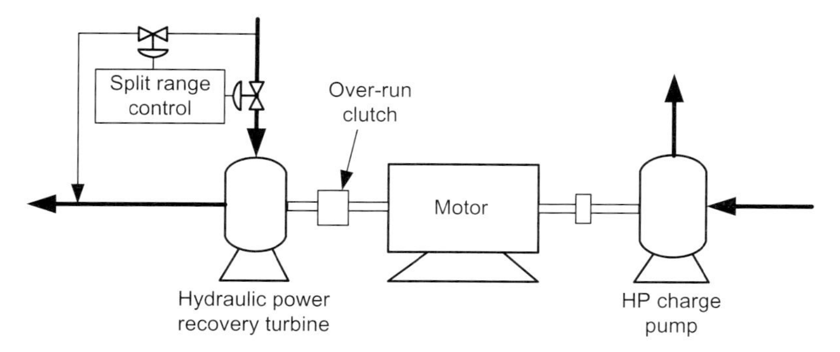

Figure 8.10 Power recovery train.

8.2.5 Steam Turbines

8.2.5.1 Factors Affecting Performance

Most of the change in steam enthalpy between inlet and exhaust emerges as shaft work. Examination of the Mollier Diagram shows that at high pressure, incremental temperature increases enthalpy proportionally more than incremental pressure. Note that lines of constant pressure are steeply inclined and closely bunched whereas lines of constant temperature have shallow inclination. However, at low steam pressure, it is the pressure lines that have shallow inclination.

Therefore in order to provide the maximum differential between inlet and exhaust enthalpy, inlet temperature should be maximized and exhaust pressure minimized. The steam line from boiler to turbine should well insulated, but even then there is likely to be temperature loss depending on the length of line. It is better to adjust the boiler de-superheater set point so that steam arrives at the turbine at the turbine's maximum safe temperature.

Many steam systems are designed at nominal pressure levels, for example, 40 bar, 10 bar and 3 bar. These pressures may not necessarily be the best for the plant concerned and so slow adjustments while monitoring fuel usage can often result in a net reduction.

Condensing turbines should be avoided if possible since even the best reject up to 60% of the steam energy to the turbine's condenser. For an existing condensing turbine, the most important way to reduce steam consumption is to achieve the best possible exhaust vacuum. The only limit is the amount of moisture in the turbine exhaust which, if too great, risks LP blade erosion. Obviously, higher inlet temperature not only improves performance but also reduces this risk.

8.2.5.2 Single Stage Turbines

Work is defined as force exerted multiplied by the distance this force causes an object to move. So, the maximum force steam can exert on a turbine wheel occurs if the wheel is held stationary. As the wheel speed increases, the force falls until finally it reaches zero when the turbine blades are moving at the same speed as the stea0m. Distance moved is zero if the wheel is stationary reaching a maximum when the force falls to zero. Multiplying force x distance thus produces theoretical maximum work when the velocity of the turbine blades is approximately half the velocity of steam from the turbine's nozzle(s). This is illustrated in Figure 8.11. In practice, the number (known as velocity ratio) turns out to be about 0.47 rather than 0.5.

For typical applications, this means that the turbine should rotate at around 9–10 000 rpm. However, most single stage turbines are driving pumps etc. and for existing installations there is rarely a gearbox to step down turbine speed due to concerns about reliability and initial cost. Therefore, where power is 50 Hz the turbine is set to 3000 rpm and at 60 Hz (e.g., USA) the turbine runs at 3600 rpm. Isentropic efficiency is severely reduced at 3000 rpm to around 35%, compared with a possible 75% or more at the ideal correct velocity ratio. In other words,

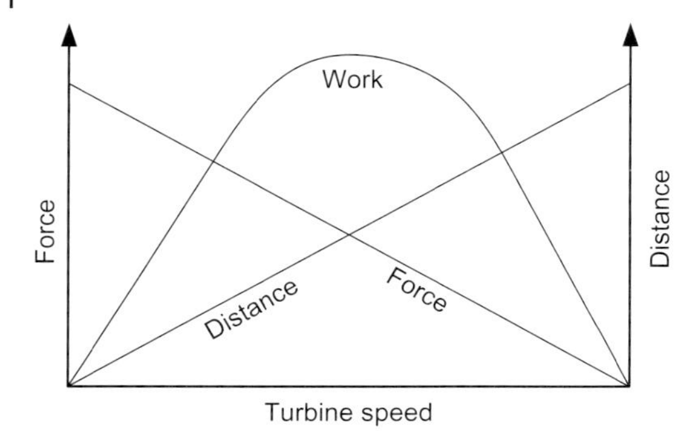

Figure 8.11 Turbine work vs. speed.

without a gearbox, steam consumption is doubled. Some argue that since the exhaust steam is used in the process, efficiency is not important. This is not correct since energy produced from the turbine is made up by energy from a boiler – usually at over 85% cycle efficiency. This is far better than the efficiency of marginal electrical energy, and so shaft work extracted from steam should be maximized.

For new installations, it is possible to purchase small single stage (or compound stage) machines with an integral gearbox. Isentropic efficiency then rises to 75% or higher.

A new generation of electronic governors is beginning to supersede the older style 'flywheel'-hydraulic governor. Electronic governors have a major energy-saving advantage in that they can be coupled to adjust turbine speed to match the required process variable. For instance, turbine driven boiler fans can run with wide open dampers and the speed controlled to match the set point for flue gas oxygen content. Boiler feed pumps can be run at a speed required to maintain steam drum level. (See also Section 8.2.2.1).

The majority of single stage turbines have simple drilled nozzle plates. If the pressure ratio from inlet to exhaust is supercritical (and it usually is), then maximum steam velocity equals sonic velocity and increasing upstream pressure only increases mass flow through steam becoming denser at the choke. It is possible to retrofit convergent–divergent nozzles which accelerate the steam to supersonic velocity, thus increasing output power without increasing steam flow.

Many single stage turbines are fitted manual overload valves (also called auxiliary valves or hand valves). These are intended for use only if the power required is higher than produced with the governor valve wide open. Unfortunately, many operators do not realize that this can waste steam. The effect of leaving a manual valve open is illustrated in Figure 8.12.

8.2.5.3 **Multistage Turbines**

Most multistage steam turbines are engineered to suit each application. There are two main types – impulse stages and reaction stages. In the impulse type, all pressure drop is taken across the stationary nozzle diaphragms whereas the reaction

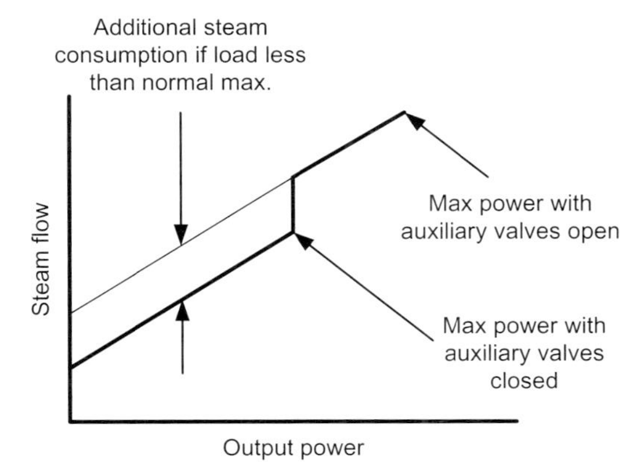

Figure 8.12 Effect of turbine auxiliary valves.

the reaction type shares pressure drop equally between the moving and stationary rows of blades. Both types usually have an impulse 'control stage', mainly to limit axial thrust, but also to allow for multi-valve operation whereby steam is fed progressively to sections of the nozzle ring as the valves open in sequence. Reaction turbines have higher isentropic efficiency than impulse machines, but 'pure' impulse is rare and impulse blading with a degree of reaction is nearly as good for medium sized machines. Manufacturers' catalogs rarely quote efficiency and when buying new this should be investigated.

One potential problem with multi-valve machines is a dip in efficiency when a valve starts to lift off its seat. The initial steam flow is insufficient to generate high nozzle velocity and so contributes little power from the control stage. Operating at 'valve points' is therefore to be avoided if possible.

8.2.6 Gas Turbines

Inlet air filtration is important for all gas turbines since the most common cause of reduced output (and falling efficiency) is air compressor fouling. Also, pressure drop across the filter can have a significant impact: 100 mm H_2O pressure drop reduces output by 1%. Choice of filter type depends on location.

Virtually all gas turbines except extremely small units use axial air compressors to compress the air prior to introduction of fuel in the combusters. It is common to carry out compressor cleaning on a fixed routine but monitored experience has shown that fouling is not a linear progression but occurs in steps depending on ambient conditions–weather, dust storms, etc. It is therefore better to monitor the air compressor efficiency and clean on as an 'as needed' basis. For efficiency monitoring see Section 8.2.1.3.

Downstream of the filter it can be attractive to use 'fogging'. As the same implies, this is ultrafine mist which produces evaporative cooling within the compressor section. The result is up to 2.5% efficiency improvement and up to 20% increase in output. As well as reducing CO_2 emissions (less fuel), the system also gives lower NO_x emissions.

All gas turbines have high exhaust temperature typically 500 °C–750 °C plus 15% or more oxygen in the exhaust. They are therefore well suited to combine with a heat recovery steam generator (HRSG). If duct burners are included to increase steam production, then the incremental steam is produced at close to 100% efficiency.

8.2.6.1 **Frame Engines**

These are heavy industrial units.

Advantages

- robust;
- relatively long intervals between maintenance downtime;
- fuel flexibility and can also be dual fuel;
- fuel pressure required generally lower than aero types, which can be an advantage if using gas requiring compression;
- moderate sensitivity to air compressor fouling;
- able to accept large step changes in load.

Disadvantages

- cost per kW is high for the size range used in process plants;
- cannot simply exchange engines so major maintenance involves extended downtime;
- single shaft machines have poor part load efficiency;
- peak efficiency less than aero-derivative types – but newer engines are getting better.

8.2.6.2 **Aero-Derivative Engines**

As the name implies, these are derived from aircraft engines with the jet exhaust driving the output power turbine. Pressure from the airline industry has resulted in larger and more efficient engines with the latest models reaching ~42% thermal efficiency.

Advantages

- suitable for service exchange (as done for aircraft);
- installed cost per kW low for the size range used in process plants;
- high efficiency, good part load efficiency – means less CO_2 per kW output;
- able to accept medium step changes in load.

Disadvantages

- engineered for specific fuel(s) such as natural gas or kerosene;
- high fuel pressure required;
- sensitive to fouling.

8.2.7 Electric Motors

Motors run most efficiently near their designed power rating. It is therefore good practice to operate between 75% and 100% of full load rating.

Older installations that have been re-vamped or de-bottlenecked can often be found to have significant line loss. If the voltage at motor terminals is less than design then output power is reduced. Current must then be increased to reach the desired output and the combined I^2R loss for motor and supply cables can be significant.

Unbalanced three-phase voltage affects a motor's current, speed, torque and temperature rise. Equal loads on all three phases of electrical distribution helps assure voltage balance while minimizing voltage losses.

The value of power saved by purchasing high efficiency motors can pay out in under 12 months in some cases. They can improve efficiency from 3% for typical process pump applications to eight percent for small units. Efficiency is improved by:

- high grade steel for armature and stator laminations;
- thinner laminations;
- heavier gage copper windings;
- better bearings;
- improved cooling fan design.

The odd shape of the curve in Figure 8.13 reflects typical current industry products. This is a guideline only since there a 2–3% scatter between different manufacturers.

Synchronous motors are more efficient (and more expensive) than induction motors and are mainly used as drivers for ratings in excess of ~3 MW. However, when buying new, it is worth checking whether energy saving outweighs the extra

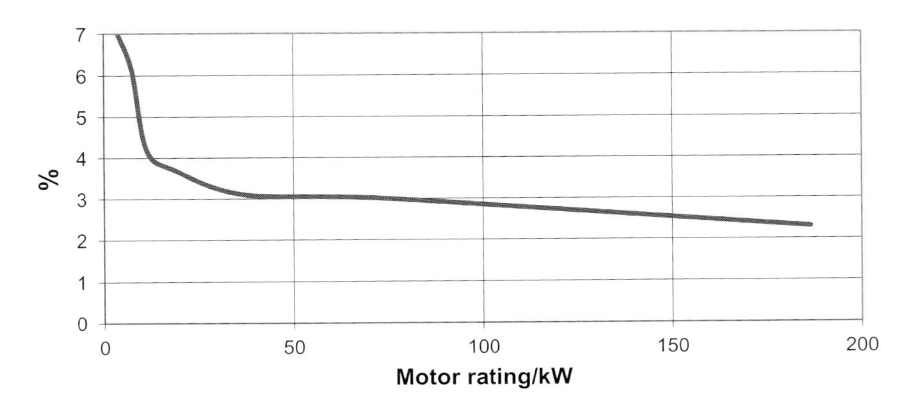

Figure 8.13 Typical advantage of high efficiency motors.

cost. Multi-pole low speed synchronous motors are a good choice for driving reciprocating compressors.

Synchronous motors can also be used for power factor correction. When fitted with suitable controls, they can maintain system power factor at a pre-set level. This is better than fixed capacitor banks since induction motor power factor changes according to load, with power factor worsening as the load reduces. By generating VARs into the system, power factor is improved and supply current (with its associated I^2R loss) is reduced.

8.2.8
Air Coolers

Air coolers are used where the process temperature is too high to use cooling water (risk of scaling) or the load on the cooling water system would be excessive. There are many possible problems with air cooler performance including:

- fouled or loose fins;
- air leakage through gaps between tube bundles and side walls;
- excessive clearance between fan blade tips and the plenum chamber. (Tip air recirculation);
- poor fan hub design (center hub air recirculation);
- fans running stalled (excessive pressure drop across the finned tube bundle due to fouling coupled with high angled blades);
- lack of air inlet 'bell-mouth' to assist air flow.

Figure 8.14 illustrates 'bad' versus 'good' design.

On the left there is hub recirculation, tip recirculation and uneven velocity profile. The fan on the right has a correctly designed center disk, zero tip clearance

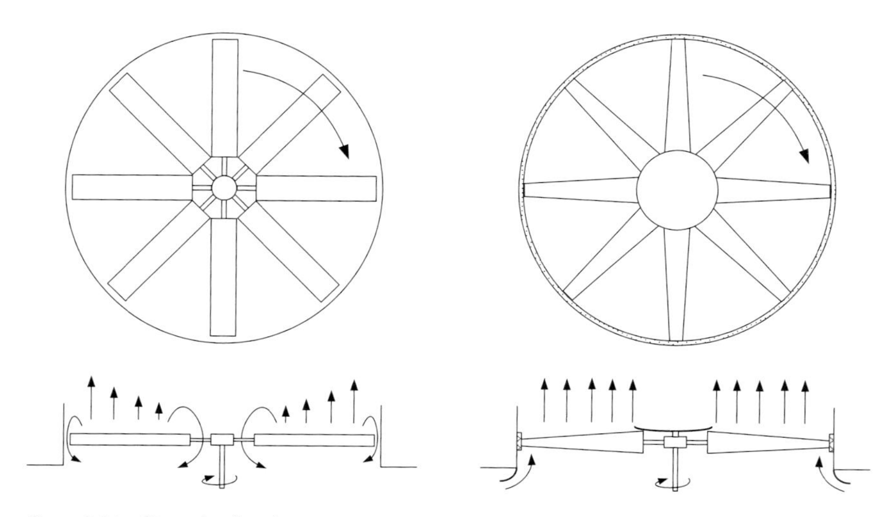

Figure 8.14 Air cooler fan design improvements.

(due to aluminum honeycomb tip seals), air inlet bell mouth and correct aerodynamic blade profile along the blade radius.

On the left, there is recirculation of air at the fan hub and the blade tips. The blades are simple parallel aerofoil sections, generating a higher pressure ratio towards the tip than the root.

On the right, there is a center seal disk, abradable tip seals and variable chord width blades with decreasing attack angle from hub to tip – giving constant pressure ratio along the blade length.

Figure 8.15 shows the typical effects of:

'A' lack of air inlet bell mouth
'B' inadequate center seal disk
'C' excessive tip clearance
'D' total impact A + B + C.

Note that there are wide differences in air cooler design. In particular, some use relatively small diameter fans with high velocity pressure and other use large diameter fans with lower velocity pressure.

For existing fans, it is worth checking the velocity pressure to see whether modifications are justified.

If we define velocity pressure (V_P) as kinetic energy per unit volume of air passing through the fan, then this is:

$$V_P = (0.5\, m\, V^2)/(\text{Volume})$$

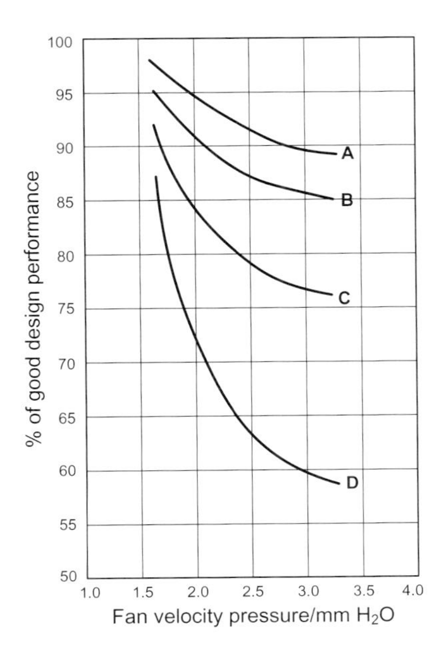

Figure 8.15 Effect of air cooler fan improvements. A: lack of air inlet bell mouth; B: inadequate center seal disk; C: excessive tip clearance; D: total impact A + B + C.

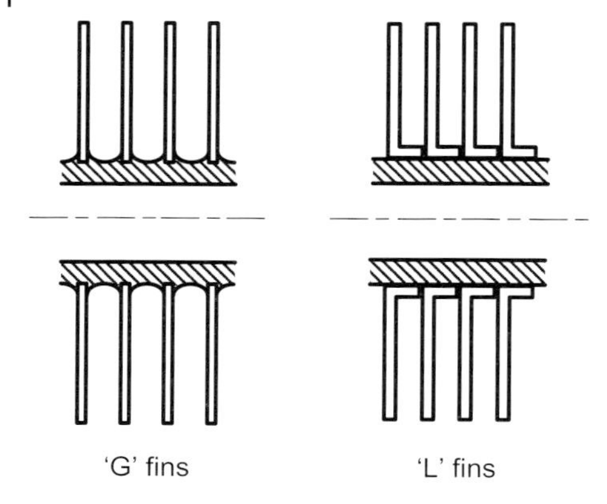

Figure 8.16 Air cooler fin types.

where m = mass
V = velocity of air entering the fan
mass/volume is density (ρ).

$$V_P = 0.5\,\rho\,V^2$$

where the units of V_P are pascals; ρ is $kg\,m^{-3}$; V is $m\,s^{-1}$.

Air density is ~$1.2\,kg\,m^{-3}$ and 1 pascal = 0.1 mm H_2O so the formula simplifies to:

$$V_P(\text{mm } H_2O) = 0.06\,V^2$$

The velocity of air entering the fan can be measured using a hand-held anemometer.

8.2.8.1 **Air Cooler Fin Types**

There are basically two types of fin attachment. 'G' fins and 'L' fins. The 'G' fins are embedded in grooves in the cooler tubes but 'L' fins are wrapped around the outside. There is a drawback with the (cheaper) 'L' type: corrosion under the base of the 'L' can occur if the cooler is out of service for even a few weeks, and this greatly reduces heat transfer. Figure 8.16 illustrates the types. 'G' fins are recommended for any future project.

8.3 Fixed Equipment

8.3.1 Fired Heaters

Fired heaters (including boilers) are unavoidable in many processes. Efficiency is generally best with the lowest possible stack temperature and oxygen content.

The amount of fuel required can be reduced in a number of ways including:

- The process fluid requiring heating should use waste heat as efficiently as possible to maximize the heater coil inlet temperature and minimize heater duty.
- Forced draft burners can generally be controlled to lower stack oxygen than natural draft burners.
- Preheat of combustion air.

8.3.1.1 **Configuration and Design**

There are a number of heater configurations: vertical cylinder; vertical tube cabin type; horizontal tube cabin type; etc. The type chosen depends to some extent on the service.

One aspect sometimes overlooked is 'box loss'. This refers to heat lost through the external metal skin. It is common to find older heaters with external skin temperature approaching 60 °C. Some idea of the loss is that 10 m^2 of exposed skin at this temperature in a 15 kph wind will lose ~90 000 kcal h^{-1}. It is therefore worthwhile specifying that internal refractory should not produce a skin temperature in excess of a named value – depending on ambient conditions for the location. Also, where appropriate, a cylinder design has less exposed skin area than a cabin type of the same capacity.

Older heaters often have firebrick or castable refractory. There is a wide variety of alternatives available which are far superior. Instead of costly re-lining, it is often possible to apply a coating to the existing refractory. Some of these coatings improve emissivity and thus improve heat pickup, especially in the heater's radiant zone.

Combustion air preheat is invariably attractive to improve efficiency. There are a number of systems but the best and most reliable is simple heat exchange between flue gas and combustion air. Care is required when retrofitting such a system since the reduction in fuel tends to move the heat load from the radiant section to the convection section, possibly limiting heater capacity. Note also that it is possible to take energy from a large heater stack and share the preheated air across a number of heaters. This avoids the cost of multiple individual systems.

8.3.1.2 **Fuel System**

The amount of fuel fired is usually set to control the process fluid exit temperature (or steam pressure in the case of boilers). A simple control system risks 'overshoot'. This is due to the time lag between increased fuel firing and the increase in process fluid temperature. There is therefore a period when excess fuel is fired. Proper tuning of the controls avoids this problem.

Apart from hydrogen, the fuel producing the least CO_2 is methane – the main component of natural gas. Unlike most fuel oils, acid gas dew point corrosion is not a problem for natural gas and so there is no imposed bottom limit to stack temperature.

8.3.1.3 Burner Design

NO_x formation rises exponentially with increasing temperature and linearly with time. Old style burners produce undesirable levels of NO_x since air and fuel enter separately producing high temperature hot spots at the fuel-air boundary. Modern burners pre-mix fuel gas and air to avoid this problem.

8.3.1.4 Instrumentation and Control

Assuming that an air preheat system is installed, there are two control loops required. The induced draft fan should be used to control draft at the heater arch (immediately below the convection section) at ~2 mm water gage negative pressure. The forced draft fan should be linked to stack oxygen. Actually, the reading is not taken from the stack but at the arch. This is for safety reasons since any air leaking past the convection section maintenance access doors will give a false high reading in the stack and risk incomplete combustion. Peak energy efficiency is achieved by using VFD on both fans, with dampers wide open.

8.3.2 Flares and Flare Systems

Obviously, any flaring is undesirable and ideally a flare release should only occur during emergency upsets. However, there are often many hundreds of valves connected to the flare system – control valves as well as pressure relief valves – and it can be difficult to keep these leak-free, especially since maintenance access to the flare system is limited.

There are two main groups – elevated flares and ground flares.

The traditional elevated pipe flare requires a continuous purge as well as permanently lit pilot flames. Fitting a molecular seal greatly reduces the purge rate required. For instance, an open 500 mm flare tip requires ~700 $nm^3 h^{-1}$ purge rate (depending on gas type) and fitting a molecular seal reduces this to around 20 $nm^3 h^{-1}$ or less.

Smoke suppression is generally achieved using steam. Often, the steam flow is controlled manually by control room operators watching the flare on closed circuit television. This is wasteful since steam is regularly left at high rate, and often

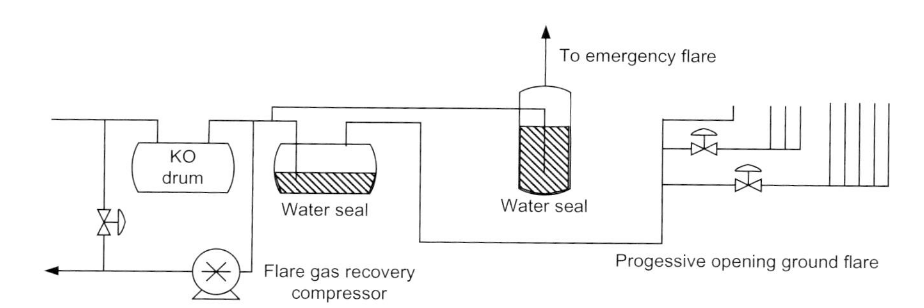

Figure 8.17 Energy efficient flare configuration.

excessive when actually needed. There are now systems that control steam automatically to the minimum required.

Ground flares are generally 'multi-point' with small individual burners designed to avoid the need for smoke suppression. They are usually surrounded by a wall which limits radiation and noise. Typical design has the burners arranged in banks of progressively larger numbers. Controls based on header pressure ensure that the minimum number of burners necessary are in operation. Since there is always the possibility that the control system could fail, for safety reasons an elevated flare is normally also included, fed from a water seal pot with a deeper seal than the seal pot feeding the ground flare.

For large systems where leakage to flare is difficult to eliminate, a flare gas recovery compressor offers a solution to avoid flaring, generally feeding the fuel gas system. The compressor should be a positive displacement machine due to highly variable gas molecular weight and flow, with spillback control set to maintain the knockout drum pressure below the water seal pressure. Figure 8.17 shows a typical complete system.

8.3.3 Piping

Over the lifetime of a process plant, friction losses in piping systems are a major energy consumer.

8.3.3.1 Capital Cost versus Running Cost

As an example, take 100 $m^3 h^{-1}$ of water flowing through a pipe where by using a larger line diameter, the avoidable pressure drop is 0.1 bar/100 m. The pumping cost using a 60% efficient pump is ~0.27 kW per 100 m equivalent pipe length. This amounts to ~2.3 MWh per year.

Figure 8.18 illustrates the effect of line size on pressure drop for this case. Whereas traditional design practice would almost certainly select 150 mm pipe,

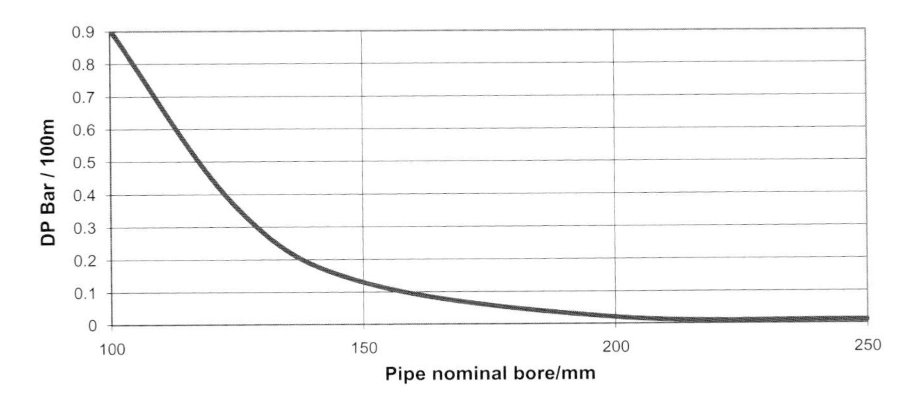

Figure 8.18 Pressure drop vs. pipe diameter.

there may well be a case for choosing 175 mm based on lifetime running cost. Note also that compared with traditional practice, unusual diameter pipe is becoming available so that there is a wider choice with smaller cost increments for increasing line size.

8.3.3.2 Design for Low Line Loss

As noted above, line diameter is the key variable. There are a number of other ways that line loss can be reduced. Table 8.1 below illustrates the equivalent line length in meters for swing check valves, 'T' junctions and long and short radius elbows.

Among the measures for reducing line loss in new installations are:

- Optimize line size based on running cost.
- Use VFD pump motors and full line size control valves for back-up if thought necessary.
- Use swept bends or long radius elbows.
- Consider expansion joints to avoid large expansion loops.
- Venturi meters offer less pressure drop than orifice plate flow meters. Where accuracy is not critical, use ultrasonic flowmeters.
- Select pipe with smooth inner surface.
- Use a 'Y' to split flow to the suction of paired pumps.

For existing installations, it is rare that any modification is worthwhile. Nevertheless, where plants have been de-bottlenecked, it is recommended to carry out a single gage pressure survey to identify where modification may be justified. Rerating orifice plate flowmeters (larger bore) and re-sizing control valve trim can help reduce loss.

For pumping applications with low head/high flow such as cooling water circuits, balanced check valves are preferred to swing check valves which often have significant pressure drop. Swing check valves in vertical lines should be avoided or fitted with external balance weights.

Table 8.1 Equivalent line lengths.

Line Size mm	Check Valves	'T' Junctions	Elbows r/d = 1.0	Elbows r/d = 0.5
100	6	5	3	5
150	9	8	5	7
200	12	11	7	9
250	15	14	9	11

8.3.4 Insulation

8.3.4.1 Economic Insulation Thickness

Heat loss depends on:

- the range of ambient conditions at the site location;
- the type of insulation;
- insulation thickness;
- cladding emissivity;
- Wind speed.

It can generally be assumed that the amount of energy lost is replaced by fuel used at ~80% efficiency. The installed cost of insulation should be used, not just material cost, for comparison with the energy cost.

On existing installations, it is good practice to carry out a thermal imaging survey in order to calculate the economic incentives for insulation and re-insulation, and also to prioritize the work.

8.3.4.2 Mis-Use of Insulation

A common observation is that lines are often insulated which should be left bare and are left bare when they should be insulated. Some of the more obvious offenders are listed below.

- **Compressor discharge lines:** These lines mostly lead to inter-coolers or after-coolers. During process design, it is common for engineers to mark them on the P&I Diagrams for 'personnel protection insulation'. But, when the plant is built, the majority of the hot lines are well out of reach and so should be left bare to reduce load on the cooling water system or air coolers. The lower resulting process temperature saves horsepower.

- **Lines to air coolers:** These often have the same problem. Headers and drop lines are insulated, yet the cooler header boxes (closest to walkways) have to be left bare to allow visual checks for leaks.

- **Refrigeration circuits:** It was common in the past to rely on ice build-up to provide insulation. It is far better to fit properly sealed insulation to reduce 'heat in-leak' which increases refrigeration compressor power demand.

- **Corrosion under insulation:** This is a fairly recent concern. There have been cases where poor sealing of insulation cladding coupled with metal temperature below 100 °C have resulted in serious corrosion due to rain penetration. Some operators have therefore chosen to leave metal bare when the surface temperature is less than 100 °C. This is false economy and insulation is still recommended – with special care taken on sealing the cladding.

There is a widespread belief that the flanged connection of the heads of heat exchangers should not be insulated since this risks leakage when the bolts expand. Provided the bolts are correctly specified and tightened, this risk is minimal. The key is correct tightening. Experiment has shown that bolt tension can vary up to 50% for identical torque settings since this is affected by the degree of friction between nut, studbolt and washer. Ideally, tension should be determined by bolt stretch (measured with a micrometer: ~0.1% of the distance between nut faces) or by simultaneous hydraulic tightening of the bolt circle. Flanges which *should* be left bare include high pressure hydrogen lines since any leakage will auto-ignite.

8.3.4.3 **Insulation Types**

There is a large variety of insulation methods and materials, some of which are still being used after 100 years or more, asbestos being the notable exception.

In recent years, a range of new products have become available. One of these, known generically as 'aerogel' is especially noteworthy. Thickness required is considerably less than traditional materials and it is extremely light. This means that in many cases, it can be applied over existing insulation along pipe racks for example.

Factors to consider when choosing insulation type are:

- temperature of the surface to be insulated;
- durability;
- need to remove and replace during maintenance activities;
- installation cost;
- space limitations to applied thickness;
- purchase cost.

Mineral wool is a better insulator than pre-formed calcium silicate, but is more time consuming to install and awkward to remove and replace for maintenance.

Padded blankets covered with waterproof cloth are recommended for items such as hot isolation valve bodies (often left bare). These can be fixed in place using stainless steel wire and so are easily removed and replaced during maintenance.

8.3.4.4 **Sealing**

For refrigeration circuits, the most important detail is to ensure that the external vapor barrier is intact. Without this, atmospheric moisture forms ice under the insulation which eventually bursts off. Although ice itself acts to some degree as an insulator, there is still considerable energy loss reflected in increased load on the refrigeration compressors.

Corrosion under insulation has recently been highlighted as a problem. This can occur if the external cladding allows rain to leak through. Also, wet mineral wool is a poor insulator so correct application and sealing of the external cladding is again important–as for low temperature insulation.

8.3.5
Tank Farms

It is unfortunate that the vast majority of installations fail to take advantage of local topography when it can be beneficial. Some of the largest pumps in a coastal plant are used for ship loading. It is therefore energy efficient to mount export tanks as high as possible to reduce pumping cost. In most cases, the pressure at which product leaves the process on the way to storage is high enough easily to fill elevated tanks. Surprisingly, elevated tanks containing flammable product are safer than when in a flat tank farm. This is because leakage can be routed away from the tank to an impounding area rather than have the risk of the tank sitting in a potential sea of flame.

8.3.5.1 Tank Gas Blanketing

Fixed roof tanks containing volatile flammable liquid, even with an internal floating roof, will develop an explosive gas/air mix above the liquid level unless the oxygen content inside is kept below the lower flammable limit (LFL). Some operators rely on flame arrestors fitted to the roof vents but a safer option is achieved by gas blanketing. Without a vapor recovery system, this results in venting the gas mixture to atmosphere during tank filling.

Many operators use nitrogen as blanket gas, but nitrogen manufacture is energy intensive. A better solution is to use low pressure fuel gas, feeding all blanketed tanks from a common manifold running at ~5 mbar positive pressure. If the pressure rises then gas can be re-compressed into the fuel gas system. For safety, in case the compressor trips or is otherwise unavailable, excess gas can be routed (on split range control) to a dedicated low pressure ground flare.

8.3.5.2 Tank Heating

Heated tanks lose temperature faster as storage temperature increases. In one case involving a large tank farm, operators turned steam on until a pre-set temperature was reached then turned the steam off. This 'sawtooth' temperature profile consumed almost double the steam compared with controlling the steam to maintain a constant tank temperature. When the new controls were implemented, the net saving was a staggering 50 t€/h.

Stored heavy products tend to produce a colder layer at the tank walls with a hot core. It is therefore important for temperature control that thermowells penetrate at least 450 mm into the tank to get a representative reading.

Depending on the product, it is possible to stop heating altogether until shortly before the tank contents have to be moved.

Bayonet type steam heaters can be more effective to provide low pump suction viscosity than heating the whole tank.

8.3.6 Steam Systems

8.3.6.1 Header Pressure and Temperature Control

Most steam systems have nominal HP, MP and LP headers. Typically, HP steam is ~40 bar, MP steam ~10 bar and LP steam ~3 bar. These pressures, chosen at the design stage, are not necessarily the most energy efficient. Rather than run at 'design' pressure, it is generally better to experiment by monitoring fuel consumption as the HP/MP/LP pressure set points are *slowly* changed to find the optimum combination.

Turbines use less steam if inlet temperature is increased. This is especially important for condensing steam turbines (See Section 8.2.5.1). Thus, the HP system should run at maximum allowable temperature since the majority of HP users are steam turbines. Conversely, steam reboilers can be bottlenecked if the steam supply is superheated since in this case, the heat transfer coefficient is low until the temperature falls to the point where the steam-side of the reboiler tubes is wet. However, rather than lower the header temperature (if it also feeds steam turbines), it is better to install local de-superheaters for steam reboilers.

8.3.6.2 Boilers

Utility steam boilers rarely have poor energy efficiency. Areas to check are:

- Is the blow-down excessive? (An automatic system is preferable with the minimum practical volume of water in piping between the steam drum and the measurement point).
- Are the fans using minimum energy? (See Section 8.2.3.1).
- Is there a case for using LP steam for air preheat?

8.3.6.3 De-Aerators

Many sites have excess LP steam, and one of the large users is the boiler feed water de-aerator. It can pay to increase de-aerator pressure which will then require more steam to raise the water to the higher steam saturation temperature.

De-aerator vents (for inert gas removal) are mostly fitted with manual valves. Rather than adjusting a valve, it may be preferable to use a gate valve with a hole drilled in the gate, sized to release the correct volume at the de-aerator design operating conditions. Invariably, with a manual valve, the vent rate is set conservatively high, sometimes excessively so. Modern de-aerator design should include a vent condenser to minimize the steam released along with the oxygen, nitrogen, CO_2 etc.

Depending on energy cost, it may be economical to retrofit a vent condenser to existing de-aerators – typically an exchanger in the cold make-up water.

8.3.6.4 Steam Traps and Condensate Recovery

There are many varieties of steam traps and for each application they should be chosen with care. There are advantages and disadvantages for each type.

Thermostatic traps do not open until the condensate temperature has fallen below steam temperature. For this reason, it is common to see 1 to 3 M of bare line upstream of the trap – giving a continuous energy loss to atmosphere.

Thermodynamic traps rely for closure on discharging until all condensate is gone and steam enters the trap. There is therefore some steam leakage into the condensate system (if condensate recovery is installed) on every drainage cycle. This can give rise to water hammer. Ob the positive side, these traps can operate with up to 80% back pressure.

Mechanical traps such as the ball float type only discharge condensate. They are only suitable for large flows, such as steam reboilers. They are also relatively expensive.

There are trap manufacturers' web sites which give considerable detail on trap design, performance and selection.

Condensate recovery is always a challenge. If the condensate has to be pumped, then it should be recovered to an elevated drum giving adequate pump NPSH for return to the utility de-aerator. For condensate recovery from the HP steam system to operate satisfactorily, the drum pressure has to be below the HP steam system pressure and so the drum generates flash steam. In a well designed system, this flash steam simply passes into the MP steam header.

Low pressure condensate also loses flash steam at most sites, especially when fed from thermodynamic traps. Depending on the location of the collection drum, it may be worth adding cold water which would otherwise be used as part of the de-aerator make-up. This eliminates the flash steam, reduces the chance of water hammer and retains all the energy apart from insulation loss. An alternative is to install a long, bare, self-draining vent above the drum in order to condense the majority of the flash steam.

8.3.7 Cooling Water Systems

For most applications, it is energy efficient to have the water as cold as possible (except in cold winter conditions when 15–17 °C is the safe minimum).

For example, energy is saved by improved compressor suction cooling and inter-cooling since the resulting increase in gas density reduces compressor power requirement. Product run-down cooling can also be important in reducing tank farm fugitive emissions for existing installations. In many cases, recovery of C_3 and C_4 from fuel gas is also driven by cooling water temperature.

8.3.7.1 Cooling Towers

Process plant cooling towers are available in a range of designs. The most common are cross flow and counter flow with air drawn in by induced draft fans. Figures 8.19 and 8.20 below illustrate the main differences. For some reason, cross flow towers are common in North America and counter flow towers predominate in Europe and Asia.

In a cross flow tower, air is drawn through the falling water from the side, over the full height of the packing. In a counter flow tower, air enters underneath the packing and then rises vertically through the packing in the opposite direction to the falling water.

Cross flow towers are usually wooden construction. Counter flow are mostly concrete. Note that wooden towers are a serious fire risk when they dry out during turnarounds. A well designed and maintained tower should be able to cool the water to within 3 °C of ambient wet bulb temperature.

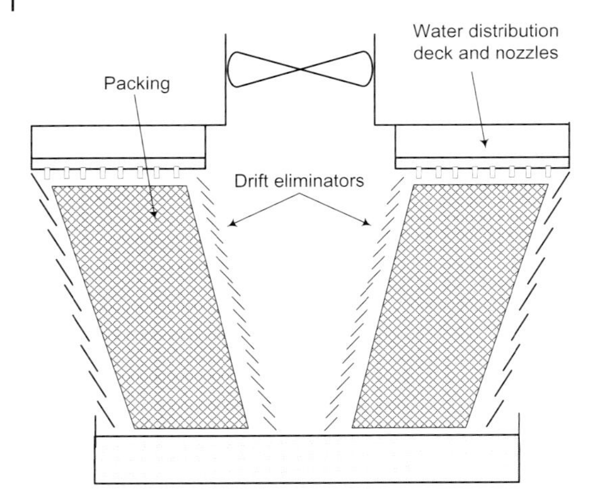

Figure 8.19 Cross-flow cooling tower.

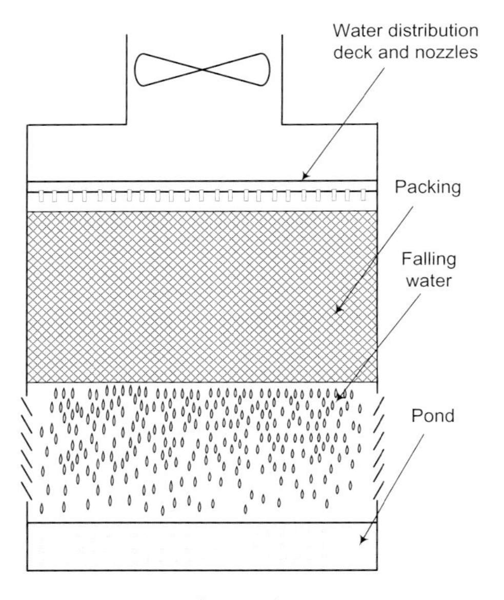

Figure 8.20 Down-flow cooling tower.

One attraction of the counter flow type is that it requires less plot space than a cross flow tower of the same rating. Cross flow towers also generally require a higher elevation of the water return lines – increasing pump power required.

There are many potential problems that can reduce cooling tower performance. For all types it is important that the water distribution is evenly spread through the packing. If there are areas with reduced water flow, the air sucks through the packing preferentially in these locations, reducing the cooling where water is flowing more densely.

Cross flow towers have open decks so that operators can check whether distribution nozzles are plugged with debris. However, the current environmental regulations mean that certain chemicals for preventing build of algae are now banned, and with sunlight playing directly on the open decks this is difficult and expensive to control. As a result, it is necessary to add sunshades over the decks. Sun also shines down the fan shafts promoting algal growth on the drift eliminators.

Depending on local conditions, it can be attractive to use a hybrid counter flow cooling tower. These towers route return water through finned tubes above the drift eliminators to achieve sensible cooling before the water is passed through the packing for evaporative cooling. In the past, the main reason for adopting the hybrid design was to eliminate the visible plume above the tower. However, the hybrid design also reduces water make-up rate; in some cases up to 30% water saving has been reported.

8.3.7.2 Cooling Tower Fans

The fans that suck air through the packing require the same attention as described in Section 8.2.8 – seal disks and minimum tip clearance. In addition, it is preferable to have velocity recovery stacks (VR) above the fans – see Figure 8.21. Depending on the tower design, this can have a significant effect on energy consumption.

When velocity pressure approaches 10 mm water gage, the potential saving in fan power reaches 20% (Figure 8.22).

8.3.7.3 Circulation Pumps

These are some of the larger pumps in any installation and so it is worth ensuring they are as efficient as possible. There are two types that are normally selected. Due to the low NPSH available, they are either vertical turbine type or horizontal with double suction impellers. These pumps operate without control valves against system friction back pressure.

It is difficult to monitor pump condition since, for example, wear ring erosion results in loss of output but the resulting internal recirculation means little or no

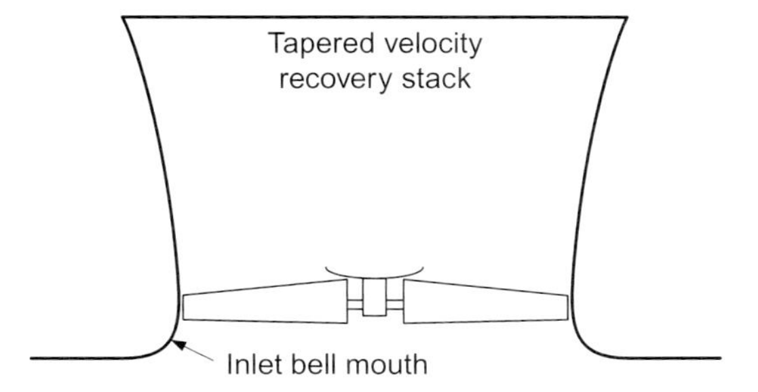

Figure 8.21 Cooling tower velocity recovery stack.

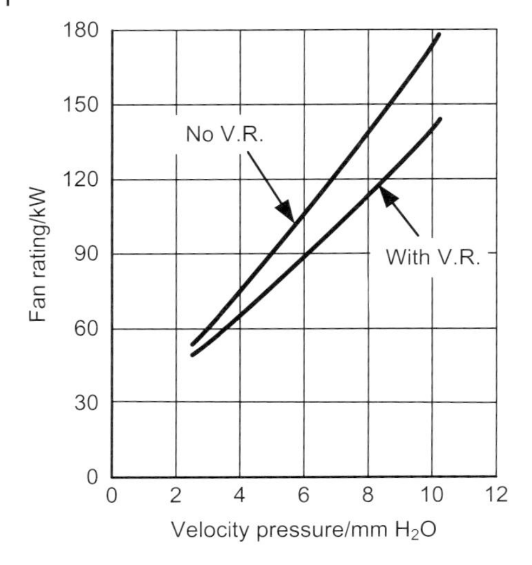

Figure 8.22 Effect of velocity recovery stack.

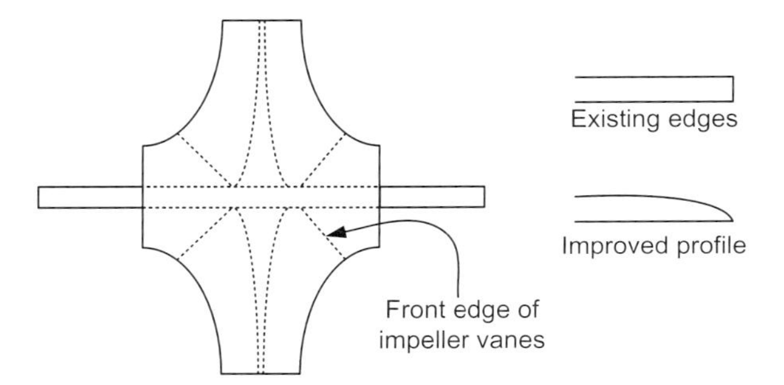

Figure 8.23 Improvement to water pump impeller.

change in input power. The cost of individual pump flowmeters is not justified, but by using a portable ultrasonic strap-on flowmeter, pumps can be compared to identify poor performance.

Double suction impellers are often supplied in rough 'as-cast' condition and simply machined and then ground for balance. It is possible to improve efficiency by improving the blade profile, especially at the suction as shown in Figure 8.23. This also reduces the risk of cavitation which is common for cooling water pumps. The grinding should be done on the convex side of each impeller vane.

These pumps are also much improved by impeller coatings. Application of a suitable coating reduces power requirement of a typical cooling water pump by ~6% – a considerable saving. See also the last paragraph of Section 8.2.2.1.

Double suction pumps should not have a horizontal elbow in the suction line since this can push flow preferentially to one side of the impeller – almost certainly

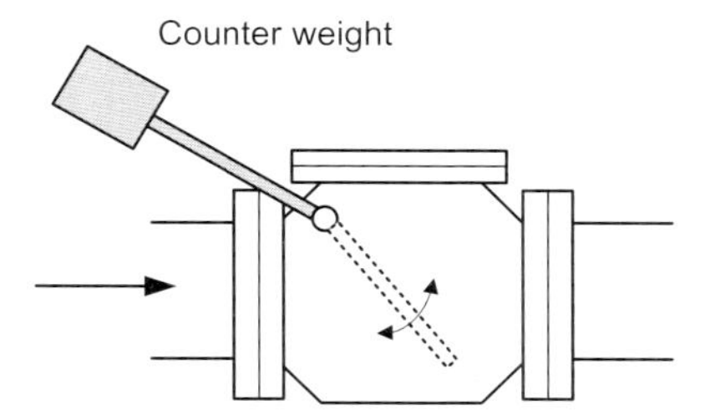

Figure 8.24 Reduction of check valve pressure drop.

leading to cavitation damage as well as increasing thrust load and reducing performance.

Surprisingly, there is often significant loss across the pump discharge check valves. Cooling water pumps are generally high flow, low head and so balanced check valves should be used and swing check valves avoided unless they are fitted with an external counter weight as sketched in Figure 8.24. Pressure drops of 0.2 bar have been observed in this service. Another advantage of the external weighted arm is that operators can physically move it to check that the valve is not seized in the open position (and also that it is not stuck part open).

8.3.7.4 **Tower Packing**

Tower packing (or 'fill') improves evaporative cooling by exposing a large surface area of water to the air that is drawn through the tower. Packing is available in a range of styles and designs, from old style wooden slats to modern structured blocks.

The performance of older towers can often be improved by installing modern packing, though care is required to ensure that the possible increase in packing pressure drop can be handled by the fan(s).

There is a particular concern in countries where winter temperature is very low. In this case, without remedial action, the packing can become part blocked with ice and the extra weight risks mechanical damage. Various strategies have proved successful, including partial blocking of the air intakes, reversing fan direction and isolating and draining a percentage of the tower cells. Hybrid towers are attractive in extreme low temperatures since the packing can be by-passed completely and evaporation loss eliminated along with the cost of chemical dosing.

8.3.7.5 **System Tuning**

It is common to find the cooling water isolation valves fully open at all water-cooled equipment. This is not good practice since conservative design generally allows for flow rates in excess of what is required.

If the temperature rise between water inlet and outlet of an exchanger is small – say 1 or 2 °C – then almost certainly the water flow rate is higher than necessary. In this case, the water exit valve should be throttled to the minimum safe flow rate, consistent with achieving the desired cooling and avoiding the risk of scale and fouling. Generally, exit water temperature should not exceed 40–45 °C to avoid scaling, and exchanger tube velocity should not be lower than ~0.8 m s^{-1}. A portable strap-on ultrasonic flowmeter can give a reasonable guide to water flow rate.

There are a number of benefits from system tuning. In most cases, overall water circulation rate is reduced. The resultant higher supply pressure means that flow is increased through critical service exchangers. In existing plants this can result in significant energy savings where, for example, improved cooling of a steam turbine surface condenser gives better vacuum and reduced turbine steam requirement.

Reduced water circulation rate through a cooling tower increases cooling residence time and also the lower water flow rate through the packing results in increased fan air flow. Both of these factors tend to give a better approach to ambient wet bulb temperature – colder cooling water supply.

9 Energy Efficient Refineries

Carlos Augusto Arentz Pereira

What is a refinery? A huge energy consuming and transforming machine that uses heat and power to separate and recombine atoms. An immense thermal engine with lots of flows coming in and out, and in midway frantically crossing with each other; a few main pieces of equipment and a bunch of many auxiliary ones, spending, changing and moving energy. Refineries need energy to convert crude oil into transportation and heating fuels, chemicals, and other products. Enormous amounts of heat are required to separate crude oil into its components, such as gasoline and diesel, also to crack hydrocarbon molecules into smaller ones, which generate lighter and more valuable products. All this should be coordinated, in order to provide appropriate conditions, especially pressure and temperature, for these core processes, at their best economic point. This economic point depends on many items, like the refinery technological design stage, cast of crude oils it processes, portfolio of products, and also the way that is operated. All these issues have been changing a lot over the years, influenced by each other and by economics. And one of the great drivers of these changes was energy conservation.

9.1 Historical Evolution from Energy Conservation to Energy Efficiency in Refineries

Energy conservation emerged during the oil crisis in the 1970s as a matter of survival for refiners. Since a refinery is an energy converting machine, that uses energy to perform this task, reducing its consumption means fewer costs and more products to sell, leading to better overall results. And at that time, as the industry faced pinched margins, that was a matter of survival. Because of this urge for survival, energy conservation evokes energy rationing and usually means restrictions and discomfort.

Energy efficiency, on the other hand, may be considered the maturation of energy conservation. It is a more complex and sophisticated concept. Instead of being born during a crisis, it was bred by consolidation and endurance of energy conservation activities. It impels better use of natural resources and, consequently,

Managing CO_2 Emissions in the Chemical Industry. Edited by Leimkühler

ISBN: 978-3-527-32659-4

human and economic development. Through technology and investments, the same processes should use less energy, which means more productivity, less operational costs and environmental impacts. It is a new point of view that must be considered since the conception of the design. Energy efficiency means evolution, doing things right and better in each cycle. This evolution from conservation to efficiency concept was brought about in refineries for many reasons.

9.1.1 Global Scenarios and Impact on the Oil Business

A refinery is a complex and integrated system designed to separate and transform hydrocarbon molecules contained in crude oil into specified products for diverse final purposes. It may vary from a simple refinery that runs a basic physical separation of crude oil by distillation, to a complex one that can also perform a series of chemical reactions to convert petroleum into a wider array of products. This process configuration is determined by the variety of different crude oils that are to be used and target market products. These many processing options intend to enhance yields of specific products that meet that target market specifications. All refineries use significant quantities of heat and power. For economic reasons, they are designed to use any lesser cost residual energy form, in particular those that don't match market standards, like refinery fuel gas, heavy fuel oil or coke produced in their own operations. Due to natural gas production associated with crude oil and environmental restrictions, many refineries have been adapted to include natural gas in their energy mix. But regularly, a typical refinery fuel supply is composed of 50 to 70% of refinery gas, petroleum coke (either from a coking unit or catalytic cracking regeneration) and other heavy oil by-products. On average, 10 to 20% of an equivalent barrel of crude is required as energy to refine the rest into products. Power demand may be augmented by external electrical energy supply or by self-generation. This is an economic and security option, because it reduces energy costs and guarantees operational continuity, making its operation independent of external power sources. Energy is the largest cash operating expense in any refinery ranging from at least 30 up to 60% total manageable costs.

There are nowadays approximately 700 refineries in the world, with an overall refining capacity around 85 million barrels a day. Most of these refineries and capacity lay in USA (almost 23% of the refineries and 22% of capacity) and Western and Central Europe (18% refineries with 18% of capacity). Asian Pacific countries together sum 27% of refineries and 29% of capacity. China and Japan alone respond for one third of the refineries and almost half of the region capacity. Most of the other countries have smaller numbers and capacity share. In general terms, the refining business is dominated by USA, Europe, East Asia and Latin America, which together account for almost 80% of global refineries and capacity.

The usual capacity of a refinery ranges from 1000 to almost 1 000 000 bbl/day, with average size of 150 000 bbl/day, using something like 9% of its volume throughput as energy. They are normally integrated with logistics facilities, like an oil terminal with a harbor or a pipeline complex and some other industry, like a petrochemical, a power plant or any other energy demanding plant. Generally, refinery energy consumption is a function of its size and complexity, bigger and more complex plants tend to consume more energy per input barrel and in absolute terms, simple and small refineries consume less. And of course their CO_2 emissions follow the same pattern.

For the last 40 years, refineries have progressively increased their complexity, investing in processing units to upgrade oil products, reducing fuel oil yield and enhancing gasoline, kerosene and diesel type products yield, matching products to tighter quality standards, especially regarding environmental law. The energy conservation movement has made refineries around the world reduce their energy intensity, meaning that the average refinery nowadays requires 25% less energy to refine one barrel than the average one did four decades ago.

The trend over recent years has been towards more complex and energy intensive processing for many reasons. Existing oil reserves are depleting, enhancing competition for light and sweet crudes and at same time, pushing refiners to process cheaper heavy and sour crudes. On the way around, more rigid product standards especially related to pollution reduction, together with lesser quality crudes, concentrate emissions and hazardous components in the refinery, demand auxiliary processes to treat and dispose them. These factors together are leading to refineries demanding more energy in absolute figures, although they are more energy efficient in terms of productivity. Concerning most of the energy demand forecasts, petroleum will remain the main world's energy source for the next 50 years. But the greenhouse effect seems to be here to stay and law makers and governments are doing something about that, like emission restrictions for fuel end user over sulfur, nitrogen, particulates. And not to forget CO_2, there is the Kyoto protocol and whatever instrument that replaces it.

Considering this path that refiners have followed, and previewing challenges for the future, energy conservation or energy efficiency activities will continue to be included with good management practices. Reasons? First of all, (the old and good reason): money! Energy saving is and will always be driven basically by return rate and cost reduction, either operational or investment. It will depend on energy prices, but remember that there should be extra gains! Whenever you accomplish any energy efficiency action there is a good chance that you will also be reducing some other operational expense, like water for cooling and of course water treatment costs and emission reduction when you burn less fuel. And when you demand less water, fuel, steam, you can postpone the investment for new facilities, and that means more money in hand, less interest rates to pay and more to receive. And you might also be helping to cope with environmental regulation. So, energy efficiency has been and still is, an important part of refinery management. Now we will discuss some ways and points of view of how to deal with it and make it valuable and practicable.

9.2 Good Practices for Energy Conservation Programs

Concerning all the lessons learned and results achieved, it is possible to trace a route to energy efficiency, not only as a short-term goal, but as process, an improvement cyclic process.

Such a process should have three distinctive phases:

- awareness and education;
- good housekeeping;
- upgrading and technological improvement.

There are two main objectives. First is to alert all involved personnel about the potential values of savings. A crucial first task is measurement. Without a good energy balance, it is impossible to know where, how and what is consumed. Such data reveal losses and where opportunities for better energy use lie. Audits and benchmarking with other similar players provide targets to be pursued. Sound economics is also needed; measuring wasted energy without taking the relative costs into consideration provides no return.

After you have raised their attention for the problem with data, the second objective is to show that there is a prize and it can be achieved. With the numbers in hand it is possible to prioritize actions and set some goals. Although they will demand work and money, they are attainable; something can be done to get that energy back. And results can be accessed by operations and investments.

Day-to-day work or good housekeeping means making the most of existing assets. This, in turn, means running and maintaining facilities at maximum efficiency, as well as learning, developing and using best practices. Developing these practices or procedures is a trial cycle. It involves learning about what is the primary energy consumption situation, investigating where the losses and opportunities for better use are and setting some goals for improvement. It is necessary to study, design and experiment new ways of performing the same operations, to try to avoid these losses. If losses can be avoided, you have begun improving the process energy productivity. In case it is a failure or results are below the expected, it is time to go back to the drawing board, and another trial cycle.

All this cycle gives feedback on education, the list of 'non achievements' and 'wrong doings' that is very valuable. After some time, running this repetition, a maximum possible efficiency should be reached and another lesson is learned, the limits of the actual system. In this point it is necessary to find out if that result is enough. If it is not enough, the industry in general, competition or recent technical developments announce that it can be done better; the time has come to consider investment.

But to begin to run this proceeding, a foundation stone must be established. This is the energy balance.

9.2.1
Energy and Material Balances

Energy use monitoring is basic for an energy efficiency program and its management. Lord Kelvin said 'To measure is to know.' and 'If you cannot measure it, you cannot improve it.' Measurement allows tracking and control, so that it is possible to learn about how much and what source of energy is used, and for which purposes. It is the foundation of any energy efficiency program.

A comprehensive site energy balance must include all energy sources used by refinery units, especially fuel, steam and power in all its forms. Only after a good energy balance has been generated and regularly checked can the system be effectively evaluated. The current status can be determined and a baseline established for energy consumption of the facility. With this tool it will be possible to:

- Track historical energy consumption, learning which production events impact it.
- Forecast energy consumption with production.
- Prioritize points of attention and consequent actions to reduce obvious losses.
- Determine the success of those actions by comparing change from expected and effective consumption.
- Compare with other typical refineries and units to help set realistic targets for improvement.
- Get data to spread energy opportunities and generate effective participation in the program by company staff and commitment from management.
- Together with accounting data, generate costs for each energy source.
- Evaluate impact of investment by demonstrating energy savings with future projects.

To set a good energy balance, the first step is to gather all available data and determine what additional information is needed to estimate energy use. It is necessary to prepare balances for every energy form used, normally including fuels, steam and power.

A reasonably accurate fuel balance is of main importance because a refinery usually relies by more than 80% on fuel consumption. Commonly, fuel is the main direct heating source for the major processes on radiating furnaces, and steam generation is mostly done in boilers using fuels. Because a significant amount of fuel used as energy comes from the refinery's own production, this balance influences the complex throughput and production. These flows must be checked with overall refinery production mass balance before being considered in the energy balance. But some fuel is imported, especially natural gas, and must be included in the balance.

The most difficult part of the energy balance is normally the steam balance. Steam is consumed at different levels of pressure and temperature and they are closely related to each other. Usually the highest pressure level is generated in boilers and the others lower levels are produced by steam turbines and valves. After consumption, steam may become condensate that should be returned whenever possible to the boilers for economic reasons, and the rest of it most certainly will end up in the wastewater system either as drained condensate or as oily water. So a good steam balance should be based on an equipment-by-equipment sum of steam flows for each pressure level. The water balance is a consequence of the steam balance. Since you cannot measure many of the steam uses that end up as losses, measuring the inflow of water is a valuable help to close accounts on the steam balance. Because of a natural tendency for refineries to be attached to other facilities that tend to either consume or produce steam, there is a good chance that imports or exports of it are likely to occur and must be taken into account. Another mass balance appears since steam energy balance is related to water and condensate, and they all together add up in the water consumption balance.

It is not uncommon that a large proportion of power demands in a refinery, meaning electricity, is purchased from an external provider, although for economic reasons, the rest of it is usually generated inside the refinery by means of fuels or steam from waste heat recovery. The power balance tends to be the easiest one among all other energy sources because it is metered directly in energy. The balance is done by comparison of the total power supplies, purchased and self-produced, with the total load, the whole sum of all electrical demands. The tricky part is relating total electrical demand to overall energy demand, because the electrical power rate over steam and fuel will depend on generating system efficiency. This isn't a big issue, and monitoring and tracking of energy balance of the utility generating unit will do this job. Nor does it influence the overall energy balance, but it may become an issue when the aim is to share electricity costs among process units.

The energy balance is a regular procedure that must be kept, registered and improved. Remember that this is the main pillar over which the energy efficiency program is erected and maintaining it is a basic daily work. To accomplish it, a lot of data is required, and some standardization of metering, laboratory analysis and basic assumptions are needed.

9.2.1.1 Measurements and Basic Units

The energy balance deals with two basic forms of energy: heat and power. Fuel and steam respond basically for the heat demand and electricity and steam deliver power. Eventually, some refineries that have gas turbines may have a small amount of fuel directly linked to power, whenever it is used as a driver to impel any equipment other than a power generator.

This means that the balance has to work with two basic measurements, one linked to heat and the other to power. Electrical power can be measured directly in energy by a powermeter or wattmeter, the preferred units regularly being MWh or kWh. Heat has to be measured indirectly by mass flow of steam and fuel and converted to energy on its heat content.

For fuel, since it is burned, the heating value is the energy involved. For each fuel stream, it can be determined through laboratory analysis. If the composition of fuels produced changes frequently with different crude slates or operational mode, it is wise to analyze them regularly, using a calorimeter. For gaseous streams, this is demanding, but the method is a composition analysis, chromatography. With the basic gas components it is possible to calculate its heating value. Two heating values can be assigned to fuels depending on the amount of hydrogen, which forms water in the combustion. It depends, whether the latent heat of water formed is to be included or excluded. If it is included, this is called the fuel's high heating value or HHV. If the latent heat energy is not included, it is referred to as its low heating value or LHV. High and low heating values can both be used to calculate combustion efficiency. Depending on gas sources, especially refinery gas, generated from different process units and external natural gas supplies, it is possible that fuel gas composition may vary along the distribution system, so it is good practice to perform daily analyses at some significant points, like near to the natural gas connection, near the main producers and near high-consuming equipment, like boilers and huge furnaces.

For steam, the available energy it possesses is expressed as enthalpy that can be determined by pressure and temperature, measured over the various equipment, headers or process stages, through the whole plant. The best approach is to recognize the main steam levels that are used and assume standards of pressure and temperature for each one of them. This will simplify steam energy balance, resuming it basically to mass flow metering. To guarantee that this standard enthalpy will remain valid, just keep track of pressure and temperature at different points of the facility. If it is discovered that there are significant variations over time, operational conditions, or regions of the compound, corrections or diverse standards should be used.

After establishing these basics, it is possible to perform all the necessary calculations that generate the energy balance and some skill is in demand for reaching a solid result.

9.2.1.2 Calculi and Approximation

After converting all measurements in energy, a basic calculation on an energy balance is matching production and consumption of each energy form. Ideally, with all energy inputs and outputs measured, the difference between the sum of productions minus the sum of consumptions should be positive and would represent the losses. And the next step would be tracking these losses to try to avoid or reduce them. But this fortunate option scarcely happens most of the time in the real world. There are many reasons for this to be rare, like inaccurate instrument readings, arithmetic and recording errors, wrong or inappropriate use of conversion factors or of standard conditions. There is also much lack of metering, which eventually leads to poor estimates of quantities.

The challenge lies in selecting the essential measurements to reach a level of accuracy that suits the needs of the energy efficiency potential. Good practice is to guarantee that precise meters are installed on the main production and consumption energy streams, including imports and exports points, like:

- refinery gas producing units;
- boilers and furnaces;
- electric motors, gas and steam turbines (big drivers);
- big steam heaters;
- catalytic cracking unit flue gas (CO gas) (when it is the case).

In general, it is imperative to measure all sources and if possible to address something like 70% of the consumers. When it gets down to process unit level, around 60% of energy use should be related to very few pieces of equipment and the other 40% spread among many smaller ones, not worth metering individually. In these cases it is indispensable to have a unit battery limit meter for each individual energy stream. With the consumption details of the main equipment and the overall measurement it is possible to make good tracking of energy; what is consumed by whom, where and when. In this trail, losses, wastes and poor efficiencies will be discovered.

One should record not only the results, but the concept that helped built this methodology; make it simple to understand. If it is a continuing and evolving job, it has to be traceable and reconcilable, to give every technical person that runs into it the ability to comprehend, check its accuracy and improve it. Explain assumptions underlying the calculations, like what is the time period considered, which unmetered flows are estimated, and how this estimative is done. Detail all functions that evaluate any dependent consumption, like steam for stripping in a distillation column, explaining how it was built and how it is expected to be used. The refinery scenario changes over time, for seasonal operational reasons like turnaround maintenance, or new projects and adjustments will be needed. This registry and improvement process will give the balance accuracy, consistency, reconcilability and legitimacy. All these values are very important to reliability in estimating the energy efficiency opportunities.

9.2.1.3 **Analysis and Basis**

Analysis of the energy balance is a powerful tool for achieving an overall view of site energy flows, its losses and efficiency improvement opportunities. It enables a specialist or the person in charge of energy management to learn about the system, its flaws, strengths, weakness and sensitivities. With this knowledge it is possible to establish a structured approach to improvements.

The analysis itself has two stages:

- analyzing data collection and calculation methods; and
- analyzing results and inferring actions.

The first stage should be structured and systematic and its benefit is to avoid errors and advance in data availability. This is accomplished by checking the consistency of energy consumption variations for each period with operational logs. Energy flows must follow certain trends and ratios with production rates. If the figures don't show coherence between them, it is time to verify meters, review formulas, assumptions and eventually consider more metering.

The second stage is an investigative and screening phase. Once the previous one has been approved, energy balance can be analyzed over historical results, benchmarking with data from similar plants and units, to determine energy consumption patterns and forecasts. Comparison with the actual consumption opens the way to find out gaps between them. These gaps can be explained either by inefficiencies or better, energy efficiency opportunities or lack of knowledge of the system that turns the wheel of the learning cycle and improves the energy balance. Benchmarking can be accomplished by some different approaches. One is surveying similar processes, tabulating results, and choosing the best performing processes as the iconic best practice standards to which refinery units and processes must strive to meet. Another option is to generate a benchmark from the plant's own past performance, analyzing operational conditions when that point was reached and comparing it with current performance. With this in hand management can set increasing improvement goals, like being 10% better than the previous best, concerning differences of operation at each moment. A third option is building a performance model based on a software and feeding the equipment's operational data into models that attempt to simulate the potential performance of each equipment and by instance of the whole refinery. This kind of comparison allows better identification of opportunities for energy efficiency. Such a model can also be used to identify which part of the plant is degrading in performance, and loss of energy efficiency can be prevented by operational or maintenance intervention. It allows the effect of changes in operating procedures, or capital projects to be forecast, and the generation of a dynamic energy efficiency target for the site, taking into account operational mode, throughput, seasonality etc. In large refineries, petrochemical and complex plants this approach is nowadays almost indispensable.

It begins to be possible to determine subtleties, like how much energy really is variable with production and the amount that is fixed, that will always have to be spent regardless the operational mode. The limit from manageable and non manageable consumption is traced. Relating these to processes can provide a detailed scope of areas which offer significant benefits and produces a list of priorities, ranking the size and places where there are better options for improvement, from the so-called 'low-hanging fruit' to the 'hard-cracking nut'.

Assuming that data and calculus methodology are available and consistent, the most important requirement for performing this analysis is practical process knowledge in refining processes. Delegating this procedure to seasoned personnel helps the learning cycle and in the meantime allows maximum improvements in energy efficiency. These specialists will be able to trace the plant itinerary for better results encompassing short, medium and long-term operational actions and investment projects, guaranteeing consistency between them. This knowledge can be translated into simulation tools that can be used dynamically to forecast results and help spread the energy efficiency culture, while the refinery develops and grows over the years. This will ensure that future decisions whether operational, maintenance or investments will be more concerned with energy efficiency options.

9.2.1.4 Standards, Averages and Deviations

It is good practice for any energy balance to assume some standards for units like BTU, kcal, KJ, or MWh, any basic unit that is of common use for the personnel involved. It many cases it is good practice to comply also with some other physical significant package, like direct steam flow of tons per hour. For fuel, establishing a kind of currency, like an equivalent, in terms of mass or volume, of some common usage product, like a standard fuel oil barrel or natural gas 1000 cubic meter or feet. This helps communication and simplifies general evaluation of results and amounts for daily use on operational decisions.

Another standard that should be established is a kind of expected energy operational range. This means, considering the balance data series and its relations with processes and throughput for a certain operational mode, an expected reasonable range of consumption for each energy form should be complied with. An example, for a distillation column with a specific crude and campaign, for instance, maximization of diesel, the stripping steam flow must stay in between a determined minimum and maximum. Another sample would be establishing expected time patterns for pump operations and maintenance check. Reasons for noncompliance from that are to explained and recorded.

These values come from those same analyses performed over energy balances and benchmarking. It should express the proper manner in which the unit should be operated, and is a way to try to standardize operational and maintenance best practices. This procedure helps to spread knowledge and gage practices through the whole facility. Needless to say, now and then some argument about the practice, numbers and motivations of it will arise, but that is a good thing, if this movement shows that the standard might be either wrong or old-fashioned, and pushes improvement and awareness for the project.

As a natural consequence of instituting better practices by average numbers and model procedures, we must learn how to deal with deviations. Deviations can be tracked by historical comparison with usual dispersion of values on energy balance, and it is possible to classify in two ways. One is the permitted and conceivable deviation, like numbers a little higher than expected average. An example: for steam consumption being higher than usual because of an asphalt campaign; since dealing with viscous oils demands more heating, an extra steam demand is justified. Relating operations and updating the records for these numbers increases knowledge about the process. The second one is trickier, when numbers are normally very far away from those expected with no apparent operational justification, but measurements confirm it. The chances are that it was a mistake and some improper operation was done. Trying to track down in which area the episode may have occurred is a good option. Investigating the reasons, but not pursuing the guilty. Once it is possible to determine where it might have happened, communicating the fact and its possible consequences, in terms of losses and lack of understanding, will eventually bring about the causes. Registering again on the balance all the facts will help the learning process and avoid its repetition in the future.

9.2.1.5 Usual Figures

Concerning some of the previously mentioned numbers, an average refinery should be consuming something like 9% of its crude oil throughput as energy. Energy is one of the biggest controllable costs in the refining business going up to 50%. Ninety percent of energy demands in this average refinery are provided by fuels either from fuels produced or imported. The remaining 10% of that energy demand is imported electricity. The final energy needs are divided between heat with 80% and the remaining 20% being power demand. Heat is released in fired heaters for direct process use and these account for almost 50% of overall energy demand, and boilers supply the other 30%. Burnt fuels from self-production that provide the basic energy for heat are mostly refinery fuel gas, FCC coke and some residual fuel. Natural gas is also a major fuel supply for most refineries, considered here as an import. Adding all typical fuels, refineries are mainly gas consumers, summing refinery gas and natural gas flows, going up to 70% of energy demand. Power is supplied by electricity, purchased or generated inside the refinery and steam, and these three power sources share almost evenly the demand.

Steam is usually distributed at least three pressure levels, mostly superheated. The highest commonly drives electricity generator steam turbines. The second high pressure, is used in steam turbines alternatively with electricity, to move most of the sensitivity equipment, like the FCC blower, distillation units main pumps and others. The lowest pressure usually has the lowest superheating quality and normally matches the required temperature for general heating and stripping demands. Actually, there can be more steam levels even up to ten, many of them ranging only in some areas and units, directly related to minor uses as stripping or heating for some special process. Generally steam represents something like 20% of final energy use, but is responsible for most of the scattered energy losses, particularly in the distributing system.

These figures indicate that a good energy balance must be accurate around fired heaters and boilers, because they deal with the bigger share of energy released in the refinery. Precision on steam balances, specially on the main pressure levels cover another significant amount of energy. Any loss detection and recovery on these systems should be significant. But these are just normal figures, an overview of a general refinery is fundamental to learn about the installed processes and that depends on each refinery's particular configuration.

9.2.2 Process Units

A refinery is composed of a number of different process units for separating and transforming crude oil into products. Any refinery may be configured with a certain combination of some of these process units and this arrangement may vary significantly between refineries. These units can be classified according to

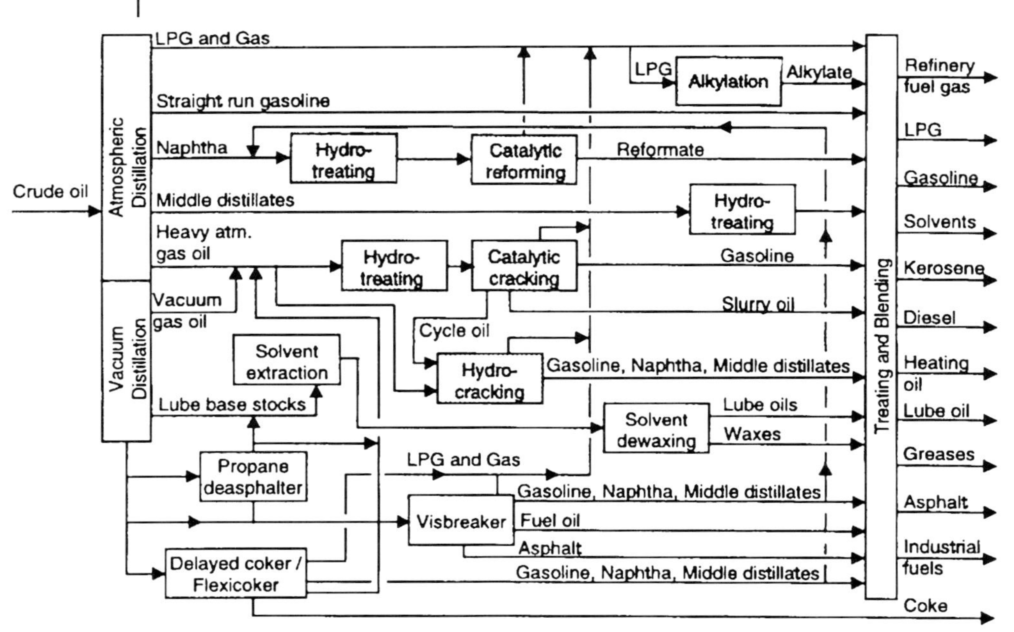

Figure 9.1 Simplified refinery flowchart.

whether they perform physical or chemical processes. Here they are nominated transformation and separation units. Figure 9.1 shows an almost complete refinery flowchart addressing all process steps that are described in the following section.

9.2.2.1 Transformation Units

The majority of typical process units perform some kind of chemical reaction. We can include in this class conversion units, like thermal cracking, as a coking or a visbreaking, cracking processes, like fluid catalytic cracking, steam cracking and hydrocracking, catalytic reforming, isomerization, alkylation and polymerization. In this same category, we can include the treatment processes, hydrodesulfurization and hydrotreating.

To break and recombine molecules, high pressure, temperature and consequently energy are required over relative small amounts of matter. So, generally, these processes are very energy intensive per mass. On the other hand, they usually run over small shares of the crude oil input of the refinery, so that their energy requirements may not be the most significant in the overall refinery energy balance.

Fluid Catalytic Cracking Catalytic cracking is the most commonly used process to convert heavy oils, like light and heavy gas oil from atmospheric or vacuum distillation, coking and deasphalting units into more valuable lighter products, especially gasoline. It requires both heat and catalyst to break large hydrocarbon molecules into smaller, lighter molecules.

Catalytic cracking reactions occur on moving fluidized bed reactors. In this process, oil streams are in contact with a hot zeolite type catalyst in a so-called 'riser reactor', where the cracking process takes place. The catalyst and the reaction products separate mechanically in a cyclone system in the top of this riser. Any oil remaining on the catalyst is removed by steam fed in a stripping section of the vessel that contains the cyclone on top of the reactor. The catalytic cracking reactions produce coke that is deposited over the catalyst surface, reducing its activity and selectivity. To allow the process to continue, catalyst should be continuously regenerated, by burning off the coke from it, at high temperature in the regenerator. This regeneration demands huge amounts of pressurized air, producing a large flow of a high temperature gas. Regeneration burning can be performed in a complete or partial mode, meaning that this flue gas, although mainly composed of nitrogen, may have approximately 10% of either CO or CO_2 gas. Oil products are separated by means of a fractionation train. On average, these units are responsible for processing and producing up to 25% of the overall volume output, so that they are among the most demanding energy units in a refinery, requiring great amounts of heat for the feedstock and a large power demand to achieve the appropriate process pressures.

Common measures to improve energy efficiency in these units include addition of a waste heat boiler to recover heat from the catalyst regeneration flue gas and from completed burning of its CO content, depending on the regeneration process. The installation of a power recovery gas expander turbine for the same hot flue gas is also recommended, to produce shaft power. To avoid hot hydrocarbon vapors flowing inside the boiler furnace, for safety reasons, a certain pressure drop is needed. The huge flow with this pressure drop delivers more than enough power to drive the catalyst regeneration air blower. With this equipment, depending on the unit size and feedstock, the catalytic cracker may become nolonger a net energy consumer, but an energy producer for the refinery. For opportunities in the fractionation train, refer to the analysis of the distillation units.

Reforming, Hydrocracking, Hydrotreating and Similar Units Catalytic reforming uses catalytic reactions to transform low octane distillate naphtha into high octane aromatics like benzene, toluene and xylene. Feedstock is in contact with a platinum-containing catalyst at elevated temperatures and hydrogen pressures ranging from 3.5 to 35 atm. Four kinds of reaction: dehydrogenation of naphthenes to aromatics; dehydrocyclization of paraffins to aromatics; isomerization; and hydrocracking may occur. The process might be either continuous, using moving bed reactors, or cyclic and semi-regenerative, using fixed bed reactors. Dehydrogenation reactions are very endothermic, requiring the hydrocarbon stream to be heated between each catalyst bed. All hydrocracking reactions release hydrogen, which can be used in the hydrotreating or hydrocracking processes. Feedstocks must be hydrotreated first to remove sulfur, nitrogen and metallic contaminants that cloak the catalyst. The permitted amount of benzene in gasoline has been reduced for environmental restrictions, and the use of catalytic reforming as an octane enhancer is decreasing.

Catalytic hydrocracking occurs in a fixed bed catalytic cracking reactor under high pressures ranging from 80 to 140 atm in the presence of hydrogen. Feedstocks are distillate fractions that cannot be cracked effectively in catalytic cracking units, like residual fuel oils. Hydrogen reduces the formation of heavy residuals streams, while increasing the yield of gasoline by reacting it with cracked products. Depending on products and unit size, it might be a single stage or multistage reactor process. The catalyst is similar to that of the catalytic cracking process, being a silica alumina crystal with scattered rare earth metals.

Hydrotreating and hydroprocessing are similar processes used to remove contaminants such as sulfur, nitrogen, oxygen, halides and trace metal impurities. Hydrotreating also converts olefins into paraffins to reduce gum formation in fuels. These units are usually needed to treat the feedstock of processes for which sulfur and nitrogen are poisons to the catalyst, like catalytic reforming and hydrocracking. It uses a fixed bed reactor catalyst in the presence of high pressure hydrogen and temperature. It produces the treated streams, fuel gases, hydrogen sulfide and ammonia. The treated product and hydrogen-rich gas are cooled after leaving the reactor and before being separated. Hydrogen is recycled to the reactor. Since these processes have high hydrogen consumption they require a hydrogen plant. This unit usually has a steam reformer and produces syngas (synthetic gas), a hydrogen and carbon monoxide mixture.

Isomerization promotes molecular rearrangement to increase the octane number, that means paraffins being converted to isoparaffins. Reactions occur at temperatures between 100 to 200 °C in the presence of a platinum-based catalyst. This catalysis demands a hydrogen atmosphere to minimize coke deposition; but hydrogen consumption is negligible.

Alkylation is the reaction of propylene and butane olefins with isobutane to form higher molecular weight and high octane number isooctane. It is a low temperature reaction conducted in presence of very strong acids, like hydrofluoric acid or nonfuming sulfuric acid. Hydrofluoric acid alkylation produces a residual acid-soluble oil that is burned in a furnace by a special burner. Sulfuric acid alkylation produces acid sludges which are burned for sulfuric acid regeneration.

All the above processes, with the exception of alkylation, run over contact with a catalyst bed usually under high temperature and high pressure, but most of the reactions involved are exothermic, and products leaving reactors have high energy content. Opportunities for energy efficiency lie in improving heat recovery over products leaving the reactor. So achieving better heat integration through preheating the fired heater feed with the products is essential. Because of high pressures, improving compressor efficiency also gives good return. This recommendation applies in particular to olefin gases in alkylation and hydrogen in hydroprocessing.

Although catalytic reforming is becoming old-fashioned for environmental reasons, these same restrictions are pushing the need for hydrotreating and hydrofinishing processes while, because of its production flexibility, hydrocracking is partially substituting pure catalytic cracking units. This indicates that hydrogenation processes will become more significant in refinery energy consumption profile in the future.

Visbreaking, Coking and Similar Units Visbreaking is a thermal cracking process that uses heat and high pressure to break big hydrocarbon molecules into smaller ones. It aims to produce more gasoline and reduce the viscosity of fuel oil. Heavy gas oils and residue from vacuum distillation are the feedstock for this unit. Feedstock is heated in a furnace up to 480 °C and then fed to a reaction chamber at a pressure of about 10 atm. High temperatures and long residence time result in a high cracking severity that boosts gasoline yield and also produces a low viscosity residue that can be added to the fuel oil pool. After the reaction step, the process stream is mixed with a cooler recycled stream, which stops the cracking, pressure is then reduced in another chamber, and lighter products are vaporized and withdrawn. Products are sent to fractionating, and the distillation bottom residue is partially recycled to cool the process stream leaving the reaction chamber; the rest is blended into residual fuel oil. It is difficult to control reaction pace, so excessive cracking happens and generates unstable compounds. Because of that, visbreaking has been substituted in many refineries by catalytic cracking.

Coking is a more severe thermal cracking process used to convert low value residual fuel oils and transform them into lighter products such as gasoline and gas oils. It also produces petroleum coke, solid carbon, with impurities and approximately 5% of hydrocarbons. There are two types of coking process: delayed coking and fluid coking. The flexicoking process is similar to fluid coking, but it gasifies the fluidized coke to produce coke gas. The vapors from the coke vessels are lighter-cracked hydrocarbon products containing hydrogen sulfide and ammonia, which are sent to fractionating.

Thermal cracking processes demand great amounts of heat at high temperature and products leaving reactors have high energy content. Opportunities for energy efficiency again are on improving heat recovery from the products.

9.2.2.2 Separation Units

Nevertheless, the very first process in oil processing historically is still the core process of any refinery, the physical separation. All distillations fall into this class, atmospheric, superatmospheric and vacuum distillation, as also any light ends recovery or specialty separations, like LPG specification or propane–butane separation. In this rank are included any solvent extractions, like crude oil desalting and dewatering.

Distillation and Fractioning Units Petroleum is a mixture of different hydrocarbon compounds with diverse boiling points ranging from 35 °C to over 450 °C. Separating it into its various fractions is the first operation in the refinery, and is accomplished by distillation. It is performed in a column equipped with internal contact elements, like fillings and trays to improve contact and separation between the liquid and vapor phases. Hydrocarbon molecules vaporize and leave the liquid mix, as their boiling point is reached, being then withdrawn from the distillation tower, condensed and collected in streams, the oil fractions. Lighter fractions with boiling point lower than 420 °C are separated in an atmospheric pressure distillation. Superatmospheric distillation is usually a primary step whenever the crude oil is very light, to favor the separation of gases and lighter products and reduce atmospheric tower duty. Heavier fractions with boiling points higher than 420 °C must be distilled in a vacuum tower at subatmospheric (vacuum) pressure, because at temperatures above this point under atmospheric conditions, hydrocarbon molecules degrade by thermal cracking reaction. To enhance separation of the molecules, steam is injected into side columns to reduce partial vapor pressure that helps lighter fractions to flow to the vapor phase. So the distillation might be described as a process that performs a huge heat exchanging, receiving and withdrawing energy to perform continuous and correlated evaporation and condensation processes.

Regardless of the feedstock, any distillation or fractioning process works basically in this way, either for crude oil, propane–butane mixtures, catalytic cracking products, whatever the liquid mix composed of different boiling temperatures compounds. The following approach addresses crude oil distillation.

Heat is primarily delivered to crude oil in a fired heater that is then fed to the column. Before being sent to the heater, crude oil is preheated in a series of heat exchangers that cool down distillation product flows, recycling energy to optimize efficiency. Additional cooling is required to allow distillation products to reach ambient temperature and it is usually supplied by water from cooling towers or air heat exchangers.

Vacuum for the secondary distillation column is provided by means of steam ejectors and vacuum pumps. Steam requirements for this unit are usually obtained through steam turbines that move internal equipment like pumps and blowers. Some medium and low pressure steam might be generated in this unit and used as stripping steam.

Regular measures to improve energy efficiency in these units are mostly related to heat containment and recovery, like heat integration between atmospheric and vacuum tower product flows in the crude preheat exchangers train. Since heat exchangers are an important piece of equipment, fouling control may be a significant option. The vapor–liquid contact element inside the distillations towers also plays an important role in heat transfer and conservation. Distillation column energy optimization can be done by tower pressure reduction when applicable, because low pressure diminishes the temperature profile to evaporate the same fluids. Reducing reflux rates and pumparound duties are important opportunities, less reflux decreases overall energy flow through the column. Pumparound is an operation where liquid is removed from the tower, cooled by a heat exchanger, and returned to the tower. Both intend to reduce the duty of the top condenser while providing enough reflux for optimal distillation, but if used in excess, they withdraw too much energy from the tower, implying a bigger fuel demand in the fired heater. Improving tray design or even substitution of traditional trays by packing is another possibility. The use of heat pumps should also be considered, since heat is removed from column top and injected at its bottom, there is an opportunity to recycle the heat rejected at the top to the bottom by means of a heat exchange fluid and compression energy, reducing net energy demand. Adjusting and controlling the steam stripping flow is also a good practice, which not only saves energy but reduces the load on the wastewater treatment system, since less oily water will be produced.

Since the distillation unit is at the heart of the refinery and receives the biggest feed in the whole complex, these units are the main energy demand and deserve special attention at any time in a refinery.

Solvent Extraction and Similar Units 'Deasphalting' removes asphaltenes from heavy vacuum residue to prepare feedstock for catalytic conversion units and for the production of lube oils. In conversion units, catalyst performance is impaired by heavy metals and high residual carbon and both are related to asphaltene concentration. In lube oil production the extracted light liquid phase makes excellent base lube oils, called 'bright stock' that can be further processed to meet specifications for lube blend stocks.

Vacuum residue, normally straight from the vacuum distillation tower, is heated and fed to the top of a trayed extraction tower at a pressure around 30 atm, while liquid propane is loaded from the bottom. Propane solvent moves up counter current to the asphalt, which is removed from the bottom after extraction. This stream, after being heated in a fired heater, enters the top tray of a baffle trayed separator. Side flashing and steam stripping towers separate deasphalted oil, asphalt and recover propane. After cooling, asphalt and deasphalted oil are ready for storage and blending. Recovered propane is compressed, cooled and drained free of the water from the stripping steam in an accumulator drum. Dry propane is recycled to extraction tower.

Other significant solvent extractions are strictly related to lube oil production. The furfural extraction process is designed to produce base lube oil with high viscosity indices and other desirable qualities, like color and stability. This process removes aromatic and naphthenic compounds from feedstock. Liquid SO_2 or phenol may be used as a solvent for this purpose. Feedstock enters an extraction tower, that can be either a trayed packed tower or rotating disc contactor, from below the bottom tray and a dry furfural stream enters from the top tray flowing counter current to the oil feed. Aromatic compounds are removed from the bottom as a rich furfural stream and treated base lube oil, the raffinate, leaves the top. Stripping towers that use inert gas like CO_2, recover furfural from the raffinate and extract. The entrained inert gas is flashed off from recovered furfural and can be either vented to the atmosphere or returned to inert gas system.

Lube oil dewaxing is necessary to ensure proper viscosity at low ambient temperatures. In solvent dewaxing, oil feedstock is diluted in solvent to lower its viscosity, chilled until the wax is crystallized, and then solvent is removed by vacuum filtration. Commonly methyl ethyl ketone (MEK) with methyl isobutyl ketone (MIBK) or MEK with toluene are used as solvents. Solvent recovery from oil and wax is performed through flashing and steam stripping.

Basically all these processes consist of blending, mixing, distilling and stripping stages.

A special extraction is the crude oil desalting, which is a process to wash the crude oil with fresh water at high temperature and pressure to dissolve, separate and remove the salts and solids. Crude oil and fresh water are mixed together to produce washed desalted crude oil and contaminated water. The main energy input is power for pumping, while mixing is provided by design. Although electrical energy is injected directly on the desalting drum to promote the coalescence of dispersed water droplets in the crude oil, its effective consumption is negligible compared with power pumping demand.

Good practices to improve energy efficiency in these units are again heat containment and recovery. Attention to solvent recovery stages is very important because solvent recycling reduces cost and improves heat conservation. Some of the same considerations for distillation apply here, specially adjusting and controlling of steam stripping for same reasons and additionally interference with solvent recovery.

Crude oil desalting has a particular relation with energy efficiency, not by its own energy consumption but for implications on the distillation tower. Poorly desalted crude oils when distilled liberate hydrochloric acids that attack the top condenser and produce rust and fouling, reducing heat transfer and energy efficiency. And the top condenser is a very sensitive piece of equipment for energy performance of the distillation unit, a main energy consumer for the refinery.

Generally extraction units represent a small share of the whole refinery energy consumption, but they deal with a small volume share of crude oil throughput, that happens to be among the higher value products of the refinery like lube oils. Any improvement in energy efficiency in these streams may be valuable.

9.2.2.3 Storage and Transport

After processing, finished products have to be stored and delivered to clients, or semi-finished products stored to be finished. Although these operations are not intensive energy consuming, the significant amount of volumes involved and some specialties can be observed.

Tankage Storage tanks are used to store crude oil and intermediate process streams for further processing. Finished products are stored in tanks before transportation off site. As a matter of fact, approximately 50% of the products in a regular refinery are prepared and specified at the storage tanks by means of blending. Many tanks deal with low viscosity products at ambient temperature. However, heated storage tanks are common in refineries. They are used to store products whose viscosity properties restrict flow at normal ambient temperatures. Products heavier than diesel oil, such as heavy gas oils, lube oils and fuel oils, depending what is ambient temperature in the part of the world that the refinery is, can be stored in heated tanks. Usually tanks are heated by heating coils or bayonet type immersed heaters. Steam is normally used as the heating medium. It is common also, wherever there is immersed heating, that the tank needs to be agitated by side-located propeller agitators. Eventually external circulating heating is used for tanks, and products are mixed by means of jet mixing, where the hot return stream enters the tank through a jet nozzle. External heating is used when there is a potentially hazardous situation, if steam from the immersed heater were to leak.

Measures for energy efficiency in the tank farm are insulation of heated tanks, accurate heater design and steam feed. In the heater design of course the best fitted steam trap and best steam pressure must be selected. Needless to say that condensate recovery is also included. Use of an agitator for mixing is reinforced, instead of the common procedure of pumping around the liquids to promote the blending. This last practice spends much more energy than necessary for the mixing and is less effective, compared with what is obtained in a good blending of the products by an agitator.

Pipelines The flow of fluids is an essential and rather common daily operation in a refinery. Since all fluid flows pass in a pipe to be directed or withdrawn from processes and storage tanks, it is possible to estimate that a refinery regularly has to pump more than twice the volume of the distillation throughput. Considering crude oil, all products and necessary side flows, it is easy to find out that flow power demand represents 10–15% of whole refinery energy demand. And 85% of this energy demand is supplied by electricity. Energy consumed for flow depends on fluid properties, its flow rate and also on size, length and physical characteristics of the pipe. Pipework must be designed for optimal performance and should be selected balancing some factors. Adequate flow velocity to minimize erosion in piping and fittings, pipe routing to avoid many direction changes with the presence of too many bends and curves, and an internal diameter complying to plant standard pipe diameters. Inadequate pipe sizing can cause pressure losses and increase in leaks, but while increasing pipe diameter may save energy, on the other hand, it raises investment and maintenance costs for system components. Pipe diameter doesn't influence installation costs by much, but one thing is always true, undersized pipe diameter results in unnecessary energy consumption. And once the system is installed, these costs become a fixed energy cost of the refinery.

Pipeline design and installation are strictly linked to pump selection and matching piping and pumping system is what makes energy efficiency for the set.

Another important point on pipeline is related to heating and heat recovery. As mentioned previously, oil fractions with viscosity at ambient temperature, higher then 100 cP or 400 SSU, need heating to reduce its viscosity and achieve better flowing condition. To move these fluids around it is critical to keep their temperature, otherwise more friction losses appear and more energy is spent. Maintaining the pipeline at an appropriate temperature is achieved by thermal insulation and steam tracing. Tracing is included inside the insulation; this choice, while bringing extra costs to installation, makes pipe replacement or repair simpler. Steam temperature is necessarily hotter than the temperature to be maintained and all materials should withstand that temperature. Energy efficiency opportunities here are proper design, selection of adequate steam pressure and a steam trap for the tracing. The best economical thickness is calculated by a balance between energy loss and cost of insulation, including maintenance and expected life endurance of the insulation. Thicker insulation conserves energy but uses more material and installation is more expensive. Fiberglass or calcium silicates are the most usual insulation materials. An outer cover of aluminum or stainless steel provides mechanical resistance to minor damage and weatherproofing.

Ships and Barges Since a refinery deals with huge volumes of liquids, either of crude oil and products, and a significant share of costs is related to transportation options, moving bulk volumes at low cost is demanding for economical operation. Because of this, many large complexes are located on coastal or riverside sites, allowing shipment by barge or ships. Tankers and barges can be loaded and unloaded at special jetties or docks for handling petroleum products. Not unusually, particularly big tankers, the very large crude carriers known as VLCC, use submarine pipelines at deep water anchorage with floating loading facilities.

Different from pipes, that are commonly used for flows inside the refinery, ships are just meant for sending flow out to the battery limits. Compared with other means of transportation, shipping is often more energy efficient than trucks or trains.

Opportunities for energy efficiency related to the refinery in this area are linked to ship consumption while docked and the timing of loading and unloading operations. Most of the ships while docked are somehow linked to energy system of the dock that might be supplied by the refinery. The more time in overall transferring operation, the more energy is demanded. And timing is related to refinery production scheduling, ship scheduling and product loading and unloading operations. These are also impacted by details such as berth availability, and delays or constraints on the product loading system. Neglecting the coordination of this scheduling implies that more energy is likely to be spent. And this is actually just waste, because any time above the expected necessary operational time means an extra energy requirement that doesn't produce more.

Of course there are other energy efficiency opportunities over ships operation, like regulating ship engine, proper voyage speed etc., but these are beyond refinery operational ground.

Rail and Truck As crude oil is refined and more specified streams are produced, smaller volumes need to be moved. This requires other transportation means, less bulky and with more flexible delivery capacity. The common method for shipping products with these characteristics is by road or rail in suitably designed tankers.

The same comments may apply as previously on energy efficiency opportunities on ships, related to scheduling and coordination but tank storage play a decisive role on smoothing consequences since the volumes when it is dealing with tankers are much smaller than ships or barges. But another point must receive some attention when loading viscous fuels to trucks and trains: loading facilities for these products also require heating to guarantee proper viscosity for the flow. If the loading dock is located too far away from the product heated tank and from the heat source, like the steam header or boiler that feeds steam to it, there may be reasonable heat losses in this process. Installing the truck and train loading dock near to the heated tank farm and a steam source is an opportunity to reduce energy consumption.

Actually energy requirements for supplying and distribution before and after refinery boundaries represent only 2 or 3% of all cycle energy demand. That doesn't mean it is not worthwhile because it may represent a relevant share of logistic costs for certain products.

9.3
Awareness and Motivational Work

One of the most important steps in any energy efficiency program is getting people committed to it. And to get commitment, it is critical to show them, through the whole organizational structure, the opportunities raised and then tell them they can be achieved, by their work.

9.3.1
Communication

It is demanding to get management and employee support for energy efficiency efforts, and that is achieved through a continuous, consistent, and open dialog. The organization must explain to employees the reasons and the goals it wants to accomplish through its energy efficiency programs. This aims to gain support and to indicate to them what their role is concerning involvement and requirements. A direct link must be drawn to the company strategy, vision and business plan, otherwise it may seem weak among many other corporate demands on the workforce.

The conclusion is that communication must be designed innovatively in order to continually captivate the appropriate audience. This is achieved by developing and marketing key messages to motivate these clients to engage in the energy efficiency program. And the message must show clear commitment between energy efficiency and corporate strategy, using the right promotional channels at the right times.

9.3.1.1 Target Clients

Clients for energy efficiency communication can comprise a wide range of the public, especially some key employees. Actually it is important to get all employees' attention but this is accomplished by getting key personnel involved early and often. These are decision maker managers, engineers, technical and operational staff.

Senior management commitment is fundamental, if concrete actions and projects are to be driven from any energy efficiency program. One of the most valuable aspects of managerial engagement is trustworthiness, so the program may not be seen as a one-off initiative or something that may pass.

Energy efficiency activity is in essence a technical issue and commitment from the technical ranks is also demanding. They are the ones who provide energy monitoring and targeting from data collection, energy balance, analysis, interpretation and benchmarking. From these, standards and expected performance is also set, and the improvement procedures and projects. Technical personnel also perform most of the basic actions to improve energy efficiency, from operational and maintenance best practices to project design and installation. Among them, people should be picked to play critical roles in the energy efficiency program – the energy experts and the program champion or best person. They are meant to be a focal point for communications, resources, actions, programs and training to create conditions for success. Considerations on this position are described later.

General staff and employees' support at all levels of the organization is important to reach its goals. Neither management nor technical commitment alone will likely be enough to sustain the effort over time. Remember that it is a company-wide program and needs cross-departmental cooperation. And each department should have their own first priority, although aligned to the main business strategy, and that might not be linked to energy efficiency. Raising their awareness may at least avoid internal barriers.

The external public must not be neglected. Showing them the company plans, goals and achievements is always a precious occasion to demonstrate willingness and the contribution of the company to environmental and sustainability.

Energy efficiency is a subject related to culture, and the organization should institutionalize it, turning energy improvement opportunities in business as usual and as public image enhancer.

9.3.1.2 **Language**

Good communication tells the audience what they need to know about the current situation and what they should do differently in the future. Some energy data must be provided, so the audience can understand their role and course of action, where they fit and what they can do. For that reason, pertinent information must be delivered in a language that is appropriate to their position, background and usual tasks. This helps to support their engagement with energy efficiency. The message must be clear, in a concise style, as in a one-on-one conversation. Simple, understandable language is better for general staff than technical jargon. But technical slang may fit in just fine for engineers, operators, maintainers, designers and draftsmen. For managers, it must follow the same tone that strategy and the business plan uses.

The main goal with language is to break the ice and build a link, almost a partnership with your audience. Once this is done, a great step is accomplished and the following work becomes much easier.

9.3.1.3 **Media**

Energy efficiency activities should be accompanied by comprehensive publicity to maintain momentum. This may include regular publications in either corporate magazines or newspapers and when suitable external publications for specific sectors. At the beginning or for particular moments, a campaign newsletter might be appropriate. Appearances and occasional sponsorship at trade fairs and conferences linked to the theme enable the campaign to address relevant audiences, from company employees to public opinion. It also promotes networking with other companies and extends knowledge about best practices from different points of view. Internal seminars help to share program results and experiences, and enhance awareness by advertising information about company efforts.

Placing posters and other exhibits in passage areas or in places where activities are being accomplished, kill two birds with one stone. Spread the word and provide recognition for the area personnel since they become an example to be followed, it is a kind of 'Employee of the month' picture. The next step is usually that someone is going to ask them how they did it, and they proudly tell their story about the energy efficiency program.

Contests and prizes are options but must be strictly attached to some measurable goal and it is very important to avoid turning them into the main purpose. The targets are the energy efficiency awareness and results; the prize is just a mean to reach them.

Development of a marketing campaign might be an option but beware of collision with other public relations activities. Trying to integrate with existing motivational regular campaigns, like safety, environmental or cost reduction, is always a good call, as it gives muscle to the ongoing actions and usually reduces energy efficiency awareness costs.

A website containing information, free advice, guidance, resources and identification of training opportunities is a modern approach and can be easily updated. Establishment of networks between energy professionals and staff engaged in the program, assists promoting and disseminating best practices.

A successful campaign is not done simply by distributing a few posters and stickers or regularly publishing a brief article in the company newsletter. Such a naïve approach cannot be expected to have a major and continuous impact. Like any advertising campaign, it is important to keep the awareness campaign fresh to maintain its impact. Regularly update and change the material being used to keep people interested in your campaign. Outdated old-fashioned news and figures, hanging on a poster or displayed in a website, spread a very different message. The message is that actually nobody really cares.

9.3.2 Education

9.3.2.1 Speeches

Spreading knowledge of energy efficient techniques and exhibiting ongoing results is another regular task to keep the program alive. And speeches should be tailor-made for each audience, complying with their language as mentioned previously. But every speech must address some points to make it hit the spot.

It is possible to summarize the points that any speech to any audience has to address in some basic points:

- What are current performance levels?
- Which and where are the potential savings?
- What is in it for that audience?

First present the findings from energy balances and benchmarking processes. Placing the figures of the refinery among industry peers, nation, state, region, city and comparisons to familiar devices or physical measures helps to evaluate the size of what is being addressed. Secondly show the opportunities. It is important to separate and stress results that may be achieved with simple actions and behavioral change, from those that require investment. This draws a line between the individual effort and the company resource provision. Of course, human resources is always the driver but separating the 'what is up to you ' from 'what is up to us' calls people to their own responsibilities. The third point is specific to the regular audience work. What is that area accountability, where are the opportunities and what can they do about it. At this point display and promote the successful cases, which prove that it is achievable and gives return. For the external public rather than those relating to work, take the chance to advertise the gaps and what the company is planning to do about them.

Focus on the essential for their comprehension; if further details are needed they will ask. Be transparent on both positive and negative aspects, don't try to hide misfits, because if they come out, this ruins all the positive aspects of the message. Use graphs and pictures to make points that would take too many words; remember that a picture paints a thousand words. Launching images, figures and metaphors of common objects and daily standards which the audience know about, gets the message across more easily.

9.3.2.2 **Courses**

Communication is a good first step, but long-term engagement of staff is ensured through training; this is critical for the success of the program and requires the people involved to be trained and educated in energy efficient principles and techniques. Formal training is proposed as part of the program, encompassing he essential components of the energy efficiency program including application to real situations, and it has to be designed to meet the needs of different staff groups. Energy efficiency training should focus on improving energy awareness and reducing energy consumption. Generally courses should comprehend three approach levels:

- awareness;
- technical;
- management.

The awareness approach means involving employees in general and should deliver some basic training in energy. It might simply teach the difference between energy sources, usages inside the plant, related costs, unit conversion, basic conversion equipment function etc. Some directions over energy conservation at home also apply. This is a level equalizing training, aiming to bring all potential engaged personnel to a basic understanding of the subject. Remember that employees place a high value on training, so this could be understood as motivational work. It might be a good idea to promote short training sessions on energy into other ongoing in-house training for employees, such as safety. If possible, it should be delivered by a qualified energy expert of the company, or even the energy manager if there is such a position. It enhances networking, helps break down departmental barriers and demonstrates the importance of the theme.

The technical staff, like engineers, operators, maintenance and other workers whose jobs can affect energy use in the plant, need to be trained in the principles and technologies that are related to the systems and tasks that they work with, concentrating on the practical aspects of their tasks. The first option is to concentrate training on specific systems for which they are responsible. There are specialized courses for fired heaters, boilers, mechanical and electrical equipment, heat exchangers, pumps etc. Syllabuses must address thermodynamic principles applied to the systems involved and be delivered on a level consistent with the audience's background. These courses may aim to upgrade job qualifications in the long term and can bring positive performance effects. They may be delivered by university teachers, specialized private enterprises and in-house technical staff. Secondly, management training in energy systems can be broadened. It is helpful to have a wider understanding of the systems operation altogether. Integrating energy management into the whole company matrix is essentially a cross-training to extend employee capabilities and view of the system. When workers acquire this broader view and range of skills, they tend to interact better and more effectively with peers throughout the facility. They get more motivated towards energy efficiency and it becomes easier to recognize potential system problems, opportunities and the implications of each other's acts in overall efficiency. Courses can be supplemented by ongoing professional development and study.

A set of skills is needed to manage an energy efficiency program and among them can be included leadership capacity, planning, delegation, communication, negotiation, technical process knowledge, and economics. These abilities and the specific training will be defined in the job description in Section 9.6.4 on leadership.

Once you have educated and motivated the staff, it is time for accomplishment, to put all the messages and learning into practice to justify the effort.

9.4 Saving Energy by Operation and Maintenance

Energy efficiency is a multifunctional and team task. Although individual actions influence and might be decisive in reaching some targets, without a coordinated plan, it is very difficult to achieve continuous improvement. From this point of view maintenance and operation are strictly related activities that together make the good housekeeping step.

9.4.1
Scheduling and Maintenance

Maintenance may be seen as a task required at two separate times, one during ongoing operation and the other when the operation stops. Ongoing operational maintenance includes tasks that don't impair or require stopping of the main process equipment that significantly reduce refinery production. Some routine tasks are included in this category, like lubrication, examination, monitoring, inspection and cleaning of equipment, calibration of instruments, repairing insulation, checking and tuning electric systems, repainting, changing lightbulbs and steam traps, and fixing leaks. Leakage happens to be the single biggest area of energy wastage that can be addressed by what is normally a simple and low cost maintenance service.

But many other maintenance activities require some slowdown or stop in production. Among these it is possible to quote: procedures to repair burners and draft control equipment such as fan and blowers, for fired heaters and boilers; cleaning of heat transfer surfaces in heat exchangers, fired heaters and boilers to reduce fouling; maintenance of motors, turbines and drivers to recover optimal energy performance; maintenance of pumps, fans, compressors and blowers to restore flow rate; replacement of worn impellers, bearings and sealings, and checking and adjusting alignment. Refurbishing and replacement of expected and unexpected worn parts and some special repairs, like leaks from awkward places, electric connections, and other auxiliary equipment such as safety checks on valves. These services might be planned in a predictive or preventive manner and depending on their burden on operational costs or safety may be postponed for some time, but they have to be done at some moment.

A good maintenance program has to contemplate all planning, scheduling and monitoring for the maintenance assignments to be executed in a predictable and well organized way. The benefits of such a maintenance system are reliability, safety, availability and confidence in the operational assets that improve team work and employee morale. And it must match production planning and scheduling. Actually they must be structured together and when this planning and scheduling are well established, they can contribute to energy efficiency with substantial reduction of wasted energy and lost production by avoiding or diminishing effects of equipment breakdowns. But this effort begins well before any tools are seized or spare parts are changed.

9.4.2
Pre-Maintenance Work

Maintenance work actually begins in the operating phase and on the drawing-board and this issue will be addressed in Section 9.5 on upgrading. During operation all monitoring that is executed to collect data for energy balance must be completed by further monitoring, observation and recording. This information can be used to detect problems, investigate causes and determine solutions to improve the efficiency of the system. This monitoring can indicate damage, wear, inadequate installation, equipment limits, exceed or unexpected operational conditions etc. It should include wear monitoring, vibration analyses, flow and energy condition monitoring (effective pressures and temperatures and differentials head and temperature equipment and systems).

Good record-keeping of events and incidents during operational procedures, like change of burners in a furnace with damaged tips, unusual behavior of pressure during a pumping procedure, presence of water or any other contaminant, or a product in an unexpected location may help the investigation into causes that may lead to some inspection or maintenance tasks either in routine maintenance or at the next turnaround.

Especially for energy efficiency targets, a long list of items that need attention, that may affect maintenance for better results, can be built. Below is a small sample addressing only furnaces:

- fuel supply and burner pressures;
- fuel temperature;
- burner damper settings;
- windbox to furnace air pressure differential;
- air flow indication;
- air preheater inlet gas O_2;
- stack gas analysis O_2, CO, CO_2 etc;
- flue gas and air temperature and pressure;
- air temperature to air heater;
- unburned combustion indication;
- stack appearance;
- flame appearance;
- forced draft fan discharge;
- furnace boiler outlet;
- economizer differential;
- air heater and gas-side differential;
- soot blower operation;
- general condition of burners like erosion, deposits, missing or damaged parts;
- atomizing steam pressure;
- refractory condition.

This list partially comes with the equipment itself, but is in constant construction. During operation a regular interface and exchange of ideas between operational and maintenance crews is demanding. It promotes learning about the equipment, the system, what to look after, strengths and weaknesses. It is another step of the knowledge process that leads to better results in energy efficiency.

9.4.3
Conditioning and Testing

Conditioning and testing generally mean preparing lines and equipment for operation for initial use or after maintenance for start-up. There should be regular procedures for clean-out and conditioning of each kind of equipment, following the manufacturers' documents and internal best practices. All safety procedures should be performed for pressure, temperature, leaks, refastening joints and so forth.

Addressing energy efficiency some points deserve special attention. Before lighting any fire or turning anything on, do check if all insulation is properly installed and refractory is cured. If they are not attended from the beginning, energy is wasted to the environment at a higher rate than designed. Especially if the refractory is not cured, it breaks at the start of operation, and does not performs its tasks so more energy is lost. On initial pressurization, check if any leaks listed to be repaired really are secure. Verify thoroughly all repairs and equipment parts replacement that are related to energy performance, if they were accomplished and how effectively the renewed piece is working. Big differences from previous or expected performance should eventually postpone start-up for additional verification or at least be recorded for monitoring during the following operational cycle.

Beyond this phase, optimized warm-up procedures must be followed to reduce supplementary energy use and this is particular applicable to thermal processes such as combustion, fired heaters, furnaces, boilers, heat exchangers and steam systems.

This step tells a lot about gap reduction from the pre- to post-maintenance condition, and helps forecast what can be expected over energy performance and problems for operation.

9.4.4 Best Practices in Operation

On average about 30% of the potential energy efficiency opportunities that are found on the energy balance and analysis stage can be achieved through operational best practices. These are low cost saving energy opportunities that derive from behavioral and procedures changes. Once these targets are reached, cash flow from these savings can promote improvements that depend on investment. Implementation is also a part of the learning cycle and gives immediate return for the awareness and education stage, building knowledge and skills by making the most of existing assets while preparing room for upgrading and better technologies. Among operational best practices it is possible to detach the more significant for refineries.

9.4.4.1 Combustion

Combustion is the quick oxidation of fuel, releasing heat. Complete combustion of a fuel is then desired to allow the maximum delivery of its chemical energy content. Ideally, stoichiometric combustion occurs when fuel and oxygen in exact proportions react completely and yield maximum heat energy. However these ideal proportions of fuel and oxygen are just theoretical and cannot be applied for two reasons. First since it is a chemical reaction, to displace equilibrium towards products and to consume all fuel, an excess of oxygen is required and second, oxygen from air comes with nitrogen that reduces energy efficiency by absorbing heat from combustion and eventually reacts to form oxides. So too much excess oxygen cools the flame and increases NO_x pollutants, while too little oxygen leads to incomplete combustion producing CO, unburned carbon or sooting and leads to a delayed combustion condition, which can result in furnace explosion.

Good combustion is obtained by mixing fuel and as little excess air as possible, to promote complete fuel burning and generate energy to match process needs, coping with safety and economics. This is accomplished by controlling the burner temperature to be high enough to initiate and maintain fuel ignition, turbulence at burner tip to promote intimate mixing of fuel and oxygen, and time to allow complete combustion. These parameters are provided by burner design and careful operation and maintenance of furnaces and burners.

The main energy efficiency opportunity for operating furnaces is controlling excess air. Combustion controls should be used to decrease the amount of excess air to a point where the amount of excess O_2 is set to the optimal minimum, almost at the limit at which actual waste of energy occurs. Searching for this maximum energy efficiency point is a critical operation; small reductions in excess air, still guaranteeing complete fuel burning can represent significant energy savings. But if the air inlet is slightly less than the minimum oxygen requirement, there is not enough oxygen to complete the reaction, CO and smoking can appear. While less energy is released per mass of fuel and more fuel is demanded to cope with the absolute heat requirement of the furnace, heat receiving tubes can become fouled, and the equipment performance is reduced. Since combustion conditions vary through the operational period, finding a setting point a little higher than the minimum avoids these occurrences. Excess air can be measured by the amount of O_2 in the stack gas. It can be done periodically by sampling flue gas with an Orsat analyzer or continuously, with a recording gas analyzer. Of course continuous readings give more chances for operational improvement and better efficiency results. Eventually the presence of continuous CO and CO_2 analyzers make this procedure more accurate. Excess air control is managed by draft control.

Draft provides air for the combustion process while exhausting flue gases to atmosphere. When induced by a stack alone it is called natural draft. Flue gas behaves similarly to air because both have nitrogen as the major component, and since its temperature is higher than ambient external temperature it is much lighter than the outside air. This pushes the flue gas up making the outside air flow in through the burners into the furnace. This kind of draft can only be controlled by hand-operated dampers in the stack and by flaps around the burners. Control is not precise and air flow depends mainly on heater load. But draft can also be controlled more accurately when air or flue gas flows are produced by fans. It is possible to blow air to the furnace through a blower that is called a forced draft, draw flue gas from the stack by a fan, inducing air to enter in the furnace in an induced draft or to use both to balance the air fed to the furnace, known as a balanced draft. When air is forced in, furnace pressure becomes higher than the external atmospheric pressure, when induced, slightly below and when balanced, it may vary from positive to negative relative to atmospheric. Whatever form of fan arrangement is used, a mechanical driven draft enables more accurate control.

Openings in the furnace can also lead to significant heat losses. If furnace pressure is below atmospheric, undesired cold air may infiltrate into the furnace and depending where it happens, can either increase excess air in a non controllable mood or simply cool heating surfaces and hot flue gas streams before they leave their energy in the proper spots. The other way around is if the furnace pressure is above atmospheric, hot streams leak out of the furnace, increasing heat losses to the ambient. The opportunity is to prevent and track openings in the furnace while ensuring that furnace is operated at a pressure slightly positive when the draft system allows.

9.4.4.2 Heat Transfer

Together with combustion, heat transfer is a complementary and critical process for energy efficient operation. Combustion is just the heat-releasing phase and its good performance just guarantees maximum availability of this energy. Good heat transfer is the other side of the process, guaranteeing maximum absorption of liberated heat. In a simple thermodynamic approach, to achieve maximum yield in a heat machine, it is necessary to minimize heat escape. If a refinery is treated like a huge heat machine, avoiding heat losses by containment and recovery is paramount for energy efficiency, and it is obtained by controlling heat transfer.

Heat is transferred by three ways: radiation, convection and conduction, and all are dependent on temperature from bodies exchanging energy. For convection and conduction, physical contact between bodies is essential, but not for radiation. So in a heat exchanger like a shell and tube or tube bundling, heat is transferred by conduction and convection, while a fired heater is in fact a heat exchanger in which the major heat amount is transferred by radiation. Conduction transfer depends mainly on the constituent material of the equipment, and convection, on the physical characteristics of fluids exchanging heat and equipment design. For radiation, like in process fired heaters or in boilers, the interaction between combustion flame, flue gases and absorbing heat tubes must be addressed at the design stage, to achieve high heat transfer rates by radiation and convection, the last for economic purposes, to increase overall energy efficiency of the equipment.

During operation either in furnaces or heat exchangers to guarantee maximum heat transfer rates, avoiding fouling is a big energy efficiency opportunity. Fouling occurs due to film deposits, such as oil films or scaling due to corrosion. It can be avoided in fired heaters by proper combustion, as mentioned in the previous section, avoiding low air conditions with soot formation and averting flame incidence over heat absorbing tubes. This can be accomplished partially by the same procedures for good combustion, with monitoring of tube skin temperature inside the furnace. Keeping heat absorbing surfaces clean is one step for quality heat transfer in furnaces. Heat containment is another, obtained by proper refractory and insulation. Refractory is any material that can withstand high temperatures and abrasive or corrosive conditions, can tolerate sudden temperatures changes, and possesses low thermal conductivity and a low thermal expansion coefficient. Also it has a reasonable emissivity, defined as the ability to both absorb and radiate heat. These properties together give more chance for heat absorption while preventing it from being wasted to the atmosphere. Monitoring the condition of the refractory is another relevant operational task.

One improvement opportunity related to this area is preheating the combustion air. Recovering heat from stack flue gas to preheat air entering the furnace is a form of heat recovery. And once heat is recycled to the furnace, energy losses are reduced and reagents mixing and reaction improves. However, higher air temperature for combustion increases nitrogen oxide formation, that is a regulated pollutant, so finding out what should be its limit and tracking it, is another significant operational procedure.

Heat recovery from all fluids processed or produced heat transfer is the most important operation for energy efficiency in a refinery. This is achieved by proper design and sequencing of heat exchangers. Fouling is taken into account usually by appropriate flow velocity design inside the heat exchangers and some extra heat transfer area known as 'fouling factor'. This factor should not be exaggerated otherwise an oversized exchanger may signify lower velocities and ends up increasing fouling. The optimal sequence of exchangers is also a design task where methodologies like 'pinch' apply. For operations, when running a train of heat exchangers, it is important to maintain flow rates and track temperatures of fluids passing through the exchangers, to guarantee the expected heat transfer exchanges to occur. If they don't comply, fuel demand on fired heaters or steam demand on exchangers along the process rises, revealing a low energy efficiency point. Recording these facts indicates a maintenance intervention spot for next turnaround.

9.4.4.3 **Cooling**

Cooling is required in a refinery for many reasons, from returning products to ambient temperature for storage and delivery, passing through refrigerating equipment to guarantee their integrity to condensing or quenching processes. Depending on the application, it can be provided by water or air. Air cooling is an option especially to condensing distillation top fluids, while cooling water is almost a universal choice for equipment refrigeration, because of its thermal capacity.

Disregarding equipment refrigeration that is specified by design, products cooling and condensing duties are related to efficient operation. More recovery on heat exchangers batteries from fluids leaving unit limits, less cooling demand and adequate control of the process, like top distillation tower temperature, steam stripping flow and fired heater operation reduces demand on top condensers. In general, the better the heat supplied to the process is recovered, less cooling is needed and that is also a monitoring issue to evaluate an energy efficient operation.

9.4.4.4 **Fluid Movement**

Pumping, compressing and blowing systems account for nearly 20% of the refinery power demand. An average refinery may own more 1000 pumps, fans, blowers and compressors with more then 100 HP or 75 kW. The actual energy requirement for fluid movements is achieved in a balance between piping design, previously mentioned, and its movement driver match. In other words, matching piping and pumping or compressing system design is essential for better energy efficiency performance. There is a huge opportunity involving specially pump selection, because it is one of the most frequently used pieces of equipment in a refinery. Pumps are energy inefficient as equipment, roughly 40% of energy input is lost in conversion, and they are often improperly sized or used. Optimization of pumping systems can have significant energy impacts.

Opportunities for energy efficiency around fluid movement begin with the proper selection and sizing of equipment. In pumping systems in particular, eliminating control valve or use of by-passes to control flow are measures that increase energy efficiency. Better control of flow may be achieved by speed variation or using pieces of equipment in parallel to meet varying demand. Controlling the pumping or compressing speed is the most efficient way to control the flow, because power consumption is lower at reduced speed. The most common method to reduce pump speed is through a variable speed drive. It allows speed adjustments over a continuous range, avoiding the need to jump from speed to speed, as with multiple speed pumps. Variable speed systems can be mechanical using hydraulic clutches, fluid couplings, adjustable belts and pulleys, or electrical using eddy current clutches, wound rotor motor controllers, and variable frequency drives where electrical frequency is varied to change the motor's rotational speed. An extra advantage for variable speed control is that it can be linked to computerized advanced process control, demanding the exactly necessary energy for each flow. Replacing old and eventually improper sized equipment by newer more energy efficient and better fitted ones is always a good option.

9.4.4.5 Energy Distribution

Energy is mainly supplied to the refinery in form of heat by burning of fuels and in the form of power either as electricity or steam as previously cited. Measures addressed so far rely basically on the operation of a simple unit, but some energy distribution through the diverse processes is also an opportunity for improvement.

Steam Steam can be used throughout the refinery, supplying energy for process heating, mechanical drives, enhanced separation processes, and can also be the water source for operations and chemical reactions. It can be generated by waste heat recovery from processes, cogeneration and boilers. Considering those three usual steam pressures available in a refinery, the highest one should be generated in a boiler or in a cogeneration system, which will be addressed in Chapter 10. But the medium and lower pressures will be generated by waste heat recovery systems and steam turbines and these steam flows can be carried elsewhere in the complex for use. This is another recycling option.

Opportunities lie in the recovery of energy wastes of heating systems combustion flue gases, final cooling in heat exchanger trains, and even refrigerant water before its return to the cooling tower. They may provide low temperature or low grade heat but significant savings can be realized if they are in cascade and their heat recovered. Many high temperature furnaces have some heat recovery device to generate steam and this is another option for the distribution of thermal energy.

Efficient operation in this system is achieved by implementing actions like ensuring that the process has correct temperature control. Otherwise, instead of just reducing heat waste, it becomes a steam provider for the system, and that is not its purpose. Monitoring of the necessary process steam pressures and temperatures must be in place. Each process unit and equipment should have steam consumption standards and guidelines for regular adjustment of equipment depending on process changes, like throughput, campaign, temperature etc. General steam distribution awareness applies, like checking steam leaks, poor pipe insulation, faulty steam traps and providing their repair. Ensuring that as much condensate is returned from steam usage in the process is another step to be checked.

One particular operational feature is in generating proper amounts of low pressure steam through the use of back pressure steam turbines. This must be balanced all over the facility, matching steam and electrical power demand. Here a very important interface with electricity appears, where relevant savings can be obtained.

Electrical Power The task of the electrical power distribution system is to deliver the right amount of energy for the appropriate electrical equipment. The main uses of electricity are for heating, illumination and mechanical drive. Heating and illumination are usually very specific final consumption modes and although heat might be performed by other energy sources, when electricity is in use this normally means that other heat sources are not to applicable for some process reason. But mechanical driving builds an interconnection between diverse energy options inside a refinery.

Electric motors, gas and steam turbines can be used to drive pumping, compression and much other motive machinery. Choosing between them for each application depends on the availability of each energy form and its cost compared one to the other. But this apparently simple decision relies on the overall cost of the utility for the whole refinery. Just as an example, if in a refinery, high pressure steam is available and some process demands low pressure steam, using a turbine as a pressure reduction device while driving the fan or pump might be an economic option. But if that low pressure steam is in surplus, this choice might not be the best. Instead if the same operation can be performed by an electrical motor, using it could be better. And this equation also depends on whether electricity is generated on site, when usually it might be more efficient to use electric drivers. Constantly calculating and providing guidelines for choice between turbines and motors is a critical and sensible means of improving energy efficient operation of the electrical and steam distribution energy systems. And of course, complying with these guidelines and giving feedback about equipment availability is an important operational procedure.

Concerning electrical power planning and scheduling of large electric loads and equipment operations is another issue. An integrated approach for electrical load management demands reduction of energy usage peaks and a smoother, reliable and more energy efficient operation. Again some general care must be taken in electrical use , like reducing excessive illumination by switching, changing lamps etc; controlling lighting by awareness signs, use of daylight, skylights; or illumination devices with timers and presence sensors. When replacing electrical equipment either lamps or motors the use of higher efficiency ones should always be considered. Consider also the appropriate ambient temperature setting for air conditioning systems.

9.5
Upgrading and New Projects for Better Energy Performance

Completing potential savings in a refinery, the other 70% are upgrades that imply investment. Most of the investment in energy efficiency is based on well-established and common use technologies whose payback is guaranteed. But applying new technologies can bring competitive advantages, especially if it can significantly reduce energy costs, and as energy prices rise, these opportunities for payback can become more attractive.

Large savings may be accomplished with high return rates and fast payback. And good management of selection and implementation of energy efficiency projects can make the difference by achieving real enduring and competitive enhancements. This is accomplished by a combination of choosing the right opportunities with adequate corporate attention and some use of new technologies.

Opportunities that are prioritized for investment must follow the magnitude of potential savings, which come from energy balance analysis. So among main investments there will be waste heat recovery, fired heaters and boilers, heat exchangers, steam systems, energy integration etc. Balancing the size and impact of these projects for their selection, between the process units and systems where maintenance and operational best practices are being accomplished, shows commitment and recognition from management over personnel efforts.

Corporate attention can be easily determined by establishing a special capital funding or allowing a differentiated expected return rate for energy efficiency projects. Even so, results should be tracked based on the same built-in assumptions used for corporate strategy, like energy prices and like any regular project. Additional benefits from efficiency projects should also be taken into account, like low risk, pollution and emissions reduction or support to comply with environmental regulations. This can also bring positive impacts on corporation reputation, employee morale and productivity and all these issues should be somehow considered. Even with this set of tools aiming to push energy efficiency investments many of them cannot provide acceptable paybacks by themselves alone. It is important to regularly review the calculations and always keep track of joint opportunities, for example, when a review and enlargement of systems takes place due to process needs. Combining increased productivity with reduced emissions, and reduced energy costs can provide an opportunity to incorporate that efficiency enhancement into the project.

One warning here is about high budgets on the special funding for energy efficiency. Once you have exclusive money for it, there is a chance that anything might be forced to fit into the program. This can jeopardize the overall return rate and make serious proposals banal. A neat budget blended with some competition with business-as-usual projects, some allowance for a smaller rate of return, and some noneconomic measurements seems to be a better option.

Technology is crucial to improve energy efficiency, but it can bring risks to the expected return rate of the program. Supporting investment in newer and more efficient technologies to replace inefficient ones is an option for continuous improvement. Some provision for research and development is to be made for demonstration projects and it should be encompassed within the portfolio of solid and well-known investment solutions. This practice seems to be more effective, since it gives room to experiment with new equipment and methodologies, but has reduced impact in the case of faulty results. An example can be oxygen enrichment, that means to supplement combustion air with oxygen, making less nitrogen flow in the furnace with oxygen. This can reduce the effects of energy losses from the presence of nitrogen and may increase heater capacity, by allowing more fuel to be burned and more heat released with the same equipment.

Reaching a balanced program between best housekeeping practices and investments is an achievement that really turns energy efficiency into a continuous process inside the refinery. Awareness, best practices and investment improvements are the building blocks for an energy efficient organization and can be considered one big step towards sustainable development. Management must have the ability and willingness to perform all these tasks as daily work. Establishing an organization with this capacity is an indispensable cornerstone for energy efficiency.

9.6 Organizational Issues on Energy

Energy efficiency achievements demand human, technical, financial and organizational capacities. In particular organization can either raise or overturn barriers to this effort.

9.6.1 Initial Work

Any enduring and serious task developed inside any company must have support from top management. To achieve real and sound results an energy management program needs to be tailor-made, specifically developed for that organization, to match its culture and core business demands. The first step of initial work is to commit the board of directors and top management and it is essential to initiate and support such a program.

Normally the first priority of top management is for production, economic and financial results and indexes related to regulatory issues, like safety or pollution. If top management is not aware of energy efficiency opportunities concerning potential economic return, avoidance of pollution and increase in competitiveness due to cost reduction, no program will begin. Identifying issues that are important to management helps their understanding that it is a business opportunity and helps to obtain that fundamental commitment.

Once this commitment is obtained it is possible to ask top management to announce to the company that an energy efficiency program is underway and that has sponsorship from them. The second step of initial work is to raise personnel awareness for the program, its importance and the high profile that it carries. Now activities start.

9.6.2
Committees

Involving a cross-section of employees from diverse areas across the refinery as soon as the program is started is the best way to spread and promote short- and long-term commitment throughout the organization. So a committee must be formally designated composed of representatives from major refinery activities that use energy, people with knowledge of processes and energy usage, that must bring institutional knowledge and operational experience. A coordinator must be appointed to head the committee and it is recommended that this person holds a managerial position, in order to express that high profile and empower the committee with actual decision capability. The committee should also include a workers' union and safety representative and someone responsible for administrative and marketing tasks. It ought also to be a temporary task, at least for some of the representatives. This shift enforces participation of more staff personnel and spreads the culture.

This committee has an important leadership role in initial tasks as planning, collecting and evaluating opportunities and eventually delegating tasks, establishing deadlines, providing training, guidance and assistance. Another critical job that begins with designation is communicating the importance of energy efficiency to staff.

Even with top management commitment, the energy efficiency committee receives an assignment that occupies the void on organizational chart. It is the space between the boxes, a kind of twilight zone. Although the results of this activity can be profitable with reasonable return rates, the absolute amount of money it provides in general tends to be of a magnitude much smaller than any other core business action produces. And effectively there is a good chance that the money it renders is not actually seen, since it is mostly cost reduction, no influx of currency, so it is not that easy to connect energy efficiency with money for the staff. Another misfit is that any organizational cell has a job description, accountabilities and incentives normally linked to some company priority. Committee participants should be attached to some cell and have their primary obligations and they participate in the group as a kind of part-time job. If this dedication to energy efficiency is not considered within their conventional tasks and accounted for in their regular evaluation there is a good chance that they tend to face the committee as a secondary and low priority activity.

But if the committee has enough support and is able to launch the energy efficiency program it is also building solid bases for its continuity. This may lead to a formal organization, like an energy management or an energy assistant position, really incorporated in the culture as a core business function. Another important legacy of a good committee work is the generation of energy experts and champions who usually are brought from the components of these committees.

Committees are good and can do a great job, especially for the first moves, raising awareness and keeping momentum. But without defined goals, leadership sponsoring and expertise, these groups tend to get a little lost after some time and lose credibility. This is a technological issue and has to be strictly attached to core business. It is essential to have a dedicated team of trained and experienced experts to make it active and reliable.

9.6.3 **Task Forces**

Task forces are very different from committees. Committees are permanent, although participants may vary, but they are all drawn from company employee staff. Their goal is institutional and broadening, maintaining energy efficiency activities going, tracing opportunities, monitoring results, proposing projects and behavioral changes in a continuous workflow. Eventually this committee may be turned into an organizational cell of the company. Task forces on the other hand are by definition transitory, having an initial and final work dates. Members are designated by name and normally they remain the same until the work ends. Their target is usually specific and directly focused, linked to some deadline decision, like a project, use of a certain technology, an opportunity operation or maintenance, a permanent procedure change, audit in a facility etc.

Experience has shown that there are some key aspects for a good task force. First is the task force coordinator, responsible for the group's daily activities, the schedule keeper. It is his responsibility to ensure that all planned actions are accomplished, if milestones are being achieved and whenever they aren't, he must trigger planning and target reviews. Choice for this position is likely to be dictated by the nature of the problem, personal skills and experience added to some seniority and/ or managerial position. It is desirable that this person has negotiating skills and focus, because to comply with a tight schedule, negotiation and flexibility is demanded.

Task force composition, concerning that it is attached to energy efficiency, should be centered on the committee. Some members must be attached to the task force through the whole task development, and because it is time driven, these personnel must have this task as top priority, eventually interrupting their regular activities to give full attention to the task force goal, only resuming normal duties after its completion. But combining them with other technical personnel, who do know operations intimately but aren't directly involved in energy efficiency, can provide breakthrough performance.

There should also be room for part-time members to address certain limited aspects of the project, expert with some specific know-how. Briefly, the task team can be selected to focus on the specific problem and be made up from a mix of in-house personnel and outside specialists.

Management participation is absolutely crucial to task force success and it must be kept fully appraised of group pace. It is a management function to provide fundamental direction, resolve conflicts of priorities and ensure that everything is on schedule. Without management authority and influence little priority is given to the task force and it is unlikely that goals will be achieved. Another activity for the group coordinator is updating management about the task force achievement and asking help and support whenever needed. It is a fundamental interface for success.

Here we reach the point where skills and knowledge become basic to drive the energy efficiency work, no matter what kind of structure is supporting it. It is possible to design a specific job with particular requirements, training and development, and that makes a career.

9.6.4 Leadership

Any successful program needs a leader, a coordinator that who can take responsibility and be granted authority to develop, implement, and maintain the energy efficiency program. It is recommended that this person should be an expert skilled both in engineering and financial principles. Knowledge about company culture and good networking, internally and externally, are desirable. Effective communication at all corporation levels to learn different perspectives and ideas and negotiate. Also valuable are good administrative and project management abilities. He must show pertinacious enthusiasm, willingness to advocate for the cause, tireless commitment and trustworthiness to gain staff respect. This position is usually known as 'energy champion,' a reference figure to the energy efficiency program, empowered to give direction, monitor results and advise management about the program.

Beside him there might be some energy experts who are knowledgeable in specific areas either related to energy consumption by processes, like distillation or catalytic cracking or in energy processes, like combustion or heat transfer. They should act as consultants and trainers in the field of energy efficiency multiplying awareness and commitment from staff. Needless to say, it is mandatory that all of them are active members of the committee; they must lead at least one project or work team and regularly take charge of the committee coordination for training and experience purposes.

The energy champion and experts form the energy efficiency program core group, so a proper and thorough training program must have high priority. Continuous improvement relies mainly on them, so keeping this group updated with state of art technology and best practices is a very suitable way to refresh the program. This education is meant to specialize professionals on energy, and can begin by in-house training and self study about their own processes. Complementary formal education is an option, like short expert courses offered by associations, consultants or universities and comprehensive long-term courses as post graduations and certifications should not be disregarded. Regular participation in external events, like seminars or conferences, helps build knowledge and networking, while providing opportunities for external marketing of their own results to an external public. Subsidiary training on communication, finance, project and management should also be provided.

In smaller companies, the tendency is to turn this task into an additional assignment for an already existing staff function. But with such characteristics and potential results these occupations can easily become full-time jobs especially in large complexes.

What would be this function in an organizational structure? A kind of Energy Program Manager placed like an aide to the top management, or to the production or refinery manager. General assignments for this position are:

- build and lead the committee;
- plan and implement all activities related to energy efficiency;
- collect, organize, and disseminate information;
- delegate tasks and establish deadlines;
- advise top management and get their involvement;
- raise awareness and provide training;
- promote cross functional support;
- ask for investment and activities budgets;
- forecast, report and be accountable for energy efficiency program results.

Eventually, if a more broadening and aggressive action is to be expected this task can be merged with the utility manager, creating an energy manager, who also runs the utility system, like boilers, power generators, water treatment etc.

Whatever way of structuring from an additional assignment to a specific position, once the road to energy efficiency is taken, if this activity is to be valued, all aspects that compose any job position must be considered from planning to performance measurement and recognition. And this measurement is related to accountability for the energy efficiency program results.

9.6.5
Accountability

Considering all that has been exposed so far, if company leaders are really committed to energy efficiency to bind resources as time, talent and money, derived results and compromise with them on a continuous and manageable mood seems fair. Assuming a fundamental principle that energy is a direct and controllable cost that can be monitored and controlled as any other production cost, implies that someone must control it, and be responsible and accountable for these results.

Accountability in energy efficiency has to comply with physical and managerial boundaries, like process units or group of process units for which energy monitoring is unique. Units that retain correlation between them concerning operation might also be an option for some accountability bundling. Managers of these units or units groups are the ones to be responsible and accountable for energy use. This means being accountable for putting forward that energy usage in their control area is within expected energy patterns, that operational crew is trained and actively engaged in the program, that the crew follows standards for each sensible operation, that energy efficiency projects and procedures are on track as expected and scheduled and so forth.

From production plans an energy consumption forecast can be derived and this can become the 'energy budget' for each unit and period. Complying with this budget, like with the financial budget, can be an accountability index that can be easily divided between units and even pieces of equipment. Defining these levels of measurement and the implications for each organizational cell, function or person, it is possible to determine responsibility for each one, targets and capability limits. Supporting functions may also have their accountability measured. Maintenance can be evaluated over expected energy consumption due to standards for equipment efficiency before and after intervention and allowable energy losses related to stoppage time. Engineering function can be measured over accuracy and availability of data and information for optimal operation, installation of projects that allow energy savings etc.

Personnel directly linked to the program as the committee and the experts should be accountable for the feasibility and execution of plans, staff awareness level based on marketing and training, participation in work groups, task forces, projects, audits etc. In summary, they must be responsible for the implementation and success of the program and accountable for its effectiveness.

Management should be accountable for providing conditions to support all these actions, like maintaining and enforcing energy efficiency as a company value, standing for the engaged personnel, supplying money and actually placing energy usage as corporate goal. This apparently simple aspect might be the one isolated issue that may guarantee the actual permanence of the energy efficiency program.

9.6.6 Corporative Goals

Energy savings alone can represent a lot of money and might be a strong incentive for energy efficiency activities and priority. But, that may not be enough for some corporate managers. The chief executive and director's board are held accountable to the owners and shareholders. Their responsibility is to create and grow the company capital value. In the refining business this is accomplished by obtaining revenue from products that exceeds the investment and operational cost of owning and operating it. Any set of equipment or system is an asset that must generate economic return. Earnings gained from the sale of products generated by these assets, the refinery itself, divided by the value of the refinery, gives the rate of return on assets, a key measure by which corporation officers are evaluated. And whatever decision these officers take, they seek investments that lead to the highest achievable return on capital employed allied with fastest possible payback.

Those responsible for the energy efficiency program need to understand how their work supports and is supported by overall corporate goals. The proposed activities to run energy efficiency provide better control over refinery assets that reduces losses and energy consumed per unit of production. This can provide reliability and this can also mean less down-time, which derives more production. So, energy efficiency is not just about reducing costs, it is also achieving greater productivity and revenue.

One approach to insert energy efficiency on the corporation priorities is to relate it to the rate of return on assets. Energy savings contribute in this index either in cost reduction as in revenue rise. This is a simple and rather coherent view and although it fits well with the company's agenda it has the inconvenience of placing energy efficiency at the same level of interest as many other better management efforts. But energy efficiency has always had an extra flavor, surpassing only economics. Relating its benefits to more corporate needs is a good option for differentiating it from others and raises its priority.

The direct benefit of efficiency is reduced energy expenditure. It can be seen as a new capital source for the corporation. These savings can also be enrolled among added shareholder value attributable to the energy efficiency program. Energy savings can simultaneously reduce emissions and water consumption and potentially contribute to reduce environmental compliance costs and mitigate exposure of the refinery to penalties. The routines involved in monitoring can help identify operational abnormalities and contain dangers and threats to life, health and property. It is also important to detach energy efficiency program contribution to health, safety and environmental corporate results.

Connecting energy efficiency results to corporate goals brings many palpable benefits to the program. It places energy culture at the same level of other usual priorities for the refinery manager and all staff. It expresses the commitment of the company to energy efficiency and spreads this statement to all company officers, even those not directly attached to the subject. This eases the work of obtaining all necessary support from corporate level to guarantee program continuity, like financial resources, education and training opportunities, compensation and motivation systems for engaged personnel etc. Incorporating this into a corporate energy policy can be a formal and worthwhile step and, as previously mentioned, may really found the business basis for the program.

9.6.7 Evolutionary Organization

Once energy efficiency is established and raised to a business issue for the refinery, by a managerial or corporate decision and an energy policy is set, organizing this activity is demanding. Some approaches for this organization have already been addressed, like additional tasks to some organization cell or function or the designation of a special advising or aide position to the site management. Nevertheless the committee is always necessary, because of multitask and cross sectional characteristics of the program.

Many refineries and companies transit between some arrangements as these varing in time due to relative energy costs, availability of personnel and momentary corporate priority for the program. The creation of an energy management or an energy efficiency program management is a natural step for many companies. But how can it evolve and what can be the options for this evolution?

Concerning the actual strong link between energy efficiency and environmental, in some companies there has been a move to either create or change this assignment or formal organizational cell into the Health, Safety and Environmental area. If the environmental compliance is an actual concern for the company, this may be a good call, if it really raises attention and priority to the subject. On the other hand this can withdraw responsibility from staff in general, turning energy efficiency into an obligatory issue like safety. This may reduce visibility about economic results, making it loose its rewarding and challenging aspect.

In companies with multiple sites, if an energy efficiency management or coordination is in charge of a broadening program, there is opportunity to build a dedicated organizational cell as an internal energy savings company, called ESCO. The advantages are that the people working on it, have energy efficiency as their sole priority. All resources and activities, like management, training, budgeting, compensation etc circle around the theme. Their revenue comes from energy savings obtained in projects among the various company's locations. If this work is well organized and established there is even opportunity for turning it into a service to be offered externally and charged to third parties, becoming another business branch. Some disadvantages come with this arrangement. Best practice operational procedures usually fall outside the jurisdiction of the ESCO. It also creates a reasonable distance between the service provider and the refinery personnel, especially plant floor. In meantime, if third party services demand begins to rise, priority conflicts between internal and external demands may occur. Anyway, it is a possible evolution that has the great advantage of rising energy efficiency to a real business level for the company.

No recommended or foolproof organization exists, and no matter what structure, refineries achieving the greatest results do have:

- top down commitment to energy efficiency and its continuous improvement;
- an integrated approach across all business aspects;

- a system to track and analyze energy performance;
- a coherent system of performance goals;
- a system to effectively reward energy performance;
- an engaged and empowered staff.

9.7
Future and Environmental Concerns

Energy will always be a critical resource for the operation and profitability of a refinery. Reduction in crude oils sources added to environmental restrictions for emissions over petroleum products, tend to increase costs. And environmental restrictions on refining process may eventually disallow operation of existing refineries in some places.

In order to prevail in the future, the refining industry will have to be sustainable, meaning being simultaneously profitable and environmentally compliant. And the optimized use of energy throughout the refinery complex is the beginning of sustainability. The sustainable refinery of the future will have energy efficiency as a paramount from design and advanced operational and maintenance best practices, taking energy productivity to new levels.

Conventional energy intensive processes, like distillation and catalytic cracking can be upgraded by more efficient and less energy demanding new technologies. Heat should be supplied in common furnaces where all fuels will be burned with minimal excess air. All high temperature demands will share the radiating section of this furnace with maximum heat absorption. Since it is a unique furnace, the overall flue gas flow can be directed to a waste heat boiler that will require minimum supplementary fuel, just for adjustment of steam temperature and generating extra steam flow for balance purposes. Because of this design option, processes will be much closer and a much greater and improved heat recovery will be possible through a wider integration between diverse products and feedstock flows. In these innovative designed heat exchangers, fouling will be essentially eliminated. Placement of flows near each other, gives the opportunity to dispose most of the pumps side by side, enabling a faster and better management of energy sources in use for fluid movement. This energy recycling, allows opportunity for advanced and better cogeneration, reducing water consumption and utilities demand in general to a minimum. Advanced state-of-the-art control systems can adjust the best setting of variables from crude oil choice up to obtaining optimized productivity and sufficient energy to accomplish it.

This vision brings no radical revolution to oil refining to cope with higher standards on energy efficiency, exception for design and management point of view. Needless to say, energy consumption reduction directly implies CO_2 emission reduction, and depending on future developments regarding climate change, energy efficiency must be integrated with other options to comply with deeper restrictions over greenhouse gas emissions regulations. Some revolutionary concepts may be applied. Diversification of load from exclusively crude oil to biomass, use of biomass as fuel and CO_2 capture from flue gas may be options to reduce net emissions.

Energy efficiency is nowadays an important opportunity to reduce costs, increase productivity and reduce pollutant emissions. With a more daring approach to application and investment, well-known energy efficiency technologies and practices can help to match the challenges for a stable, sustainable and environmentally sound future for the refining business.

9.8 Approach and Literature

When writing this chapter I have reviewed extensive literature about this subject. Since it is not a brand new issue, there are many references addressing the technical aspects of this subject in depth, so I decided to report my working experience with energy efficiency from the refinery floor to managerial office. I hope that these technical references and my practical point of view help readers build or improve their energy efficiency efforts.

Further Reading

1 Asociacion regional de empresas de petroleo y gas natural en latinoamérica y el caribe (Regional association of oil and natural gas companies in latin america and the Caribbean, ARPEL) (2002) Environmental Guideline No. 33-2002 Energy Use Monitoring and Tracking.

2 Billege I. (2009) 700 refineries supply oil products to the world. *Nafta*, **60** (7–8), 401–403, http://hrcak.srce.hr/file/65010 (accessed 26 April 2010).

3 Energy Information Administration (2009) Office of Integrated Analysis and Forecasting. U.S. Department of Energy. International Energy Outlook 2009. Washington, DC, http://www.eia.doe.gov/oiaf/ieo/pdf/0484(2009).pdf (accessed 26 April 2010).

4 European Commission, Institute for Prospective Technological Studies, European IPPC Bureau (2009) Integrated Pollution Prevention and Control Reference Document on Best Available Techniques for Energy Efficiency. Sevilla, Spain, http://eippcb.jrc.ec.europa.eu/reference/brefdownload/download_ENE.cfm (accessed 26 April 2010).

5 Gary, J.H., and Handwerk, G.E. (2001) *Petroleum Refining: Technology and Economics*, 4th edn, Marcel Dekker, Inc., New York, NY.

6 IEA Publications (2000) Energy technology and climate change: a call to action, Paris, http://www.iea.org/textbase/nppdf/free/2000/clim2000.pdf (accessed 26 April 2010).

7 IEA Publications (2005) The European Refinery Industry under the EU Emissions Trading Scheme, Competitiveness, Trade Flows and Investment Implications, International Energy Agency, Agence Internationale De L'Energie, IEA Information Paper, Julia Reinaud, http://www.iea.org/papers/2005/IEA_Refinery_Study.pdf (accessed 26 April 2010).

8 Industrial Heating Equipment Association and U.S. Department of Energy Office of Industrial Technologies (2001) Roadmap for Process Heating Technology Priority Research and Development Goals and Near-term, Non-research Goals to Improve Process Heating Technology, http://www1.eere.energy.

gov/industry/bestpractices/pdfs/process_heating_0401.pdf (accessed 26 April 2010).

9 International Petroleum Industry Environmental Conservation Association (2007) Saving Energy in the Oil and Gas Industry. London, www.ipieca.org/activities/general/downloads/Saving_Energy.pdf (accessed 26 April 2010).

10 International Petroleum Industry Environmental Conservation Association (2007) The Oil and Gas Industry and Climate Change. London, http://www.ipieca.org/activities/climate_change/downloads/publications/siaf_climate-change.pdf (accessed 26 April 2010).

11 Kreith, F., and Goswami, D.Y. (2007) *Handbook of Energy Efficiency and Renewable Energy*, CRC Press, Taylor & Francis Group, Boca Raton.

12 Lieberman, N.P., and Lieberman, E.T. (2008) *A Working Guide to Process Equipment*, 3rd edn, The McGraw-Hill Companies, Inc., New York.

13 Ozren, O. (2005) *Oil Refineries in the 21st Century*, Wiley-VCH Verlag GmbH, Weinheim.

14 Pacala, S., and Socolow, R. (2004) Stabilization wedges: solving the climate problem for the next 50 years with current technologies. *Science*, **305**(5686), 968–972.

15 Silla, H. (2003) *Chemical Process Engineering Design and Economics*, Marcel Dekker, Inc., New York.

16 Spoor, R.M. (2008) Low-carbon Refinery: Dream or Reality Hydrocarbon Processing, pp. 113–117.

17 Turner, W.C., and Doty, S. (2007) *Energy Management Handbook*, The Fairmont Press, Inc., Lilburn, GA.

18 U.S. Environmental Protection Agency (2007) Energy Trends in Selected Manufacturing Sectors: Opportunities and Challenges for Environmentally Preferable Energy Outcomes. Final Report. ICF International, Fairfax, VA, http://www.epa.gov/ispd/pdf/energy/report.pdf (accessed 26 April 2010).

19 United States. Department of Energy Industrial Technologies Program (2004) Energy Use, Loss, and Opportunities Analysis, U.S. Manufacturing and Mining, http://www1.eere.energy.gov/industry/intensiveprocesses/pdfs/energy_use_loss_opportunities_analysis.pdf (accessed 26 April 2010).

20 United States. Department of Energy Industrial Technologies Program (2004) Improving Steam System Performance: A Sourcebook for Industry, U.S. Manufacturing & Mining, http://www1.eere.energy.gov/industry/bestpractices/pdfs/steamsourcebook.pdf (accessed 26 April 2010).

21 United States. Department of Energy Industrial Technologies Program (2002) Steam System Opportunity Assessment for the Pulp and Paper, Chemical Manufacturing, and Petroleum Refining Industries, http://www1.eere.energy.gov/industry/bestpractices/pdfs/steam_assess_mainreport.pdf (accessed 26 April 2010).

22 Michael, W. (2008) Estimation of Energy Efficiencies of U.S. Petroleum Refineries. Center for Transportation Research Argonne National Laboratory, http://www.transportation.anl.gov/modeling_simulation/GREET/pdfs/energy_eff_petroleum_refineries-03-08.pdf (accessed 26 April 2010).

23 World Bank (1998) Greenhouse Gas Assessment Handbook, A Practical Guidance Document for the Assessment of Project-level Greenhouse Gas Emissions, Climate Change Series, PAPER NO. 064, http://www-wds.worldbank.org/external/default/WDSContentServer/WDSP/IB/2002/09/07/000094946_02081604154234/Rendered/PDF/multi0page.pdf (accessed 26 April 2010).

24 Worrell, E., and Galitsky, C. (2005) Energy Efficiency Improvement and Cost Saving Opportunities for Petroleum Refineries, Environmental Energy Technologies Division, Berkely, http://www.energystar.gov/ia/business/industry/ES_Petroleum_Energy_Guide.pdf (accessed 26 April 2010).

10
Energy Efficient Utility Generation and Distribution

Carlos Augusto Arentz Pereira

Any industrial plant, like a refinery, demands various energy forms to accomplish its production process. They can be heat, power and auxiliary materials. These streams are called individually utility or utilities, when a set is considered. And utility or utilities is also a synonym of the service provider for that stream. This text addresses utility or utilities inside an industrial plant, but the approach and recommendations apply equally to third parties service providers.

10.1
Characteristics

Functions in industries can be broken down into core and supporting roles. Core function is the actual transformation and production, which is the business of the plant. However, a real industry cannot operate by performing only core functions. It requires an infrastructure to supply resources, known as utilities, like steam; process, service and cooling water; instrument, service and cooling air; fuels; electricity; wastewater treatment; air emissions treatment; any heating or cooling service like air conditioning etc. Here, focus is on services that interact direct or indirectly with the production function, influencing and being influenced by it. To understand how this influence occurs, it is necessary to analyze the form in which these utilities are employed by the production process.

10.1.1
Use of Utilities

It is possible to classify the demand utilities in an industry in some basic forms:

- **Process support** in which the utility intrinsically participates in the production process, mixing physically with raw material and products. Its presence impacts on the design features of process equipment. Its quality influences the final quality of products and its absence may prevent the performance of the process. In the majority of these uses, flows are measured. Examples: Steam for

Managing CO_2 Emissions in the Chemical Industry. Edited by Leimkühler

ISBN: 978-3-527-32659-4

stripping, water for dissolution, air for particles transportation, electricity for power etc.

- **Services** in which the utility does not participate directly in the process and its use is just potential. It brings no physical implications for most process equipment, and its quality hardly affects the products. Its absence is likely to bring some inconvenience to production and it is typically not measured. Examples: Water and air for cleaning, steam for smothering, cleaning and displacement.
- **Data transport** in which the utility participates externally to the production process. It determines certain design features of some process equipment and faulty quality is likely to cause disruption to production in the medium to long term, depending on how this inadequacy occurs. Its absence may prevent the performance of the process and can bring serious consequences. It is typically not measured. Examples: Air and electricity for instruments.
- **Energy conversion and exchange** in which the utility participates externally to the process. Some characteristics of process equipment are determined by these utilities and lack of quality is likely to cause disruption of production and eventually serious consequences in the short term, depending on how this inadequacy occurs. Its absence may prevent the performance of the process and can bring serious consequences, especially if simultaneous faults of other utilities with similar characteristics happen. It is usually measured. Examples: Steam for power drive and heating, electricity for power drive and lighting, cooling water.

Another common feature is that in the first three categories utilities are physically consumed, without return to the system, thus becoming typical process wastes that often require treatment before final disposal. Only in the last case, can potential recycling or return to the generation system occur.

10.1.2 Quality

The concept of quality for a utility flux is related to final user requirements, generation capability, and distribution characteristics. Properties that can be used as quality measurements, might be standard pressure and temperature for steam, maximum allowable concentration of contaminants for industrial water, maximum allowable presence of moisture and oil for instrument air, correct tension, current and frequency for electricity and so on. Attention must be paid to quality from design to operation of generating and consuming equipment, passing through the distribution system. The quality needed for final usage must be embedded in its generation and preserved on distribution.

Analyzing quality from the point of view of utility use, as described above, it is possible to conclude that the quality concept can be quite flexible. Considering the *data transport* category, non-compliance of moisture or oil measurements in instrument air can bring serious disruption to the process control in a very short time. On the other hand, in the *service* category, these same parameters do not serve as quality measures to service air, used for displacement purposes. In *process support*, taking steam stripping as an example, despite the minimum pressure needed to overcome internal pressure, small variances in temperature influence the energy balance of the tower, but don't cause major disturbances in the process.

In the *energy conversion and exchange* category, faulty quality parameters can induce dubious situations, concerning final usage performance. Lack of tension may not allow an electric motor driven pump to run, but a little lower pressure or temperature on a steam turbine, driving a pump for the same service, can make it go. The difference between these two relies on the timing for potential consequences. In the case of faulty tension, the process does not run and losses are instantaneous. In the second case, the process runs immediately, but energy consumption may be higher than projected or expected, because of utility conditions beyond their specifications. Of course, instant energy productivity is lower and costs higher, resulting in a less competitive operation. In the long term, however, the faulty conditions tend to accelerate equipment wear and may cause future process disruption. In the same example, feeding steam out of specifications to a turbine may provoke erosion, equipment performance reduction, and eventually its destruction, at least of some fundamental parts for its operation. That can reduce life expectancy and threaten projected investment return rates and payback.

A utility quality can be either a simple or complex concept, but the way it is accepted and used, can help determine costs and productivity.

10.1.3
Energy Exchange

Comparing all the categories described here, in the utility demand of an industry, like a refinery, it is easy to conclude that utilities are basically the currency for energy exchange. And considering the idea suggested in Chapter 9, about a refinery being a complicated thermal machine, utilities are the fluids that perform the energy cycle that keep it running.

This might be considered an exaggeration, but concerning how basic design is done, in essence, this much is true. Energy in all forms: fuels, steam, electricity, water are to be the blood and body fluids, and utility equipment: boilers, cooling towers, power generators etc, are the organs of this being, the plant. They might not be the main purpose of existence, but without their support, there is no life.

And the better this exchange is performed, the longer the life span is to be expected, the more useful the equipment will be and the fewer disruptions will happen. The overall venture is likely to be less costly and more profitable. Understanding and accepting this role can influence all costs surrounding an industrial plant very positively, from financial to environmental.

10.1.4
Investment and Operational Costs

Utility costs, either investment or operational, are among those expenses that actually no one wants to afford. The reasons for that rely on the simple fact that although utilities systems are vital for any industry, their usages, in most of the cases, just represent costs in the company balance. The same thing happens in households, commercial users etc. The only exception is the investment a utility company does, because in this case, utilities production is the very product of the company, its revenue source.

No matter what priority position any utility product or system may have in a company, the smaller the investment or operational cost, the better for results. On average, utility investments may comprise around 25% of the production facility initial capital cost, while utility operational costs can account for approximately one third of energy-related expenses. And the conception of production plant from the drawing-board can significantly increase or reduce the share of utility costs in the whole life span of the industry.

If the process design is more heat integrated, meaning that input and output process flows were conceived to have an optimized heat exchanging network, utilities demands tend to be minimal, either hot or cold. Generation and distribution requirements for these utilities are diminished, implying that utility investment and operational costs share tends to be reduced, in the overall industry capital expenditure. The natural consequence is a general reduction in utility costs. Taking steam demand as an example, less steam also signifies less water demand, fewer chemicals, less piping, smaller boilers and decrease in size of all ancillary equipment and systems. Smaller size means less area for equipment, pipes etc, suggesting that more space should be available for core processes. Needless to say that with smaller utility systems, their operation becomes simpler and resource requirements, like personnel and especially money, in general diminish. And these spare resources can be directed towards core business production.

It is important to emphasize one point mentioned at the beginning. Utilities systems are a priority only in a utility company, where it is the core business. In all others, it can be a sensitive part of the process, but usually doesn't pay the bills. If it is working and supplying demand, it is just fulfilling its purpose; if not, it is impairing the plant's ability to carry out its mission, jeopardizing profitability and subsistence. So, it is possible to conclude that utilities are among those basic resources that hardly anyone pays attention to, except when a shortage happens. But, if appropriately managed, they can help make the difference to results, through energy efficiency.

10.1.5
Energy Efficiency

Utilities are almost a part of the scenery for the production; while they are there playing their role, no big deal. But any failure can make the spectacle crumble. No extra value? Consider its fundamental and biggest supporting function for process, which is to be the currency of energy conversion and exchange. In this aspect, utility generation and distribution is crucial for overall energy consumption in the industry. And as mentioned in Section 10.1.4, all energy savings or consumption reductions can end up reducing immediate costs and in the long term, avoiding additional investments in enlargement of utility systems. All these contribute to current and long term economical results.

Energy efficient operation and design is the core business of a well-managed utility system. It makes the difference from just complying with its mission to excelling and contributing significantly to the industry. To fulfill this task, understanding the utility system, its components and interrelation is basic.

10.2 Common Utilities

Since most industrial applications demand heating, cooling and power, this section lists those common utilities in use which supply these requirements.

10.2.1 Steam

A good thermal fluid must have some basic features such as high energy storage capacity and ease of transferring its energy content in diverse conditions. Many commercial transfer fluids have these qualities. But steam adds to these attributes the ability of transferring heat at constant temperature and transforming heat into work. The first attribute allows the use of reduced size heaters compared with heating fluids whose temperature varies along the exchange equipment. The second feature increases the potential energy use of steam, allowing the same stream to release power and heat. So steam can be used in steam turbines to generate power and directly drive compressors or pumps or in steam ejectors to generate vacuum.

Steam can be readily adapted to a wide range of temperatures by pressure adjustment, concerning the use of saturated steam. It is easy to distribute and control and its heat releasing ability can be used in cascade from higher to lower temperature. After its heat content is wasted, condensate can be recycled to the boiler and steam regenerated. It can be generated with high energy efficiency in boilers, either fired or wasted heat recovery ones.

In contrast to commercial thermal fluids, steam is clean, odorless, tasteless and nontoxic. It requires common and average cost construction materials for distribution, mainly steel pipes and accessories. Its use is subject to some regular losses but they are easily replaced and their overall cost is affordable. And, water is a raw material of easy supply and low cost, so far, in most parts of the world.

10.2.2
Electrical Power

Electricity is the most versatile energy form available for use inside an industry. It can be used for heating, lighting and power, disregarding the generation source or means of transmission. Compared with other utilities, its generation can be considered as simple as steam generation, but its distribution is more complex.

It is usually generated as alternate current electricity, typically at high voltage. The generated power is transmitted to final users through a distribution network, which is composed of transformers, distribution lines and control equipment. Depending on demand purpose and requirements, the necessary voltage levels are adjusted at substations near to the consumption site. There is no significant conceptual difference between a transmission line and a distribution line except for voltage level and power handling capability. Transmission lines operate at very high voltages and have the capacity of carrying large quantities of electricity over great distances. Distribution lines have limited power capacity, handling lower voltages over shorter distances. Inside an industry, quite often, the distribution approach is the one that applies.

Depending on the production process, electricity should be the main energy source, being generally used for all energy needs. Although lighting systems are used everywhere and nowadays, almost all daytime, independently of daylight availability, the majority of electrical demand is for power, to run electric motors.

Concerning installation, it is the utility with highest complexity level, demanding a lot of control and safety equipment. It is the sole utility whose faults are immediately sensed, and also the one that is usually more dependent on the availability of other utility systems. Disregarding fuels like natural gas, it is the easiest utility to be purchased from third party suppliers. Being the simplest, more commonly available and flexible energy, it is applied in general manner, and its demand is always growing.

10.2.3
Water

Water is the universal solvent and always has some chemicals, gases or minerals dissolved in it, and the so-called 'pure water' actually doesn't exist. As a matter of fact, its characteristic taste is provided by the presence of these contaminants. All these impurities bring, eventually, some undesirable additional property to water, like minerals as calcium or magnesium that cause hardness; iron, that gives it color; organic materials as biomass or oil can increase potential presence of microorganisms; atmospheric gases, specially oxygen and carbonic oxide, increase its corrosive power etc.

Independent of its state of purity, water is one of the most active and aggressive solvents of nature. In so called meantime, almost everything will dissolve in water to some extent, depending on the quantity of material exposed, until solution saturation, when no higher concentration of solids in water can be reached. This solvent property allied with its extremely high capacity to absorb and transport heat, without significant temperature variations, and as mentioned in Section 10.2.1 on steam, the ability to convert energy from heat-to-power, makes water an indispensable utility.

Water is used for many purposes in processes, like equipment cooling, maintenance cleaning, scrubbing gases for air pollution control and as the steam cycle working fluid. To accomplish all these objectives, despite its corrosive behavior, especially to steel structures, it must be properly and continuously supplied and conditioned to use. The usual source for industrial water is the public water distribution system or a captive ground fount, like a lake or a river. Growing environmental restrictions and increasing costs for wastewater disposal pushes a trend to recycle water effluents, using it as make-up for cooling tower systems. Concerning the conditioning, this means water treatment and it must comply with usage required conditions.

10.2.3.1 **Industrial Water**

For industrial use, water has to be treated to reduce its corrosivity. This treatment intends to prolong equipment usable life, reduce maintenance interventions and costs, either by repair or replacement, while maintaining the reliability and efficiency of this equipment. These targets can be achieved by treating water to prevent scaling and fouling, controlling corrosion and avoiding eventual microbiological growth. All water demands on the plant for service, cooling or steam generation must be submitted to this first stage treatment.

It is usually called external treatment and aims to reduce the concentration of suspended and dissolved solids, minimize turbidity, color and remove organic material. Water produced in this step is commonly known either as industry or service water. Service water is used as pump and instrument seal water, fire water, sanitary water, make-up to ash and flue gas scrubbing systems and all general purposes. It can be also used for potable water, after chlorination to conform with drinking water standards.

10.2.3.2 **Cooling Water**

Cooling water can be provided either by a once-through or a closed-loop system. Once-through systems normally take fresh and cool water from a river, lake or seawater and pump it directly to heat exchangers. This stream is just filtered on trash racks and screens, before circulating water pumps suction side, to remove suspended bulky materials and regularly, no additional treatment for suspended solids is required. Eventually, if the presence of organic matter and microorganisms is significant, that can be mitigated by shock chlorination. This kind of cooling utility system is subject to intense fouling, erosion and corrosion, because it is not treated, and special materials, like copper alloys, ferritic stainless steels and titanium, can be used for heat exchangers. Environmental risks and regulation are making this practice almost outdated and forbidden in most parts of the world.

Closed-loop water cooling requires a much more complex treatment. Water evaporation is demanded to recover low temperature, so hot cooling water has to be recirculated across towers, producing high concentrations of dissolved solids. When these materials reach solubility limits and concentration levels are elevated, scaling, fouling and corrosion processes, over cooling circuit components, increase substantially.

A specific internal water treatment is required. Chemical products must be injected to control higher solids concentrations effects and biological growth. Blowdown is necessary to control the concentration of dissolved solids within prescribed permitted limits, to control suspended solids in circulating water and minimize sediments accumulation in the cooling tower basin.

Treatment characteristics can be classified as cleaning or scaling. In cleaning treatments, general parameters like pH are maintained at a level in which heat exchanging surfaces are kept clean, in a kind of constant mild pickling. On the other hand, scaling treatment is less aggressive to metal surfaces, allowing some rusting and fouling to form. Control limits depend on make-up water quality, which is basically service or industrial water, and internal treatment option.

Concerning energy efficiency, the choice of treatment can bring about different outcomes. In the first option, better heat transfer coefficients are achieved, due to surface smoothness and reduced fouling, but it may imply faster equipment degradation and eventual sudden stoppages occasioned by leaks. Selection of thicker piping for this condition may reduce overall heat transfer coefficients and operational energy costs, while increasing initial capital investment. The scaling option can proportionate longer life to pieces of equipment, but reduce heat transfer. Fouling allowance can determine the need of additional spare heating transfer area in exchangers, eventual maintenance cleaning might be needed. Balancing these options depends on the financial availability for investments versus operational and ongoing maintenance costs.

10.2.3.3 Boiler Water

Water is the working fluid for steam generation and boilers, which are among the most costly and vital pieces of equipment in an industrial complex. Proper treatment and conditioning of water can increase boiler performance, energy efficiency, maintain production capability and reduce operating costs while extending its operational life. Boiler water treatment avoids scaling and corrosion, insuring safe and reliable operation. Lack of appropriate treatment can cause a series of problems to develop, ranging from loss of productivity, accelerated wear and, in extreme events, its destruction.

Boilers are the heart of the steam system and ultimately, they receive all residual contaminants that remain in feed water, after external treatment. And when water enters the boiler, elevated temperatures and pressures cause water contaminants to behave differently. Like in cooling water, since there is a continuous water make-up to compensate evaporation, a concentration cycle occurs in boiler water. Under existing conditions inside the boiler, most water-soluble components reach their solubility limit and may leave solution as particulate solids, in crystallized or amorphous forms, developing scale and deposits over boiler heat transferring surfaces. These deposits have insulation properties and can impair heat transfer.

Large amounts of deposits along the boiler can reduce heat transfer and boiler efficiency significantly and, in the meantime, they may provoke corrosion and eventually tube failure by overheating at a point. The presence of dissolved gases is also common when water enters a boiler. Certain gases in solution, like CO_2 and O_2, are released when heated and react with water to form carbonic acid (H_2CO_3), greatly increasing corrosion.

In order to control these corroding processes, two types of boiler water treatment are necessary: internal and external. External boiler water treatment is usually done immediately after industrial water treatment, deeper removal of dissolved solids, particularly major participants in scale formation, like calcium and magnesium, and to some extent silica. Silica is a chemical compound that also forms scale and may bring specific problems to high pressure steam turbines and superheating areas of the boiler. Since no treatment can completely remove all contaminants and their amount keeps growing, because of the concentration cycle, so supplementary internal treatment is needed. This is done by addition of chemicals that convert scale-forming compounds into a sludge, which can be withdrawn by bottom purge or blowdown. Chemicals that are generally used in this treatment, are sodium salts of carbonate, aluminate, phosphate, tri- and polyphosphate, sulfite and special compounds as polymers. Materials of vegetable and animal origin can be also injected to remove scale, but this is an outdated practice. Application of such methods relies on some conditions, like boiler design type, steam pressure; feed water contaminant concentration, and maximum allowable operating concentration in boiler water. Usually, depending on the quality of the make-up water, low blowdown flow rates and periodicity indicate well-treated boilers. Elevated blowdown rate is usually uneconomical, because of heat and water losses.

Silica removal is supplemented on high pressure water-tube boiler designs by installation of top drum internal accessories, and a continuous skimming or surface blowdown from this drum is necessary. Heat recovery from continuous blowdown is an energy efficiency option.

Removal of dissolved gases is accomplished by external and internal treatments. Externally to the boiler, dissolved O_2 and CO_2 are expelled by preheating the feed water before it enters the boiler. This process, called mechanical deaeration, is normally performed under vacuum, in a vessel external to the boiler. Boiler feed water enters this drum passing a kind of liquid–vapor contact element, like a small distillation tower, where it crosses counter current with steam. Vacuum and heat reduces the solubility of gases, which are vented to the atmosphere. One good energy efficiency practice here is to use low pressure steam, especially if it is available in excess. Deaeration operating pressure can be specified to use this steam and help steam balance, by absorbing a stream in surplus, hence contributing to fuel savings. Efficient mechanical deaerators can reduce oxygen concentration to very low levels, but even trace amounts can cause corrosion. Consequently, supplementary removal is necessary, and this is done by addition of a chemical oxygen scavenger, such as sodium sulfite or hydrazine. When reacting with oxygen, sodium sulfite forms soluble sodium sulfate, which increases total dissolved solids in boiler water, increasing blowdown and internal treatment requirements.

Hydrazine reacts to form nitrogen and water that do not increase solid concentration and this is of mandatory use in high pressure boilers which require low solids. The approach presented here for internal boiler water treatment is traditional and well-established by use for many years. But technology is always advancing and chemical companies and boiler manufacturers update their products constantly. Disregarding the specific chemical products mentioned here, all water quality requirements remain, whatever newer product may be considered.

Well-managed boiler water quality is critical for an energy efficient utility system. This parameter must be considered in boiler design and selection. Maintaining its expected specifications can propitiate conditions to achieve maximum boiler efficiency and since the main energy conversion happens in the boiler, this efficiency can be propagated throughout the whole plant.

10.2.3.4 **Condensate**

After its energy content is exhausted, steam returns to the liquid phase. Although at lower temperatures, relative to steam, this condensate still contains a reasonable amount of energy, almost 20% of saturated steam at same pressure. It is actually distilled water at a temperature higher than ambient, almost free of dissolved solids. Returning it to the boiler, relieves a sequence of water treatment duties, while reducing boiler water blowdown and loss of energy. Lost of condensate implies that cold treated water has to be fed to the boiler to accomplish material balance. This means more make-up water with additional treatment costs and additional fuel to heat it up to boiler water temperature. And this colder feed water tends to reduce boiler steam output and energy efficiency. Collecting and returning condensate to the boiler is another fundamental efficiency measure and is the basis of steam usage option of energy and water reuse. It may seem that condensate might be the ideal boiler water, but not quite.

Condensate may be contaminated due to steam application and physical conditions. If in steam usage, condensate contamination is expected and probable, continuous treatment must be considered and a reliable quality control is essential to indicate dangerous contamination levels as fast as possible. Parameters like conductivity or turbidity might be sufficient. For example, the presence of oil and greases from heat exchanger leaks in a refinery, demands mechanical filtration in paper cartridge filters, and oil surges that cannot be handled by existingt filters, require immediate discarding. Any oil injection to boiler water may cause immediate fouling due to carbonization of these materials on heat-transferring water-side surfaces and this can provoke immediate overheating and material failure.

Another issue is related to the physical difference between steam and water. When steam condenses, the flow volume shrinks and this causes a depression that may allow ambient gases to enter the pipe and dissolve in condensate. The same potential corrosion problems, related to specific dissolved gases, appear. Carbon dioxide reduces the pH of the condensate causing acid attack, while oxygen attacks metal directly. Before the condensate is returned to the boiler, it is fed to a deaeration device but in the collecting pipes some treatment must be provided to reduce corrosion. Corrosion of condensers, steam traps and condensate piping is common. Adding chemicals, such as neutralizing amines, helps to keep condensate pH high, but can only protect the system against acid attack from CO_2. Most commonly used are morpholine and cyclohexylamine.

Film amines can protect from oxygen attack. These compounds form a very thin organic material film over metal surfaces, blocking oxygen, but they have little effect on condensate pH. Heat transfer is minimally affected because the film is thin. Commonly used amines are dodecylamine and octadecylamine.

Since condensate systems tend to be quite large, using a mixture of amines, film amines with faster and slower condensing ones, improves equipment protection throughout the system. Chemical and physical properties of amines should be chosen depending on system length, to select the best mix. Nevertheless, attack remains and condensate lines and accessories usually present undissolved corrosion products that can be removed by an activated carbon or coke filter.

Just to emphasize that condensate with high contamination risk must not return to the boiler, especially if it contains significant amounts of hydrocarbons or any organic products, acids or caustics, seawater etc. Injection of any of these products into the boiler may cause serious accidents, boiler damage and eventual production disruption. But, depending on the contaminant, condensate may be reused as reasonably clean water, for a number of services.

Condensate is a valuable resource and even the recovery of small quantities is often economically justifiable. But how much condensate return is expected to be a good amount? Huge industrial plants, with long distance between systems and a wide variety of steam usages, tend to have lower recovery, while utility companies must have a high recovery rate as business standard and design parameter. In an average industry, like a refinery or petrochemical complex, 70% of steam demand being recovered as condensate is a very good amount. But even in these places, 30% or less, whatever the reason, is pretty awful. Condensate collection is rarely metered, and this rate is measured by water and steam balance and by itself can be a good overall measurement of energy efficiency performance for the industry.

10.2.4
Air

Air should be by far the cheapest available resource on earth; it is easily grabbed from the atmosphere, and can be used either as a raw material source or as a dump. In some instances it is like water, but depending in which part of the world, water may not be available cost-free, while air is everywhere. It is composed mainly of oxygen (21%) and nitrogen (78%), with very small quantities of other gases: some argon, other rare gases and trace elements. Among these are, water as moisture, carbon dioxide, particulates, and nitrogen oxides, that are important to determine eventual treatment requirements depending on air usage. Since everything is immersed in the atmosphere, air adjoins all surfaces and due to the low thermal conductivity of air, this air film tends to reduce heat transfer capacity. On the other hand, if it is blown over a surface like any fluid, its velocity provides convection ability and promotes heat transfer and cooling.

10.2.4.1 **Cooling**

Air cooling is an option, especially for the process industry, installed in locations where water availability is constrained and it may have significant economic advantage over conventional water cooling.

Employing air cooling implies using air cooler exchangers instead of the usual shell and tube heat exchanger. Air coolers can be either induced or forced draft, the latter being one of the most common arrangements. It consists of several heat transfer sections of finned tubes, where hot fluid passes, mounted on a frame. A fan, located either above or below the tube section induces or forces air through it.

The heat transfer coefficient of an air cooler is relatively smaller than the average water cooled exchanger. To obtain the same heat load with lower heat transfer coefficients large transfer surfaces are demanded and since the pressure drop through the finned tube sections has to be small, large areas for installation are required. Performance of air cooling is also greatly dependent on ambient temperature. At operating temperatures higher than design conditions, cooling efficiency drops, while for very cold climates, air temperature passing the exchanger has to be controlled, to prevent the temperature of the fluid being cooled, from falling below freezing or pouring point. Air flow control can be obtained by changing fan pitch. For high temperature ambient conditions, humidification can help, while for low temperature, recirculation of air can be used, depending on design.

Considering all these points, air coolers capital cost can be twice the cost of water coolers for the same service. But this option may have a much lower environmental impact, considering overall capital and operational costs, by the avoidance of water use and disposal. It may be an interesting and feasible alternative for once-through water cooling systems. As a matter of fact, cooling air may not be considered a utility itself, but an option to reduce or avoid usage of water.

10.2.4.2 **Instrument Air**

Instrument air is the utility that guarantees that the plant is in control. It is used as data transport for control valves because it supplies the driving force for activating these valves. It is compressed air produced by a separate system composed of air compressors, after coolers, receivers, air dryers, air storage tanks and supply lines. For reliability reasons there should be multiple and redundant compressor units. And for this use, the air supply must be of the highest possible quality, free of oil and humidity to allow proper function of the instruments. It might be the unique utility that impairs all other utilities production and the core process itself.

Just one general recommendation applies here. This system must not be allowed to fail, because if it does, all efficiency efforts would have been in vain. This should be the last place in the plant to search for additional energy savings. All redundancies are justifiable.

10.2.4.3 Service Air

Unlike instrument air, service air usage is quite similar to service water. It can be used for cleaning, displacement, pneumatic transportation of particulate materials like catalysts or powdered chemicals, and to drive pneumatic tools. Eventually in plants where the presence of explosive gases can occur, it can be used to pressurize ambients and equipment like electrical panels to prevent these explosive gases from entering.

Quality requirements for this stream are not tight, and normally the compressing process diminishes moisture to acceptable levels for final use. It is only compressed air, but its use at large should be discouraged and inappropriate use restrained. Compressed air can be one of the most expensive sources of power in the industry.

Many efficiency opportunities are present, but one significant parameter can greatly impact air compressor efficiency and that is intake air temperature. Much less energy is necessary to compress cool air, than that required to compress hot air. Reducing inflow air temperature by moving compressor intake to outer and higher spot may reduce dramatically compression energy.

10.3 Generating Systems

After investigating the usual utilities and assuming the concept that their main purpose is to convert and exchange energy in diverse forms to support the core plant process, it is essential to take a closer look at utility systems that deal with power.

10.3.1 Power Cycles

In a simple approach, industry energy requirements can be summarized to heat and power, and as mentioned in the beginning of this chapter, they can be supplied by an external third party, a utility provider, that can sell them separately in various and suitable forms. This approach can simplify tremendously energy management of the plant, allowing owners to concentrate on the core business, but it may imply a huge lost opportunity for energy efficiency.

Since heat and power are merely diverse energy forms, it is possible to produce them simultaneously in a consolidated system, know as cogeneration. This application can significantly increase overall energy productivity and efficiency by boosting the actual amount of useful energy. To select the proper cogeneration system, a heat and power demand map of the industry is needed, describing heat and power load profiles, for different periods and typical production conditions. Information on process waste heat recovery potential also has to be considered, as well as fuel and purchased power, availability and costs. Water availability and environmental restrictions, either for liquid effluents and air emissions, must be taken into account. This data allows evaluation of economical feasibility of various cogeneration cycles.

A cogeneration or power cycle consists of some basic individual components. They are an energy supply, which in the average industry is a fuel burner, a primary driver (a heat engine) and a heat sink, configured into an integrated system. Depending on the type of cycle, the energy supply and the primary driver can be either one single piece of equipment or a set, while the heat sink can be one single piece or many different pieces of equipment.

The primary driver may be internal combustion engines, gas turbines, a set of boilers and steam turbines, micro turbines, fuel cells and others. They may use a wide variety of fuels, including natural gas, coal, oil, biomass, hydrogen and other alternative fuels to produce power. Commonly, the mechanical energy generated in this primary driver is most likely used to drive a generator and produce electricity, but it can also drive any rotating equipment such as compressors, pumps and fans. Thermal energy leaving the system can be used directly in process applications or indirectly to produce steam, hot water, hot air or chilled water for process cooling.

Ideally, the most energy efficient path is to burn fuel using the highest possible temperature to convert chemical energy into mechanical energy in a gas turbine, an internal combustion engine or a back pressure steam turbine, using subsequently, relatively lower temperature waste heat from the primary driver to match process heat demands. But the decision to follow this ideal path depends on heat and power proportion and investment in the power cycle. Many thermodynamic cycles may be employed in this task, but here the focus will be on the most commonly used ones, Rankine and Brayton and their combination.

The Rankine cycle may use a wide variety of fuels like coal, oil, gas, biomass or nuclear power as the high temperature source, but thermodynamic operation, independent of the heat source, is fairly constant. It produces work by isentropic expansion of high pressure fluid like many other cycles. The working fluid is water or better, steam. Fuel is burned in a boiler, a furnace where heat is released to be transferred to pressurized water contained within a steel structure, either tubes or a drum. The basic premise is that it is easier to make high pressure steam, starting with high pressure water and then heat this water at a constant pressure. Water is basically an incompressible liquid and little energy is needed to compress it to high pressures. The process is controlled simply by steam pressure. Steam generated is expanded in a steam turbine, which usually drives an electrical generator (Figures 10.1 and 10.2). The Rankine steam power cycle is that most commonly used at power plants all over the world and can be considered the foundation of big power generation industry.

While the Rankine cycle uses water, constantly condensing and evaporating, the Brayton cycle is an all gas cycle, using air and combustion gases directly as working fluids. Its primary driver is a single set of air compressor and combustion gas turbine that produces mechanical energy by isentropic expansion of hot flue gas. These turbines operate at temperatures approaching flame temperature around

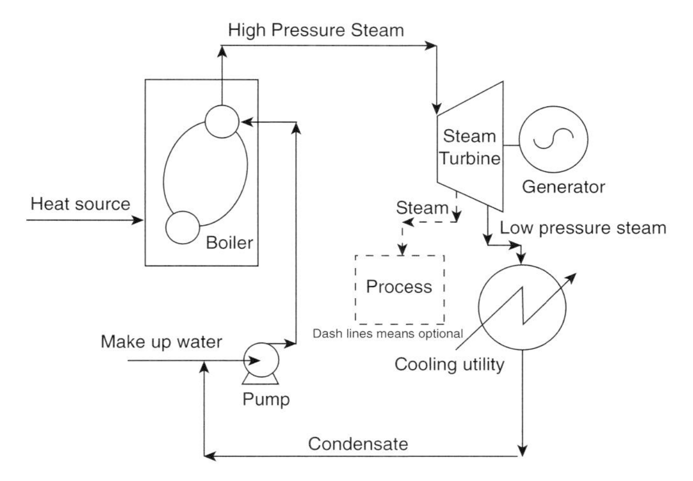

Figure 10.1 Schematic Rankine cycle.

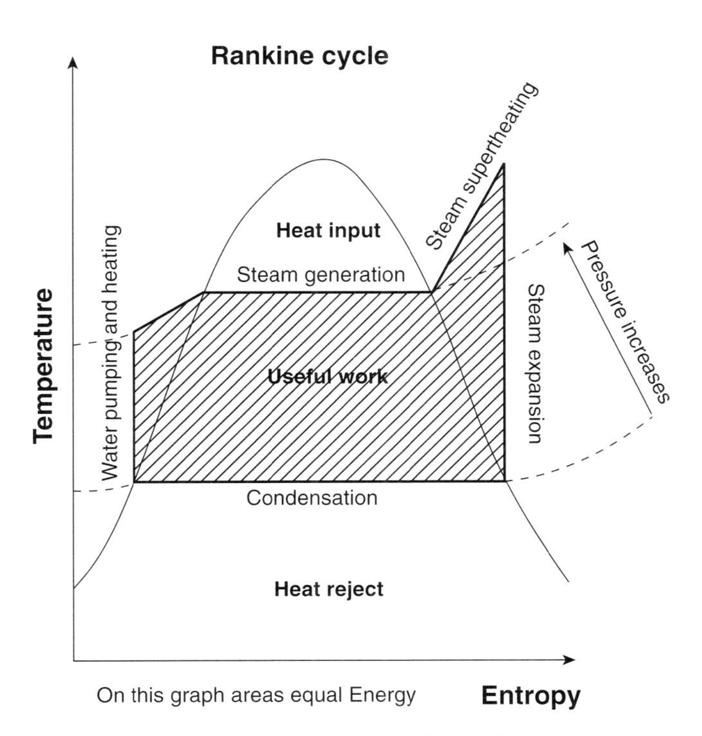

Figure 10.2 Temperature vs entropy for Rankine cycle.

1300 °C, much higher than inlet temperatures of steam turbines in Rankine cycles (lower than 650 °C). Although this difference might suggest that the Brayton cycle would have a much higher thermodynamic efficiency than the Rankine cycle, the Brayton cycle also wastes energy at a much higher exhaust temperature than the Rankine cycle. Furthermore, since the Brayton cycle uses two mechanical devices, an air compressor and gas turbine, it presents more mechanical irreversibilities, degrading thermodynamic efficiency, which means that it can only be slightly better than the best equivalent Rankine cycle (Figures 10.3 and 10.4).

The Brayton cycle can use only some fuels, mainly gases or light volatile liquids that vaporize fast. The most used fuels are natural gas, residual gases, kerosene and diesel oil. Because of these fuel restrictions, it is less polluting than an equivalent Rankine cycle. On the other hand, while the average Rankine cycle can burn almost any fuel without much preparation, a Brayton cycle may even require previous fuel treatment.

Some other different features can be pointed between Rankine and Brayton cycles, like the need of water treatment for Rankine, and the heat-to-power ratio. Rankine mandatorily produces relatively more heat for each energy unit input than Brayton. The Rankine cycle has unmatched variability of operational load conditions, due to fuel flexibility and the use of the steam turbine. Many optimal operational conditions can be achieved to comply with process heat and power demand, while Brayton, although offering more power operates in a restricted heat-to-power ratio range.

Rankine cycle plants may have a longer life span than Brayton, and all parts are reliable. A well-managed system requires relatively little maintenance. Installation

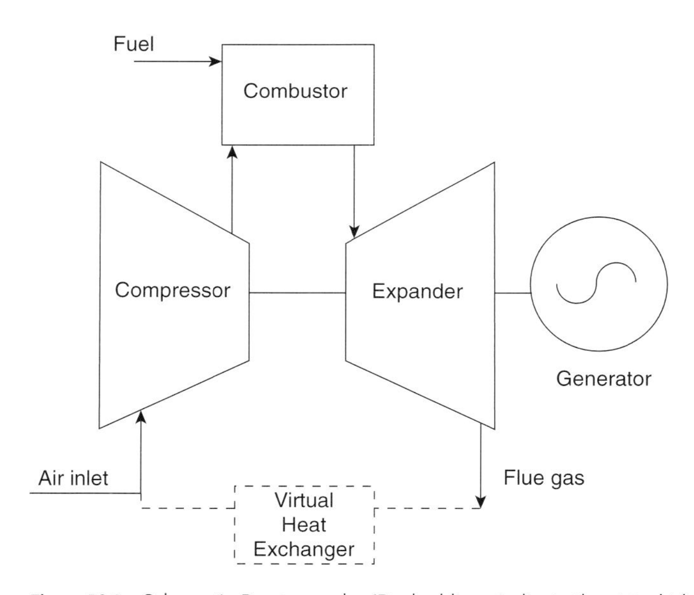

Figure 10.3 Schematic Brayton cycle. (Dashed lines indicate the virtual ideal cycle.)

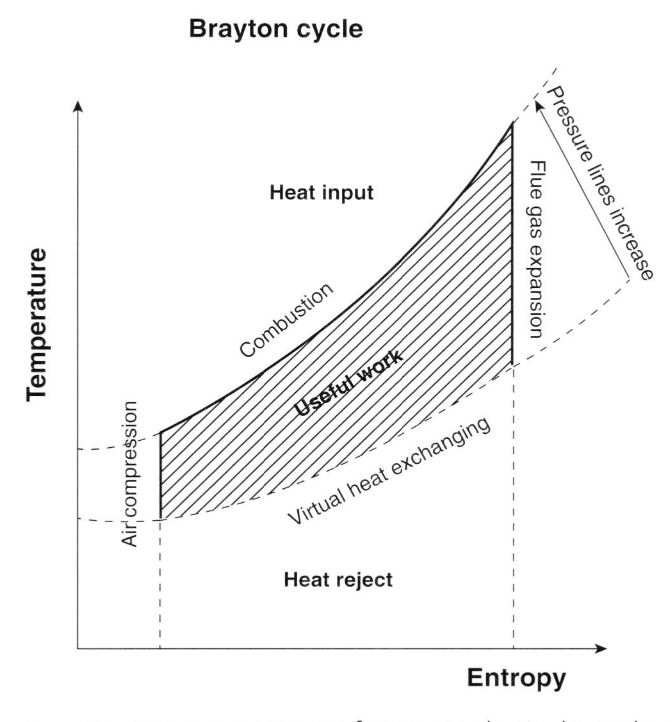

Figure 10.4 Temperature vs entropy for Brayton cycle. (On this graph areas represent energy.)

and operational costs can be quite expensive. The main pieces of equipment, boilers, steam turbines and eventually huge condensers are very expensive in terms of equipment itself and installation costs. Operational costs range high also, because of manpower and chemicals, since each energy unit produced requires a significant amount of personnel and additional materials. Brayton cycle cogeneration systems, with the same power capacity of an equivalent Rankine, may have lower capital and maintenance costs. Boilers and steam turbines will be built to order while gas turbines are normally package equipment and although with a reasonable standardization, final arrangement and installation are much likely to be specific for each location.

Concerning aspects of flexibility between heat and power demand, depending on the design and operational conditions of each industry, neither a pure Rankine nor a pure Brayton cycle may comply with its energy requirements, across the full range of possibilities, at maximum efficiency. An option to be considered is the use of a combined cycle, using the high temperature exhaust heat from the Brayton cycle gas turbine by sending it into a heat recovery steam generator to produce steam for a Rankine cycle (Figures 10.5 and 10.6). This combined cycle can achieve higher efficiencies than each cycle alone, allowing a much greater flexibility over the type of fuel used and heat-to-power ratio.

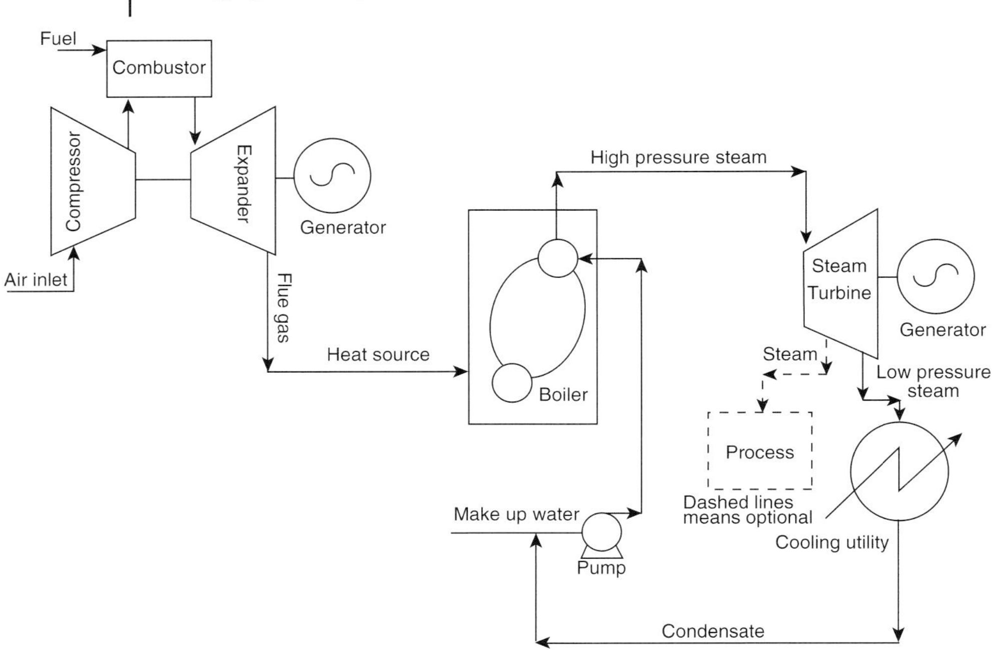

Figure 10.5 Schematic combined cycle.

Compared with purchased power and the operation of on-site boilers to provide heat, cogeneration can be considered the biggest energy efficiency opportunity for any industry. Typically, a plant with cogeneration will require 25% less primary energy compared with separate heat and power supplies. This reduced fuel consumption is the main economical and environmental benefit of cogeneration, because a plant's energy requirements are attained more efficiently with fewer emissions. The possibility of using residual products and waste materials as energy source, increases cost effectiveness while reducing the need for waste disposal or treatment.

Cogeneration is certainly not a low cost energy efficiency option, especially considering capital costs, but it can be very cost effective, particularly in the case of a grassroots project or a system replacement. Accurate planning is necessary due to capital and operational costs, design and operational complexity of cogeneration. From this approach, any medium-to-large-scale industry, that has significant power and thermal energy demands, should consider and investigate cogeneration.

Small-scale plants cannot be discarded from this consideration, and other cycles based on internal combustion engines like Otto and Diesel should be taken into account, particularly if power demand is much higher than heat requirement. These cycles have not been addressed here because of the limited power output they provide, but the general concepts apply just as in Rankine and Brayton cycles.

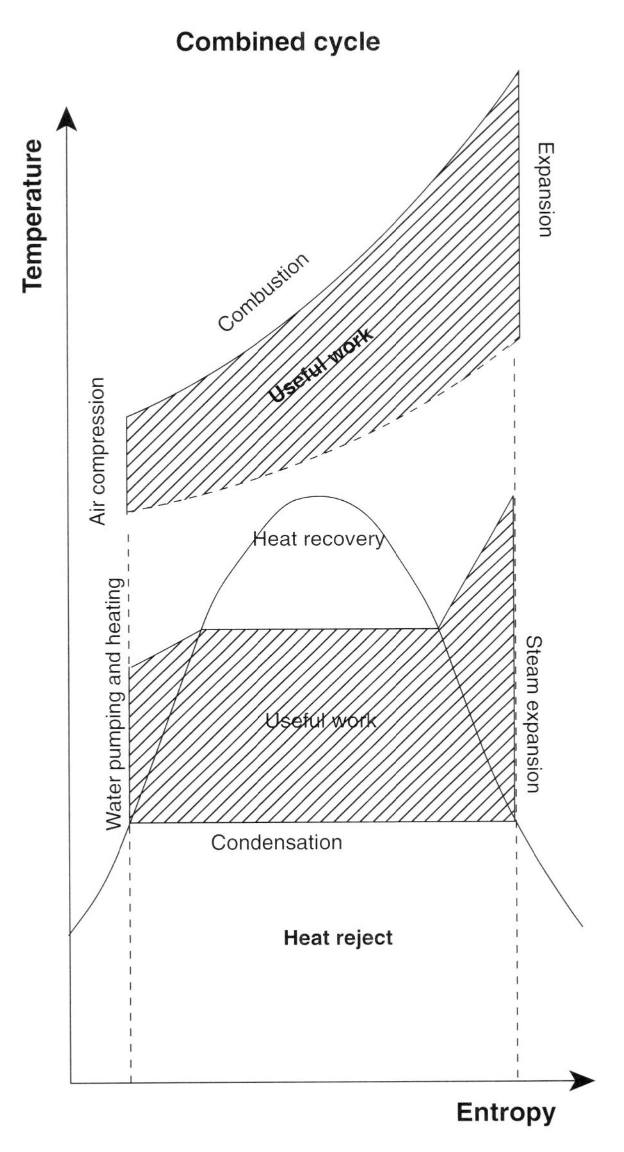

Figure 10.6 Temperature vs entropy for combined cycle. (On this graph areas represent energy.)

10.3.2 Main Pieces of Equipment

10.3.2.1 Boilers

The average equipment called a boiler is composed of a specialized radiant and / or convection tube heat exchanger linked to a drum that accumulates water and steam. Separation between steam and water occurs inside this drum that has accessories to provide humidity removal from steam. The heat source is placed internally to the tube heat exchanger and, for most designs it is a furnace, where combustion heat is to be transferred to water until it becomes heated water or steam.

The configuration may vary depending upon energy source, steam pressure and load, fuel availability, water quality, emissions restrictions, reliability etc. Concerning water passage, it can be a fired tube boiler where water is in the drum and the heat source is immersed in it, passing inside the tubes. On a water tube boiler, the heat source is wrapped in water tubes, in order to direct released heat to the water, obtaining maximum efficiency. Alternatively, the energy source may be waste heat from another process, like flue gas from a gas turbine or a process heater, configuring a recovery boiler.

For mechanical reasons, fired tube boilers have restrictions on maximum allowable pressure and load, being generally used for relatively small steam demands and low to medium steam pressures. For economic reasons, most fired tube boilers are 'packaged' type equipment, being supplied in standard sizes and ready to be mounted and connected. Water tube boilers have almost no load or pressure restrictions, being selected when the steam demand and pressure requirements are high. Usually, these boilers have some basic design standardization, but capacity, maximum operational pressure and water treatment requirements end up determining the final project customization for each application.

Normally, in the average boiler, the heat source is a furnace, where a fuel is being burned, with air supplied by a forced, induced or balanced draft system. Similarly to process fired heaters, mentioned in the previous chapter, the same warnings for energy efficiency apply. Among these points of attention are emphasized:

- Reducing flue gas losses by keeping proper air-to-fuel ratio, aiming to minimize excess air, but providing enough air to avoid unburned fuel.
- Assuring sealing in balanced draft system to avoid air infiltration or leakage. Leakage implies direct heat loss to the atmosphere while infiltration may cause flue gas oxygen content misreading, jeopardizing air-to-fuel ratio adjustment.
- Diminishing convection and radiation losses by maintaining and improving boiler thermal insulation.
- Improving heat recovery by preventing heat transfer surface fouling through soot blowing and combustion control, either in economizers and air preheaters.

On the water-side, best practices for efficient boiler design and operation are linked mainly to water treatment quality. Assuring a water quality compliant to the boiler specification reduces scale in water-side, maintains heat transfer efficiency and helps to keep surface blowdown in an acceptable range. Recovery of surface blowdown heat by exchangers and flash steam is a good option. Flash steam is explained later in this chapter.

Just reprising a issue stressed in the previous chapter, monitoring and tracking mass and energy balance around a boiler is basic, not only to identify losses. Since these are the individual main energy converters for the industry, boiler efficiency tests tell a lot about energy efficiency in the whole complex. Any inefficiency in a boiler is passed to all other energy consumers in the site and nothing can be done by process equipment to compensate these losses.

10.3.2.2 Gas Turbines

A gas turbine is a set of three pieces of equipment with different functions. First an air compressor, where atmospheric air is pressurized and delivered to the second device, a constant pressure combustion chamber, where fuel is injected and burned. Combustion takes place with high excess air. The heated flue gases are then expanded in the third piece of equipment, a turbine, producing power to drive the compressor and excess power is used for driving an external equipment. Due to the large air excess, almost 100 times the fuel mass, the exiting exhaust gas contains a high concentration of nitrogen and oxygen. And since mainly pressure energy is used to release power, this flue gas temperature is relatively near flame temperature. So, exhaust gas can be considered as heated air and ideal for heating purposes.

Gas turbines may use liquid or gaseous fuels, with the highest energy performance being achieved with liquid fuels, but lowest emissions with clean fuel gases like natural gas. Two gas turbines classes are available for power generation – heavy duty industrial and aeroderivative gas turbines. Aeroderivatives were originally developed for aviation and adapted for power generation. Heavy duty gas turbines are specially designed to match industrial use of some liquid fuels, like diesel, operate at stationery conditions such as extended operation period and load variability, reaching high efficiency.

Gas turbines normally need specialized and more frequent maintenance, compared with steam turbines. Aeroderivative types are designed to have the expander part replaced by a spare or a conditioned unit to minimize downtime, while industrial gas turbines are projected to be maintained on site, and this is generally a less expensive procedure.

Energy efficiency opportunities for gas turbine operation are linked to obtaining maximum energy conversion. Guaranteeing inlet gas design conditions like pressure, temperature and composition, prompt turbines to operate at maximum efficiency. Contaminants like ashes and sulfur compounds may result in deposits, which degrade performance and cause corrosion in the turbine expander section. Tracking and maintaining quality combustion can allow the highest possible temperature of hot gas leaving the combustors and increased temperature results in higher power output. A good specification of average local ambient air conditions, particularly temperature, helps the manufacturer provide adjustments that enhance gas turbine output significantly. Higher ambient air temperature impairs compressor performance due to lower air density, and also excessive pressure drop across exhaust gas ducts and stack. Reducing the pressure drop across air filters increases released power even more. An additional option lies here, reducing inlet air temperature by means of a refrigeration system, generated by using part of the wasted heat from the turbine.

The next great energy efficiency option for gas turbine is recovering excess heat from exhaust gas. Since this gas is in essence hot air, as mentioned before, and it can be used for any heating process, even as hot combustion air. It can be sent to a process fired heater or to a heat recovery steam generator. This option being to set a combined cycle, either in an purely unfired heat recovery steam generator, using only the sensible heat to produce steam or a fired one, using the excess air to burn supplementary fuel and raise the steam production. The supplementary fuel choice increases system flexibility to control heat-to-power ratio of the cogeneration ensemble. The next step in building a fully combined cycle, is just adding a steam turbine.

10.3.2.3 **Steam Turbines**

Steam turbines convert steam pressure energy into power, being widely used in power plants throughout industry and electric utilities. They use high pressure and high temperature superheated steam that flows through an expander, forcing the turbine to rotate, moving any equipment attached to the same shaft. After expansion, steam exits at lower pressure and temperature. The major difference of the steam turbine, relative to other primary movers like internal combustion engines and gas turbines, is that initial heat source for the process, combustion, occurs externally in a boiler, a separate device. The exit steam, depending of course on pressure and temperature, can be used for heating or can be sent to move another lower pressure steam turbine.

There are two basic steam turbines types, backpressure and condensing. The backpressure turbine produces power through expansion of high pressure steam to a lower pressure and the exiting steam is sent to an industrial process in a cogeneration system. A condensing turbine is similar to the backpressure turbine, but the low pressure side is usually below atmospheric pressure. This yields a greater heat-to-power efficiency, but the rejected steam has much lower energy content, hence reducing its value for heat recovery, so this steam leaving the turbine has to be condensed by a heat exchanger. In utility electrical plants, where electricity generation is the unique purpose, all turbines used are commonly condensing.

For cogeneration objectives, the ability to control the heat and power balance using the two types of steam turbine is helpful. This is accomplished by the extraction turbine, which simultaneously permits partial removal of steam from the turbine during expansion before condensation. This turbine can be designed to allow extraction at various pressures and flows, increasing the flexibility to satisfy both variables heat and power loads by a great deal. They are frequently used in cogeneration applications, but their average efficiency is lower than other turbine types.

The choice between backpressure turbine and extraction condensing turbine relies mainly on the balance between power and heat demands, hot temperature demand and economic factors. For same power level application, like electrical generation, compared with condensing, backpressure turbines present the simplest configuration, fewer components and lower capital cost. And, if properly projected to match process heat demand, they signify no need of cooling utility and less environmental impact. On the other hand, a backpressure turbine tends to be larger than the condensing type for same power output, because it works under a smaller enthalpy difference. Also, steam flow through the turbine depends exclusively on heat demand, which results in little or no flexibility to match its own power demand.

Therefore, to balance these amounts, there is the need either to purchase external electricity when the heat load doesn't allow enough power generation, or venting steam directly to the atmosphere to comply with power needs, a very inefficient option. In the inverse conditions, when the heat load produces more power than necessary, electricity exportation might eventually be an opportunity, but intermittent power production has almost no commercial value. All these cases present poor economical and energy performance.

Energy efficiency opportunities in steam turbines begin on selection, by choosing the appropriate turbine type that allows best control of the heat and power balance. Guaranteeing steam temperature and pressure design conditions at the turbine inlet is essential. Variations from optimal conditions can impair turbine ability to operate at maximum efficiency, regardless of its type. For condensing turbines, maintaining back pressure or vacuum is the most important factor, once deviations from optimum can reduce significantly efficiency. This can be obtained by ensuring appropriate cooling of the utility inlet temperature and the flow rate, to control fouling and scaling inside the condenser. Ensuring vacuum condition is another important aspect, because if the pressure rises, the enthalpy difference available for power generation diminishes drastically, spoiling turbine performance. Good sealing to avoid air infiltration into the condenser is mandatory.

Incorporating extracting and condensing steam turbines into industrial cogeneration systems increases the ability to control the heat and power balance. If heat demand diminishes simultaneously to an increase in power demand, steam can be passed all the way to the condensing turbine part to produce additional power, maintaining a uniform and efficient load without extra fuel consumption. As a matter of fact, industrial cogeneration produces power by steam condensation less efficiently than a large utility plant. Better overall energy efficiency can be reached if more possibilities are in hand to manage diverse optimum operation points that match process needs, availability requirements and external power purchasing costs.

10.3.3 **Ancillary Systems**

Auxiliary facilities for power cycles are supporting systems for maintaining their operation. Most important among them are water treatment facilities including external and internal treatment. Considering overall plant water demand, external treatment usually lies in a separate unit while internal treatment is by a set of equipment normally near the boilers. Electrical devices for reception, transmission and distribution of purchased and generated electricity can be included in this list.

A special and sensitive supporting system is the condensate treatment to provide recycling of condensate that reduces make-up water demand. Normally, it is attached to the boiler, in order to polish condensate to acceptable standards to be fed back to water boiler. Guaranteeing a good treating capacity is critical for a safe boiler operation, simultaneously reducing boiler water costs and increasing steam quality.

The cooling utility can be another significant ancillary system, if a condensing turbine is in use. In this case, the amount of power released might be directly linked to good cooling operation.

10.4 Utility Units

The equipment and systems mentioned above have to be grouped inside the industry, coherently with their tasks and their interface with core process and between themselves. Here follows a general description of the usual utility units.

10.4.1 Steam Generation

Steam generation can be spread around the plant in many heat recovery devices, but the main controllable production occurs in a boiler room or power plant. Here, for didactic reasons, steam and power generation are described separately.

Usually the main steam generation unit comprises the main capacity boilers, the boiler fuel system, the steam system, the condensate treatment, the feed water system and the internal water treatment system. Considering steam pressure generation, a considerable preheating water exchangers train is included. Proper operation and design of this preheating, that uses steam and hot flue gas from the boiler, is essential if the boiler is to achieve design capacities and higher energy efficiencies.

The feed water system consists of water tanks and pumps, to provide water to the boiler and automatically regulating it to match steam demand. The steam system collects and controls the steam produced, which is directed by piping to users at appropriate controlled pressure. The fuel system includes all equipment, like tanks and pumps for liquid fuel or compressors and valves for gaseous, to provide fuel for steam generation.

All the variations in steam demand are sensed in this unit by pressure drop and this demand may be cyclic or fluctuating, complicating boiler operation and control. For reliability reasons, multiple boilers are available to comply with these many and varying loads.

Managing this unit as efficiently as possible is a fundamental energy efficiency opportunity; any inefficiencies in steam generation are spread to the whole plant.

Scheduling load properly for each boiler, concerning its expected performance, helps to achieve optimum system performance while guaranteeing reliability. To accomplish this, the maximum efficiency firing range of individual boilers has to be taken into account. Efficiencies should be determined for each boiler over the full capacity range. Steady, high capacity and energy efficient boilers should hold the continuous steam load and less efficient and more flexible generators should be used at peak loads. The overall energy performance is ensured by the more efficient equipment.

10.4.2 Power Generation

The power generation unit is a section of the power plant where electrical generators devices are laid. Generators may be gas or steam turbine and eventually internal combustion engines, but whatever the primary fuel, the electricity generated must be at the same voltage and phase coupled to be channeled to final users. Also purchased energy must be connected to the system and the main reception transformer should be attached to this section, to be linked properly to site distribution lines.

Regarding all previous thoughts about energy efficiency options on power conversion, the same management cautions that are recommended for steam generation, like scheduling and control, quality and reliability, apply here. Concerning continuous power load and peak load, the same strategy used for boilers should be followed for electrical generators. The difference lies in the reliability between generated and purchased power matching priority of electrical process loads.

All electrical power generated and purchased will be distributed by high tension wiring, and segregating the feeder lines by system reliability is an option that calls for safety, continuity and in some extension for efficiency. The more critical and hazardous processes should be connected to the more reliable electrical source while less sensitive and non-potential hazardous areas can allow some sudden stoppage, without bigger consequences.

Efficiencies depend on the operational permanence of the generators, so maintaining loads close to design values surely helps optimization of the overall system. And electrical power load variations, together with heat load fluctuations, make achieving and maintaining optimal operational conditions that much more complicated. One energy efficiency opportunity here, is to decide which system assumes really sharp peak loads, either one special and reliable generator or purchased power. Normally, this choice falls to purchased power, if this system is reliable. Depending on regional transmission and distribution characteristics, other decisions may be taken.

As a matter of fact, little can be done to significantly improve energy efficiency at this point, except good operation and management of momentarily best equipment. Nearly all the decisions have already been taken when the equipment was selected and on primary mover operation.

10.4.3 Water Treatment

Usually water treatment is settled in a reasonably large area, not necessarily near the power plant. The area required is due to the volume of tanks and ponds necessary to treat and stock water. Raw water has to be conditioned to be used in many forms and each treatment stage produces one kind of utility water. A regular treatment sequence may include screening to remove big bodies like trash, leaves, fish etc. Coagulation is provided by addition of chemicals aiming to aggregate small particles into bigger ones, and a slow mixing called flocculation, provokes collision of particles, making larger ones called flocs. These flocs tend to aggregate, becoming heavier than water and settled in a slow flow tank, known as sedimentation tank or clarifier. Additional removal of small particles is provided by pressure filters using sand and gravel as filtering media. All water for industrial use is extracted from this stage, but for boiler use, further treatment is needed.

The additional boiler water treatment, to remove hardness and non-hardness salts, is called demineralization. This is achieved by using a 'cation' resin, which exchanges cations in water with hydrogen ions, producing hydrochloric, sulfuric and carbonic acid. Carbonic acid is removed in a degassing tower in which air is blown through the acid water. In sequence, water passes through an 'anion' resin, which exchanges anions with a mineral acid, like sulfuric acid, and forms water. Continuous use saturates resins with ions and regeneration of cations and anions is regularly necessary using mineral acid and caustic soda, respectively. Complete removal of silica can be achieved by correct choice of anion resin. For high pressure boilers, water quality is very restrictive, so a supplementary ions removal is demanded, called polishing, using basically the same process of resin beds, but with a higher efficiency and deeper capture capacity.

Best practices related to energy efficiency are not around the water treatment unit itself, although some significant pumping energy requirements are in this unit. The main actions to reduce energy consumption are in condensate recovery and avoiding water losses in general, even as steam. They allow lower water make-up, fewer chemicals and less energy consumption, allied to a reduced environmental impact.

10.4.4 Cooling Units

The cooling tower rejects waste heat contained in circulating cooling water utility to the atmosphere. They can be classified by the type of draft either mechanical or natural; by the way air and water flow, either crossflow or counterflow; and heat transfer mode, evaporative also known as wet, or dry. The conventional industrial cooling towers are mainly wet or evaporating ones, and for economical reasons, tend to have mechanical draft. The evaporative heat transfer process, implies high water losses, demanding constant make-up.

An industrial cooling tower is normally constructed in cell sections, each one being an almost independent tower, although sharing the same water treatment, but with separate circulation pumps and fans.

Water cooling process efficiency is greatly dependent on ambient conditions and heat rejection load. Other factors influence efficiency like heat transfer surface area, contact time between water and air streams, and air-to-water flow ratio. These factors must be balanced to select the best cooling tower for the design conditions.

Energy efficiency opportunities in water cooling units begin with selection of the appropriate cooling tower, considering site average atmospheric conditions. Selection of fillings influences air pressure drop, fan selection and sizing. Attaining cooling water quality to design conditions avoids surpassing the expected flow rate of circulating water and prevents additional tower cells being used in a certain process condition. This reduces supplementary energy consumption in pumping, besides leaving room for additional cooling demands. Staging or using variable speed drives for tower fans and water pumps is another option, especially if the tower has many cells. Turning off unnecessary cells momentarily should be considered.

10.4.5 **Ancillary Systems**

Some auxiliary systems should be placed inside the steam and power generation for monitoring reasons, like all air systems. Compressed air systems consist of compressors, air treatment and storage drums. Design conditions always adopt the premise that it will supply clean, dry and stable air at the appropriate pressure to consumers. Among opportunities for energy efficiency in air compressing systems can be cited the use of multiple stage and modulating compressors. This reduces the compressed air temperature between stages, improving compressing efficiency and allowing better control of the energy consumption. Normally, these compressors are electric power driven and operate for continuous long periods, so the choice of high efficiency motors has an attractive payout. Also recovering waste heat from interstage coolers can help save plant energy. Collecting the coolest and cleanest air possible increases compressor efficiency and reduces the need for air treatment. Constant monitoring of pressure drop across the air filter and regular cleaning reduces energy waste.

An additional and indispensable auxiliary utility system is wastewater treatment, which must comply with environmental regulations. Industry operation generates liquid effluents that are quite often discarded. Direct discharge to any water body can cause pollution and environmental degradation, so law restricts this practice to some maximum permitted parameters that must be complied with at wastewater treatments.

The treatment method is a consequence of plant processes and is divided between source segregation and pretreatment of concentrated wastewater streams. Typically it may include diverse steps depending on contaminants. If oils are present, grease traps or oil water separators are needed; if floatable solids, skimmers and air flotation are used; filtration and sedimentation is necessary for suspended solids reduction. The presence of organic matter implies biochemical oxygen demand requiring biological aerobic treatment. If waste water is to be fully recycled, eventual disinfection by chlorination may be required. And dried residual solid materials from wastewater treatment should be disposed in landfills. Not to forget that a maximum disposable temperature can be fixed to effluent streams. Needless to mention that it is not only the core process that generates liquid wastes, utilities do it too; just remember all blowdown streams mentioned in this chapter.

Concerning operation, wastewater treatment resembles water treatment units, but with a great difference. The biggest opportunity for efficiency in waste water is actually to avoid its need, by better management of the operations that produce the waste. Assuring compliance with the many recommendations cited so far, can allow an optimal operation that reduces wastewater treatment flow by a good deal. And another option is to recycle treated waste water as make-up water, for a process with less critical quality standards. It is becoming more likely nowadays to use waste water as cooling water make-up. And there is a good chance, that sometime in the not so faraway future, the only liquid effluent flow that will be allowed to an industry will be the water evaporated at a cooling tower.

10.5 Distributing Systems

The distributing system is the essential link between generation and final consumption and this step must preserve the quality included in generation process.

10.5.1 Pipes

Most of the utilities discussed in this chapter are fluids, so pipes are a unavoidable piece of equipment, with many particular features, according to the characteristics of each utility.

10.5.1.1 Steam

Steam distribution is critical to deliver required quantities of specified temperature and pressure steam to the end user. Distribution lines spread from power plant or boiler room to final use by main headers at some predetermined pressure, with numerous take off lines, close to consumption points. Boiler room location inside the industry, relative to users, is a first efficiency option, since a centrally located steam plant reduces average steam header lengths, propitiates smaller pressure drops and higher efficiency on deliverance. Final users generally operate at different steam pressures, coherent with their requirements and these are obtained by pressure regulating valves or back pressure turbines, sending reduced pressure steam into smaller local pipe networks and branch pipes, where steam can be conveyed to individual pieces of equipment.

A well-designed and efficient steam distribution piping is adequately sized, laid out and configured, presenting proper pressure balance and regulation. Diameter selection has to consider acceptable steam speed range to avoid either too slow flow that promotes high heat losses to ambient, and excessive condensate formation and very high speed that provokes erosion and tube wear. A particular parameter for the choice of small diameters is attention to avoid speed noise. For safety and efficiency reasons the option for the slightly larger pipe diameters may be more expensive, but reduces pressure drop at a given flow rate and the noise associated with steam flow. Also large diameters introduce flexibility to peak loads and allow for some increase in demand over time. Layout and alignments must be coherent with project flow directions.

Another important issue, related to overall plant design, is avoiding production and use of very low steam pressures, below 3.5 $kgf\,cm^{-2}$ (340 kPa) or 50 psig. Pressures below this level have a saturation temperature under 180 °C, offering a reduced temperature differential for average process heat demands. Specific volumes at these low pressures increase, implying bigger pipe diameters and costs for a relative low energy content flow. Also, there is a tendency for higher condensate formation, demanding more drip legs and traps with more corrosion, hence potentially more leaks.

Configuration aspects like flexibility and condensate drainage also have to be taken in account. For flexibility, piping and especially connections must accommodate expansion and contraction during start-ups and shut-downs, this prevents pipe wear and leaks. Condensate drainage must be addressed through appropriately sized drip legs, placed at regular length intervals in straight pipes and before significant equipment like turbines or valves and elevation changes. Pipe alignment should be designed with a proper pitch to promote condensate flow to drip legs, where steam traps discharge it to the return system.

Many opportunities for energy efficiency are in steam distribution; actually poor design and careless operation of this system can waste more than 10% of overall site energy demand. Even a set of efficient boilers and process plants may have its performance ruined by a badly kept steam distribution system. Installing, maintaining and improving thermal insulation is a first priority. A significant amount of heat energy may be lost for lack of insulation or for improper installation or inefficiency. Its adequate presence delivers fuel savings, better process control by maintaining process temperatures at expected levels, and safer working conditions. For safety reasons, any exposed heated surface must not surpass 60 °C to avoid skin burns and economical insulation thickness usually guarantees temperatures lower than that. Uninsulated points, damaged or wet insulation should be listed and regularly repaired to avoid increasing energy losses.

Monitoring and repairing steam leaks and steam traps also avoids considerable losses. Leaks represent energy and treated water losses allied to hazardous conditions, because high pressure steam burns can cause serious injuries, can damage nearby equipment and introduce high pitch noises, unaffordable in a healthy work environment. The lengthier and older the steam piping is, the more significant the number of leaks. They can be very costly, depending on opening size, steam pressure and period of leakage. Chasing and repairing steam leaks is a continuous and quite rewarding task. Steam traps are fundamental for steam quality and condensate recovery. Also, condensate accumulation in steam pipes is hazardous, because if a considerable amount of liquid begins to flow at steam velocity, water hammer phenomena may occur, potentially causing accidents. This subject is addressed in detail on the following section.

But steam traps can also be major contributors to energy losses in faulty operation. It is very easy for a large industry to have thousands of steam traps installed. Despite the energy loss, there is a good chance that a malfunctioning steam trap is feeding steam directly to a condensate line, pressurizing it and impairing condensate recovery, resulting in a much bigger loss in adjacent systems. Needless to say, that a blocked steam trap induces a bad heating process control, provoking extra and unnecessary energy consumption to compensate it. Regular assessment of steam trap performance is another continuous assignment that can bring about big results in energy savings, productivity and safety. Consequently, the next point to address has to be the condensate return system.

10.5.1.2 **Condensate**

Condensate is exhausted heat steam and for the reasons already expressed, returning it to the boiler results in meaningful energy savings. Actually, condensate cannot be allowed to accumulate in equipment, especially heat exchangers. These devices are designed to operate non-flooded and accumulated condensate inhibits heat transfer performance and induces corrosion. Prompt condensate removal, just after formation, is provided by steam traps. Besides heat transfer problems, simultaneous flow of steam and water in the same pipe may lead to erosion and water hammer in the pipework. Water hammer occurs when slugs of water travel down the pipe at steam speed, that is usually much higher than water velocity design. Reaching pipe accessories and direction changes may lead to disruption of the pipe and eventually accidents. Condensate is saturated water at a pipe inlet, but with flow, pressure drops and steam evolutes. If the pipe is long, there are chances that water hammer will happen.

Above all considerations over condensate recovery observed in this text, a single utmost warning prevails for condensate flow. Prevent near saturation condensate to flow just by system pressure differential. It may be not enough to avoid water hammer conditions and at the least, may not allow condensate to flow to a proper collection spot. Best practice here is to always have an extra driver for this flow. One is to take advantage of gravity, making condensate flow from higher to lower grounds. Condensate recovery lines must always be connected to pipe headers by the top. Never connect condensate from different steam pressures to the same header without good analyses. There might be moments that some streams will be facing a higher pressure ahead, and they will not flow, creating a dead spot where corrosion pops. And if there is chance that no nearby recovery header can stand the pressure of the condensate generated, flash it.

When condensate at a high pressure is released to a lower pressure, it produces flash steam at a lower temperature and low pressure condensate. This steam can be as useful as that produced by a boiler, being sent to a heating process or simply vented to atmosphere. The residual hot condensate can then be pumped to a collecting tank and eventually treatment and then back to a boiler. The pumped condensate pipe is fully flooded, meaning that water hammer is less likely to occur. These cares begins in design and system improvement and are fundamental to grant huge and continuous energy savings.

10.5.1.3 Water

An industry has several water systems, for process use like cooling, service and potable water. Independent of their function, water distribution systems tend to have similar inefficiencies and energy efficiency opportunities. Location of the water treatment unit does not have the same weight as the boiler room on steam distribution, actually the water unit is better placed closer to the main water source inlet and with a considerable available area for expansion. Water pipes are less costly than steam because they don't demand all the accessories and insulations that steam does.

Bigger opportunities are linked to good design by correct selection and sizing of pipe diameters and pumps, reducing friction losses and associated pressure drops. Best practices on operation are reducing water losses by detection and elimination of leaks, monitoring water use patterns and reducing consumption to minimum necessary by education and awareness.

10.5.1.4 Air

The air distribution systems present basically the same characteristics as water systems described in the previous section. Air headers and regulators convey compressed air from central compression plants to process units. It includes isolation valves, fluid traps and intermediate storage vessels. Pressure losses have to be compensated by higher pressure at compressor discharge. At point of use, a feeder pipe with a final isolation valve, filter and regulator delivers compressed air to processes.

Opportunities are also similar to water systems, like proper design of pipes, leak detection and repair and avoidance of improper use. Another good procedure is to regulate all uses to the lowest possible pressure, relieving compressor duty. A special aspect is to eradicate, whenever possible, use of air as energy utility in air motors because compressed air is usually the most costly utility and should be saved for specific uses.

10.5.2 Wiring

Electrical energy distribution requires cables and power transformers, implying three types of energy losses: heat in the conductor by the Joule effect; energy dispersed in magnetic fields in transformers; and energy absorption in insulating material by dielectric effect. Joule effect losses in cables may account for 2.5% and transformers losses range from 1 to 2%.

Final electrical energy consumers are in majority of industry motors, ranging up to 70% of electricity demand. Understanding the influence of basic electricity parameters delivered to consumers can help to figure out what are the best opportunities for energy efficiency.

10.5.2.1 **Phase**

The major source of mechanical power in a plant is three-phase electricity. Single phase circuits exist and have many low power uses like lighting and general electrical devices like computers and home appliances. Other single phase circuit is for instrument wiring in intrinsically safe circuits. Most plants distribute electricity by a high voltage three-phase system that is frequently used directly on large drives.

There are single phase motors, but they have an upper power limit of approximately 10 HP (7.5 kW) that does not apply to three-phase motors. Actually, in smaller sizes, three-phase electric motors may cost less than single phase motors of comparable size. On the other hand, supplying three-phase power is costly due to increased transformer distribution costs, but reduced capital costs of three-phase motors and flexibility make this option highly feasible.

10.5.2.2 **Frequency**

Frequency doesn't offer any special opportunities in itself, but it can be used to improve energy efficiency in pumping and blowing operations driven by electricity. Motors are usually limited to certain shaft speeds when power is supplied at certain frequency. Coupled to a specific service, they will deliver a fixed amount of energy and if the flow has to be controlled, energy is wasted in a throttling valve that restricts flux. However the use of variable frequency drives takes away this requirement of fixed speed and fixed power deliverance.

Variable frequency drives are attractive for use on larger motors enhancing process operations, particularly for flow control, because almost the necessary energy is delivered and valves just act for fine tuning. Concerning high inertia loads, they can provide soft start, decreasing electrical stresses and voltage sags that happen when bigger drives are turned on. Although this option may present some restrictions for use in lower power services, it is an effective and easy flow control alternative that can be used in a wide operating range. Existing motors can be retrofitted at affordable costs and can result in a more efficient operation with reduced costs. Considering its application in all big pieces of equipment, whose power consumption varies with utility demand, like boiler water pumps and draft fans, cooling water pumps etc can dramatically increase energy efficiency and controllability of the whole system.

10.5.2.3 **Power Factor**

The real electrical power used to perform a certain task is called active power. However, certain loads, especially motors, expend energy to establish the magnetic field, which is another form of power called reactive power. Although this is a virtual power, added to active power, it determines the actual demand or total power load of an electrical system. Power factor is the ratio between active power and total power and a higher reactive power implies that less active power is distributed. Theoretically, if all power loads only require active power, the power factor equals one, and the maximum power that can be transferred equals the distribution system capacity. However, if a significant number of motors in the plant are oversized by design and underloaded on operation, below 75% nominal power, higher reactive power is demanded, reducing effective energy distribution capacity. Low power factors can cause power losses in the distribution system. Voltage drops may increase, and if they happen too frequently, may provoke overheating and early failure of motors and inductive equipment. Secondary losses may occur like heat in wiring by higher current values to compensate power losses.

Energy efficiency opportunity here is to guarantee a power factor closer to 1, preferably higher than 0.85, by appropriate selection of motors capacity, matching process demand. Since this low power factor can affect third party supplier systems, penalty charges are imposed by contract, if a minimum factor is reached, increasing electricity cost. The power factor can be improved by installation of correction capacitors. This option can reduce distribution losses within the plant network while maintaining available capacity. Voltage level at consumption is increased, improving motors performance. Simultaneously total current in the system is reduced, hence reducing Joule effect losses.

10.5.2.4 Voltage and Current

Power distribution efficiency depends critically on voltage and current, since power is the product of current and voltage. Energy losses by the Joule effect are due to wiring resistance times the squared current ($R \times I^2$), so to distribute the maximum power with least losses, it is better to use the highest voltage possible.

Opportunities for efficiency are proper wiring selection to minimize energy losses. If long feeder runs are needed, the use of larger wire sizes can yield energy savings and be economically justifiable. Concerning magnetic losses on transformers, the best option is to distribute high voltage and just reduce it in a transformer close to the point of use. Proper selection of electrical equipment, guaranteeing uniform voltages at each process, reduces the number of transformers required. During operation, setting transformer taps to optimum load and keeping constant track of transformer loads, eventually eliminating unnecessary ones by regrouping charges locally. Disconnecting primary power of transformers that are not serving any active loads momentarily also reduces magnetic losses.

10.6 Design Aspects

Despite the specific features described so far about best practices, some basic decisions, that have to be taken at the design stage, will influence efficiency results for the life span of the plant. Understanding what these decisions are and how to access them, gives a great deal of chance of reaching a good balance between these requirements and energy efficiency.

10.6.1 Availability

The concept of availability here is to have access to any necessary amount of a utility, whenever it is demanded, especially in peak load conditions. On design, this can be achieved by extra capacity that can be based on maximum demand on start-up. This assumption may oversize any utility system greatly, probably making it economically unfeasible. Any extra generation or distribution capacity will result in systems bigger than the expected regular demand, which brings an operational tendency that it will be running below optimal point. This option, of course, doesn't comply with energy efficiency.

The best call is to select one most probable peak load and match some redundancy to cope with it occasionally, under controlled circumstances. Dividing this additional capacity among a number of pieces of generating equipment, also matching forecast turndowns and operational modulation of the whole system, can reduce the distance between the optimum design point and the regular operational point, allowing better performance by a good management over the life span of the system.

10.6.2 Technology

That is a very difficult call. Whenever considering utility, features like availability and reliability are paramount, while newer and state-of-the-art technology may not be considered, nor be something easy to have in hand, nor something you can trust blindfold.

Brand new technology may bring long terms benefits attached to lower costs and competitive advantage, but in the short term may signify risk. The best alternative is to mix, depending on investors' option, some seasoned and reliable technology, with some new equipment for experience. Part of that extra capacity, cited in the previous section, could be based on new technology. In the meantime there will be a chance to test it, learn and gain trust.

10.6.3 Integration with Process

This can be the ultimate opportunity for enhanced energy efficiency and least environmental impact. Instead of two separate roles, core and support, the fully integrated design approach can reveal result levels never reached before. How?

Integrating processes from concept, in an unabridged approach for energy, water and all resources, can promote the most efficient containment and recycling of all energy forms involved. This integrated industry should need much less imported energy, after start-up, and since energy transfer will tend to use processes streams, the demand on utilities is minimized. This is the same view expressed in Chapter 9, Section 9.7, but here addressing utilities in general industry.

Small-scale integration has been happening recently but what are the odds? Many barriers appear, like management and accountability, who is in charge of what and who is responsible for what. The biggest one should be complexity. The more integrated are the core business and utilities, the more energy efficiency potentially can be achieved, but in this set, no more easy decisions. The degree of integration has to be set even before design. It is a business decision, tightly attached to company values and vision.

10.7 Operational and Maintenance Aspects

After selection of the appropriate type and size of equipment, it is the turn of production. Many of the general best practices of operation and maintenance, cited for refineries, apply to utility systems. But a special burden falls on utilities, the supporting role. And exactly like at the design stage, learning what the expectations for this role are, may help match them with efficiency.

10.7.1 Stability

An operating system is considered stable, if oscillations caused by disturbances are absorbed or accommodated without further consequences for continuous operation. But, considering that the utility system deals with multiple clients and products, an immense interface with process, and a deep interconnection between diverse utilities, disturbance is a common event in daily business. Actually stability is of high concern for utilities' management. Steadiness makes easier the task of operating equipment and keeping it closer to maximum efficiency.

Best practice to avoid disturbances and unexpected transient conditions is to have information about process and management action. Utility management and crew will be involved and constantly briefed about process conditions, modifications, turndowns, potential problems etc through the whole operational period. This may be translated in other terms. Scheduling and planning for the industry has to be done with direct participation of utility personnel.

10.7.2 Safety

Energy efficiency surely walks side by side with safety. All energy efficiency best practices result in a safer working environment. Some examples are that good combustion control allows efficiency and can prevent explosions, fires, reduce pollution, and avoid presence of particulates in the air; steam leak detection and repair avoids personnel injuries, burns and noise; better insulation avoid burns; better design and operation reduces maintenance interventions, minimizing human exposure to equipment internals and hazardous conditions.

The list could be longer; the problem is that operational and maintenance crew often perceive energy efficiency and safety personnel to be acting in opposite directions, and they are not wrong in their conclusion. The option is to synchronize speech and actions between these two groups, they have a lot to obtain by working in a complementary manner.

10.7.3 Reliability

Reliability means trustworthiness in a system, which is robust enough to minimize outage occurrences and whenever they happen, their duration is short. For utilities, outages mean immediate losses and eventually accidents.

Needless to say, that any stoppage will signify a fall from an optimal operational point to scratch and the least, the waste of energy to start up again and place the operation back to optimum. Reliability is another high concern for utilities' management. Fewer stumbles mean longer runs, enduring efficiency and reduced costs. The ways to mitigate failures and consequent outages involve redundancy and standby operation of some critical equipment, or alternatively increasing the design margins of these pieces of equipment, which is similar to redundancy. All these actions resemble the approach described for availability in the design section, and have a similar noncompliance effect over energy efficiency. Option for efficiency is similar, but in the other way around. Counting on that extra capacity, included in the plant for availability, and a probability study for failures, it is possible to build a contingency plan, discarding in a coherent sequence less sensitive loads when an outage occurs, while maintaining some of the demands. This procedure will not eliminate losses, but it smoothes consequences, holds main processes and utility systems, potentially reducing downtime and accelerating pace for returning to the operational condition previous to the event.

10.7.4 Efficiency

As previously mentioned and just repeating here, the utility system guarantees the basis for the core business to perform. But energy efficiency is the way that utilities can excel in their supporting role for industry. It offers competitiveness by cost reduction and is a necessary step in sustainability by environmental compliance. It must be understood as a quality differential for utilities and pervade all aspects from design, operation and maintenance. It can simply be the first, easiest and cheapest way to address GHG emissions management, because utilities happen to be the biggest and inescapable source of CO_2 emissions in the world.

10.8 Approach and Literatur

As in Chapter 9, I have offered my impressions based on the path that I have followed in my career. For more details, I suggest some literature from a vast collection of excellent materials, to encourage the readers' research.

Further Reading

1 Asociacion regional de empresas de petroleo y gas natural en latinoamérica y el caribe (Regional association of oil and natural gas companies in latin america and the Caribbean, ARPEL) (2000) Guideline No. 31-2000 Energy Management of Steam Systems.
2 Black and Veatch (1996) *Power Plant Engineering*, Chapman & Hall, New York, p. 173.
3 Ganapathy, V. (1992) Fouling–the silent heat transfer thief. *Hydrocarbon Processing*, **71**(10), 49–52.
4 Ganapathy, V. (1994) Understand steam generator performance. *Chemical Engineering Progress*, **90**(12), 42–48.
5 Garcia-Borras, T. (1998) Improving boilers and furnaces. *Chemical Engineering*, **105**, 127–131.
6 Harrell, G. (2002) Steam System Survey Guide. U.S. Department of Energy. Washington, DC, http://www1.eere.energy.gov/industry/bestpractices/pdfs/steam_survey_guide.pdf (accessed 26 April 2010).
7 Johansson, T.B., *et al.* (1996) Options for reducing CO_2 emissions from the energy supply sector, in *Energy Policy*, vol. **24** (ed. N. France), Nos 10/11, Elsevier Science Ltd, pp. 985–1003.
8 Kreith, F., and Goswami, D.Y. (2007) *Handbook of Energy Efficiency and Renewable Energy*, CRC Press, Taylor & Francis Group, Boca Raton.
9 Lieberman, N.P., and Lieberman, E.T. (2008) *A Working Guide to Process Equipment*, 3rd edn, The McGraw-Hill Companies, Inc., New York.
10 Oland, C.B. (2002) Guide to Low-Emission Boiler and Combustion Equipment Selection. U.S. Department of Energy. Office of Industrial Technologies. Washington, DC, http://www1.eere.energy.gov/industry/bestpractices/pdfs/guide_low_emission.pdf (accessed 26 April 2010).
11 Roy, G. (1996) Selecting heavy-duty or aero-derivative gas turbines. *Hydrocarbon Processing*, **75**(4), 57–58.
12 Seneviratne, M. (2007) *A Practical Approach to Water Conservation for Commercial and Industrial Facilities*, Butterworth-Heinemann, Oxford.
13 Thumann, A. (2008) *Plant Engineers and Managers Guide to Energy Conservation*, The Fairmont Press, Lilburn, GA.
14 Turner, W.C., and Doty, S. (2007) *Energy Management Handbook*, The Fairmont Press, Inc., Lilburn, GA.
15 U.S. Environmental Protection Agency (1998) Climate Wise: Wise Rules for Industrial Efficiency–A Tool Kit for Estimating Energy Savings and Greenhouse Gas Emissions Reductions, http://www.fypower.org/pdf/Ind_EE_Wise_Rules.pdf (accessed 26 April 2010).
16 Zeitz, R.A. (ed.) (1997) *CIBO Energy Efficiency Handbook*, Council of Industrial Boiler Owners, Burke, VA, http://gasunie.eldoc.ub.rug.nl/FILES/root/2000/2043339/2043339.pdf (accessed 26 April 2010).

Part Three
Future Developments

Managing CO_2 Emissions in the Chemical Industry. Edited by Leimkühler

ISBN: 978-3-527-32659-4

11 Carbon Capture and Storage

Frank Schwendig

11.1 Background

Carbon capture and storage (CCS) is a technology for separating the CO_2 produced in technical processes and for its permanent leak-tight sequestration in repositories located well below the Earth's surface, thus ensuring that the CO_2 no longer reaches the atmosphere. This technology can give major CO_2-producing processes a climate-compatible shape.

Carbon dioxide is produced in industrial processes primarily when fossil fuels are burned in heat and electricity generation. The three fossil energy sources: coal, oil and natural gas, differ crucially in their applications and properties, and it is these differences that ultimately determine the focus of development work and the future deployment of CCS technology.

In energy terms, coal has a much higher carbon content, so that particularly large quantities of CO_2 are generated. Also, the main use of coal is in power generation, so that CO_2 emerges here in large amounts at central locations. These two facts mean that the conversion of coal into electricity is an ideal candidate for CCS. This being so, CCS is mainly being developed today for use in the generation of electricity from coal. Besides the technical considerations, the statutory environment, too, plays a role, and it is this aspect that explains why CCS is to be used, first of all, in the conversion of coal into electricity. Trade in CO_2 certificates currently extends to power generation, but leaves other industrial sectors still largely exempt. So, for power generation, the emitted CO_2 is already a cost factor, that is, the pioneering development of CCS for coal-based electricity generation also has an economic background.

The need to develop CCS for coal-based power generation becomes clear if we take a look at the expansion of coal-fired power plants especially in newly industrialized countries, above all China and India. Coal inputs have been growing dramatically since the start of the new millennium, and there is no end in sight for this development. Here, several specific differences between coal and the other two fossil energy carriers, oil and gas, become apparent. Coal is a very low-cost energy source. It is well distributed around the globe, so that its use makes a

Managing CO_2 Emissions in the Chemical Industry. Edited by Leimkühler

ISBN: 978-3-527-32659-4

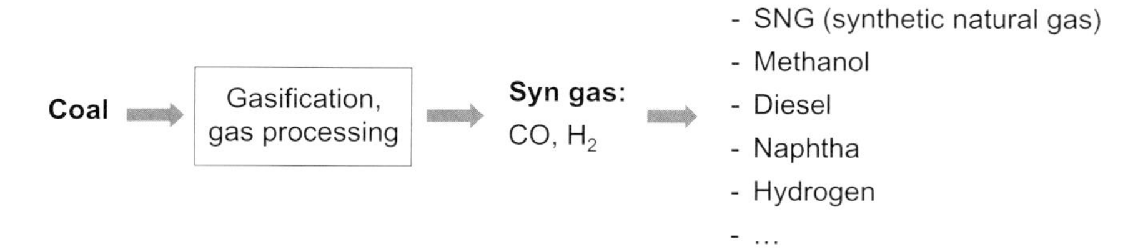

Figure 11.1 Products obtainable from coal gasification.

contribution to security of supply. In addition, the reserves are huge, so that the finite nature of coal, unlike that of oil in particular, is not a big issue as yet, nor will it be a vital question in the foreseeable future. On the contrary, numerous concepts and projects exist that envisage the use of coal as a substitute for oil and gas in the production of, for example, basic materials like naphtha, methanol, etc. – via coal gasification and synthesis – for the chemical industry. Figure 11.1 gives an overview of chemicals and products that can be obtained from coal gasification.

The dramatic rise in the use of coal is also associated with a serious increase in global CO_2 emissions. To deal with this problem, we need a technical solution that avoids CO_2 emissions, especially from coal-fired power stations, but also from other coal applications. The basic route involves raising the efficiency of power plants, which will then reduce coal inputs and, hence, CO_2 emissions while producing the same amount of power. With current developments, like an increase in the steam parameters from the present 270 bar/600 °C to 350 bar/700 °C in future or fluidized-bed lignite drying with internal waste heat utilization (WTA), the efficiency of coal-based power stations is expected to cross the 50% threshold (in terms of net calorific value). Compared with today's global average of approx. 31%, this means a reduction of nearly 40% in specific CO_2 emissions. Still, although an increase in efficiency has great potential for avoiding CO_2, it is not enough by itself. Much of the potential is swallowed up by the increase in the quantity of coal used in power generation, and even with efficiencies of over 50%, a considerable residual amount of CO_2 emissions remains, most of which must be eliminated if climate targets are to be reached. The only technology recognized today that can deliver this is CCS. This makes CCS an essential module in the strategies for lowering global CO_2 emissions.

What is more, CCS offers a special and unique feature for CO_2 mitigation: when CCS technology is used in biomass firing systems, we obtain negative CO_2 emissions, that is, in the overall balance, CO_2 is actually removed from the atmosphere. Power stations operated using biomass initially emit no CO_2 since, in a first step, the plants employed absorb CO_2 for their own growth, which they release again during combustion in power stations. Hence, this cycle is carbon-neutral. So, if the CO_2 is not emitted into the atmosphere thanks to CCS technology, but is permanently stored below ground, a biomass-based power plant is not only carbon-neutral, but is actually withdrawing CO_2 from the atmosphere. CCS is the only technology known today that will enable atmospheric CO_2 to be reduced tomorrow.

Even if CCS developments for the conversion of coal into power are very much to the fore today, other applications, too, will follow and gain in importance. CO_2 is already being separated from natural gas, for example, in order to treat the gas for further use. The captured CO_2 is stored in subterranean strata.

Besides power generation, other CO_2-intensive industrial sectors are increasingly thinking about opportunities for carbon capture and storage. Here, steel, cement, refineries, paper and the chemical industry in particular should be mentioned. In many technical processes, carbon capture is the only way to reduce CO_2 emissions to a minimum. Also, carbon capture opens up one more option: since the CO_2 is extracted in a virtually pure form, it can be used as resource or feedstock, for example, as a carbon source in chemical processes. This offers the opportunity of developing new chemical processes with a CO_2 basis.

So, CCS technology offers the chemical industry four interesting perspectives for managing CO_2:

- lowering CO_2 emissions in chemical processes;
- via coal gasification and synthesis, including capture and storage of excess CO_2, provision of an alternative to oil and natural gas as basic raw material;
- provision of CO_2 as feedstock for chemical processes;
- lowering of CO_2 emissions in power generation and process-steam generation.

Furthermore, developments and experience in the power-plant industry can yield important findings that can be transferred to the chemical industry. This is also and especially true of the two essential CCS modules: carbon storage and transportation. Here, it will be possible in future to build on the work of the power-plant industry, all the more so since it seems sensible to include the relatively low CO_2 amounts from the chemical industry in the transport and storage infrastructure of power-station projects.

11.2 General Description of the Technology with its Components

CCS consists of three main process steps.

- **Carbon capture:** In a first step, the CO_2 is captured from the process concerned (power plant, chemical plant, steel mill, ...) using special separation methods. It is then treated, so that it is available in a very pure form.

- **Transport:** If the captured CO_2 is to be sequestered below ground, it will in most cases have to be transported to a suitable storage site. This makes CO_2 transportation, too, an essential CCS component. Depending on circumstances, the CO_2 can be transported by pipeline, ship, railway or truck.

- **Carbon storage:** The third step is the permanent storage of the CO_2 in subterranean, geological formations that will prevent the CO_2 from escaping thanks to impermeable cap rock, but also using other mechanisms.

The capture methods developed today are able to remove approx. 90% of the CO_2 from the process, leaving 10% residual emissions. For each of the three main process steps, additional investment is necessary beyond that for a conventional power plant. Also, CCS technology requires considerable amounts of energy, for two reasons. First, carbon capture in the power-plant process needs energy in the form of heat and electricity. Second, storage of the captured CO_2 requires that it is available in a highly compressed form at supercritical pressures of more than 100 bar. Transportation, too, usually needs CO_2 compression. CO_2 compression is the second significant factor in the energy requirements of CCS technology. Overall, the technologies that are being developed for first-generation commercial CCS power stations must put up with an efficiency loss of over 10 percentage points.

The following sections discuss the three process steps in CCS technology.

11.3
Carbon Capture

For the capture of CO_2 from industrial processes, various concepts have been developed. Since, as was noted earlier, electricity generation and, hence, power-plant engineering were the initial focus of developments in CCS technology, present concepts are mainly geared toward deployment in power stations. All the same, some concepts use technologies that are already being implemented in other industries, though not for the purpose of avoiding CO_2, but rather on economic grounds to obtain valuable materials able to earn a price high enough to cover outlays. Moreover, these applications involve completely different, that is, much smaller, dimensions. Also, some of the solutions designed for power generation can also be transferred to CO_2 management in the chemical industry.

The tasks involved in capturing CO_2 from a power plant process can be described in greater detail and are summarized in Figure 11.2. Coal is burned in a power station, yielding a gas mixture (the flue gas) as combustion product that consists

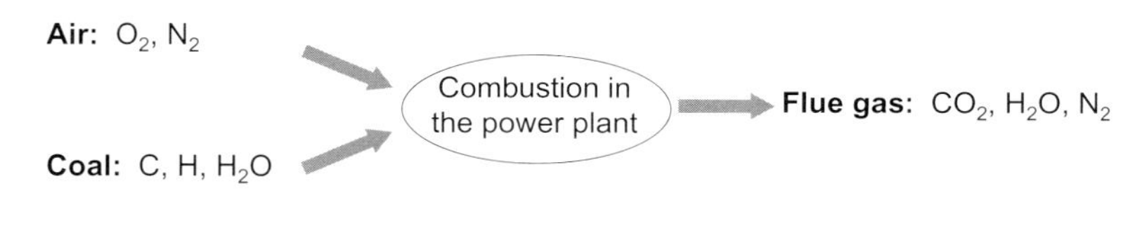

Figure 11.2 Tasks of carbon capture.

of numerous components, mainly CO_2, H_2O and, above all, N_2, which enters via the combustion air. Hence, the main task, to start with, is to develop concepts for separating the CO_2 from this mixture of three main elements. Besides these, however, other constituents, too, are encountered, like oxygen and argon as well as traces of SO_2, NO_x and others. Although taking account of these ancillary components does not affect the basic concept for capturing carbon, it greatly influences the concrete design and performance of the separation process.

To separate the CO_2 from the gas mixture of CO_2, H_2O and N_2, three basic technological routes are being pursued:

- post-combustion CO_2 scrubbing;
- the oxyfuel process; and
- IGCC with carbon capture (pre-combustion CO_2 separation).

It is expected that these three technologies will be implemented in the first generation of coal-fired power plants with carbon capture. Each has its specific properties and merits, so that parallel development of all three technologies is a sensible policy. In addition, building on these basic technologies, solutions are being investigated for the more distant future with a view to reducing the energy and cost outlays associated with carbon capture.

These three technologies for carbon capture are presented in the following sections. Proceeding from the basic idea of carbon capture, that is, splitting up the gas mixture of CO_2, H_2O and N_2, the discussion looks into specific properties, advantages and disadvantages, special development tasks and the deployment options in the chemical industry.

11.3.1 Post-Combustion CO_2 Scrubbing

11.3.1.1 The Basic Idea of Carbon Capture

In its basic concept, post-combustion CO_2 scrubbing is the simplest and most obvious technology for capturing the carbon from the power-plant process (see Figure 11.3). The idea is to target the flue-gas stream directly. Here, the CO_2 is selectively extracted from the waste-gas stream using a special scrubbing solution downstream of the power station. This is done in an absorber. The waste gas freed from the CO_2 is released into the atmosphere, while the scrubbing solution loaded with the CO_2 is further processed. It is conducted to a desorber where the CO_2 is driven out of the scrubbing liquid by the addition of heat. The CO_2 is now available

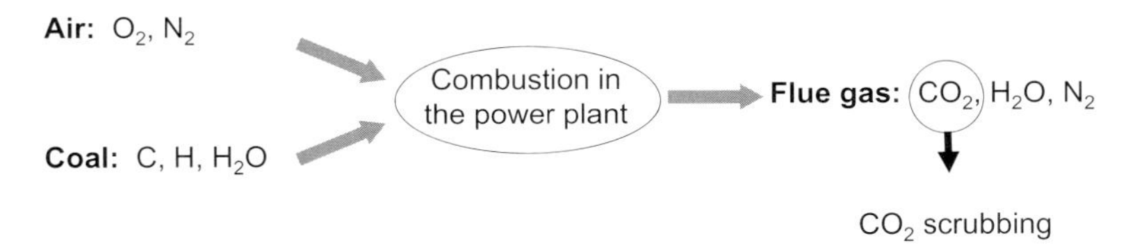

Figure 11.3 Basic idea of post-combustion CO_2 capture by means of scrubbing.

in a captured, pure form. The scrubbing solution regenerated in this way is returned to the absorber where it again absorbs CO_2.

11.3.1.2 Technological Implementation

The principle underlying this method is not new. It is already being used today, for example, to remove CO_2 from natural gas to obtain greater gas purity. However, the other uses are subject to completely different marginal conditions than in the case of CCS, in both technical and economic terms.

Carbon capture has in fact been used hitherto – within a process chain – to make either a marketable product or a product that brings some economic benefit. This means that carbon capture is currently only being used where the costs of capture are covered by the proceeds of sales. Carbon capture in a CCS setting, by contrast, is subject to different economic marginal conditions: the product, electric power, is not made possible in the first place by this process step and is not changed. The fact is that, from a cost angle, carbon capture is initially an additional outlay. This being so, the requirement that costs be kept low is much more stringent than for previous applications. On the other hand, cost savings can be obtained wherever CO_2 certificates are traded. A plant with CCS needs fewer CO_2 certificates thanks to the avoided CO_2 emissions. Ultimately, the level of CO_2 certificate costs decides whether CCS brings economic advantages. In principle, however, it must be noted that, unlike previous applications, considerable pressure exists to minimize the costs of carbon capture.

In technological terms, CO_2 scrubbing, too, faces completely new challenges. While in the past, CO_2 scrubbing was deployed wherever no technical problems worth mentioning existed, much more rigorous requirements must be met when it is used in capturing carbon from the waste gas of coal-fired power plants. This concerns, first, the necessary properties of the scrubbing solution and, second, the plant technology.

In the flue gas, the CO_2 has low partial pressure due to the atmospheric pressure of the flue gas itself, and due to the CO_2 volume percentage of below 15%. In this situation, physical absorption agents are unsuitable, and chemical absorption agents must be used. The scrubbing solution usual in today's processes under these conditions is a aqueous solution of monoethanolamine (MEA). The chemical equation for the absorption/desorption process is (Scheme 11.1):

$$HOCH_2CH_2NH_2 + CO_2 + H_2O \rightleftharpoons HOCH_2CH_2\overset{+}{N}H_3 + HCO_3^-$$

Scheme 11.1 Absorption/desorption of CO_2 using MEA.

In this form, however, MEA is unsuitable for the capture of carbon from the waste gas of coal-based power plants. The waste gas still has traces of various impurities even after flue gas scrubbing, and these have deleterious effects on the MEA solution. Particular mention must be made here of oxygen and SO_x. These components lead to rapid degradation of the MEA solution, so that much-too-frequent replacement of the solution would be necessary. What is more, energy consumption to recover the CO_2 in the desorber from the scrubbing solution is too high. This energy consumption is the crucial parameter for the overall efficiency of CO_2 scrubbing. In the case of MEA scrubbing, so much heat must be added here that the result is an excessive fall in the overall efficiency of power stations.

If CO_2 scrubbing is to be used as part of CCS, therefore, scrubbing solutions must be developed that are more resistant to impurities in the flue gas and also have lower heat requirements for desorption. These new scrubbing solutions then have to prove their fitness under operating conditions. For this purpose, pilot plants are being erected near power stations that are fed with a small diverted flue-gas stream. In RWE's Niederaussem power station, for example, a pilot CO_2 scrubbing plant was commissioned in August 2009 (see Figure 11.4). The pilot plants allow us to investigate the entire process of CO_2 scrubbing, including the performance of the scrubbing solution. They have the same building height as a later commercial-scale plant in order to provide the necessary distance for the

Figure 11.4 RWE's pilot CO_2 scrubbing plant at the Niederaussem power station.

complete absorption and desorption process, but are limited in their diameter and, hence, in their capacity to the minimum. The small amounts of captured CO_2 are usually added to the power station's waste-gas stream.

The pilot plants have one final development step still open before commercial operations, viz. the inclusion of carbon capture in the overall CCS process chain. CO_2 scrubbing must slot into the operating requirements of the power-plant process. On the one hand, this concerns the large mass flows of the waste-gas stream and the CO_2, which differ by a factor of about 10 from previous applications. On the other, the various operating modes in a power station must be tackled, including behavior in the case of failure, for example. The transport infrastructure and storage downstream of the CO_2 scrubbing system, too, are new marginal conditions, not only for carbon capture, but for the entire power station. To implement the necessary developments here, demonstration plants will be built in future, also to acquire the necessary knowledge and experience and ultimately to prove commercial deployability.

First of all, however, one major task is to develop and trial suitable scrubbing solutions in depth. In addition to resistance to impurities in the flue gas and to lowering energy consumption in desorbing the CO_2, further criteria must be taken into account. For instance, the scrubbing solution should have low steam pressure in order to reduce the loss of scrubbing agent in desorption. Reactivity with CO_2 must be high, so that the absorption process takes place at high speed. Also, of course, the scrubbing solution should be as harmless as possible and low-cost.

With these aims, three different substance classes are being considered from which suitable scrubbing solutions will emerge:

- **Amines:** The previous standard scrubbing solution MEA, too, falls under this heading. From the numerous chemical compounds in this substance group, suitable scrubbing solutions are established using screening methods and tests. Modifications and additional substances that act as activators, oxidation stabilizers or corrosion stabilizers contribute to creating the desired properties.

- **Ammonia:** Example of the chemical equation for the absorption/desorption (Scheme 11.2):

$$(NH_4)_2CO_3 + CO_2 + H_2O \rightleftharpoons 2\ NH_4HCO_3$$

Scheme 11.2 Absorption/desorption of CO_2 using ammonia.

- **Salts of amino acid:** Example of the chemical equation for the absorption/desorption process in this substance class (Scheme 11.3):

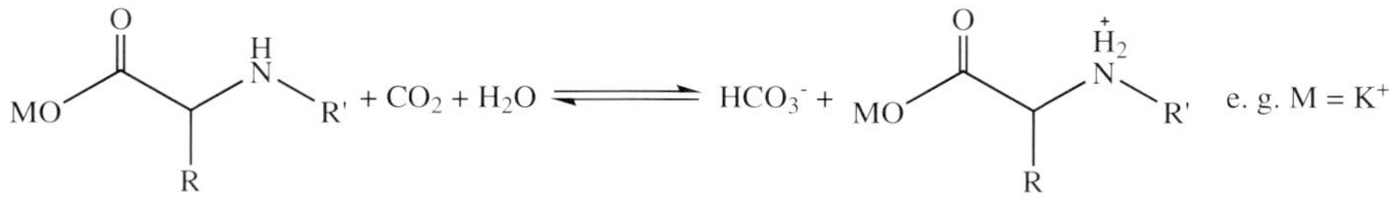

$$MOOC\text{-}CH(R)\text{-}NH\text{-}R' + CO_2 + H_2O \rightleftharpoons HCO_3^- + MOOC\text{-}CH(R)\text{-}\overset{+}{N}H_2\text{-}R' \quad \text{e. g. } M = K^+$$

Scheme 11.3 Absorption/desorption of CO_2 using amino acid salts.

For scrubbing solutions that can be commercially used in future CCS, all three substance classes offer interesting approaches and different positive properties, but also have specific drawbacks. Only long-term tests in pilot and demonstration plants will reliably show which scrubbing solutions and, hence, which process configurations will yield optimal results.

Compared with other processes for carbon capture, post-combustion CO_2 scrubbing has the interesting feature that it is downstream of the actual power-plant or production process. This means, first, that the main process is not–or hardly–impacted and, second, that existing plants can be retrofitted with CO_2-scrubbing systems. The chief point to be considered here is the fairly large space requirement for a CO_2 scrubber. A hard-coal-fired power plant with an electric output of 800 MW, for instance, needs additional space of some 16 000 m^2. Such a CO_2 scrubber separates some 550 $t\,h^{-1}$ of CO_2 from a volumetric flue-gas flow of about 600 $m^3 s^{-1}$ (standard temperature and pressure). Besides the space requirement, there are the steam needs for desorption, cooling-water needs and, possibly, further treatment of the flue gas (desulfurization) to be considered. Modern plants being erected now can be prepared for later retrofitting with CO_2 scrubbers. Ultimately, the specific measures to be taken here have to be assessed for each individual project. What is important is that retrofitting is not rendered impossible because no provision was made for this in the design and erection of the plant. Plants that are designed for retrofitting are referred to as 'capture-ready'.

11.3.1.3 Importance for the Chemical Industry

In principle, CO_2 scrubbing is a technique for capturing carbon that can be transferred to numerous processes and industrial sectors. As mentioned earlier, the technique is already being used in some processes. For more broadly-based deployment, it must be borne in mind that–in view of the multitude of possible processes–an appropriately adapted scrubbing agent is needed. The development and testing of such an agent takes time, a fact that must be taken into account in the strategic planning for CCS. For the CCS option to open, therefore, planning should start early on.

Besides actual carbon capture itself in processes used in the chemical industry, CO_2 scrubbing can also offer a new field of activity for this sector, since new scrubbing solutions will have to be developed and produced.

11.3.2 Oxyfuel

11.3.2.1 The Basic Idea of Carbon Capture

Proceeding from the above task description of capturing carbon from the gas mixture CO_2, N_2 and H_2O, the oxyfuel process starts by removing the nitrogen. This is done prior to combustion. In conventional firing systems the nitrogen reaches the flue gas via the combustion air. An upstream air-separation unit removes the nitrogen from the combustion air, and only the oxygen is used to burn the coal. Hence the process name: oxyfuel–combustion with pure oxygen.

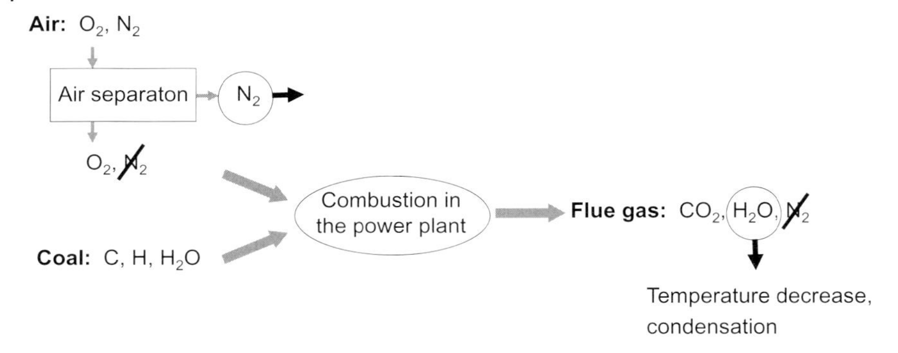

Figure 11.5 Basic idea of carbon capture based on the oxyfuel process.

So the flue gas now only consists of CO_2 and H_2O. The second step in carbon capture consists in the removal of the H_2O, which is separated from the CO_2 by lowering the temperature and condensing out. The basic idea of carbon capture based on the oxyfuel process is illustrated in Figure 11.5.

11.3.2.2 **Technological Implementation**

In a real situation, however, special heed must be paid in the oxyfuel process in particular to the other components of the flue gas, since they have a considerable influence on the quality of carbon capture. This specifically concerns gaseous components like O_2 or N_2, which can enter the process and, hence, the flue gas via various routes. Such gaseous components impair the condensing out of the water, since they lower the partial pressure of the water in the flue gas, while making it more difficult to obtain the necessary purity of the CO_2.

The gaseous substances of relevance here include above all the air components oxygen, nitrogen and argon. These gases can access the flue gas path at several points.

- The oxyfuel process ideally requires pure oxygen. Air separation produces an oxygen that still contains the other two main air components as well. To limit the investment and operating costs for air separation, an oxygen content of 95% is usually chosen. Higher purities, for example, 99%, make CO_2 treatment at the end of the process easier, but make air separation much more costly.

- The next point at which the gases can reach the process is combustion. If combustion is to be efficient, complete and low-polluting – this specifically concerns the formation of NO_x and CO as well as unburnt matter in the ash – oxygen must be added hyperstoichiometrically, that is, in excess. The excess oxygen can be found in the flue gas.

- Gaseous substances also reach the process via leaking units. The steam generator has, for example, sliding points to offset heat expansion in operations. In addition, the flue gas pass is operated in vacuum mode, so that the hot flue gases cannot escape. Conversely, however, it is possible in this way for air to reach the flue gas pass at the leaks. So, the oxyfuel process requires heavy outlays to seal the steam generator and the other components and, above all, to keep them sealed over the years of operation.

This being so, one focus of attention when designing the oxyfuel process is on the optimal balance between the outlays for avoiding gaseous substances at the front end of the process and the outlays for treating the CO_2 at the end of the process chain.

Combustion in the oxyfuel process requires special measures. Since combustion is carried out with pure oxygen, there is no atmospheric nitrogen, which would act in various ways in conventional combustion systems. During combustion, nitrogen as inert substance limits the temperature; in the flue gas, nitrogen accounts for the largest share of the mass flow. In addition, the lack of atmospheric nitrogen in the oxyfuel process shifts the entire thermal engineering. To counter these effects, cooled-off flue gas, that is, CO_2 and possibly H_2O is re-circulated from the end of the oxyfuel process chain and added to the combustion and in this way replaces atmospheric nitrogen. In this way, process conditions are obtained again that are similar to air combustion, although it must be borne in mind that CO_2 has physical properties that are different from those of N_2. For instance, CO_2 has a stronger damping effect on the reaction rate of combustion. Also, CO_2 has different heat radiation properties and a different effective heat capacity which is important for the heat transfer process in the steam generator. Determining the re-circulated CO_2 quantity, therefore, is of central importance in designing the oxyfuel process.

In large steam power plants, pulverized-coal combustion has gained acceptance for the burning of coal. With few exceptions, all new coal-fired power plants for the public power supply are based on this technology in which the coal is finely ground and combusted in a large number of burners. As alternative, fluidized-bed combustion, by contrast, is only used where special marginal conditions apply. For the oxyfuel process, too, pulverized-coal combustion is at the center of developments. Fluidized-bed combustion has special advantages for the oxyfuel process, however, so that it is attracting renewed attention. In fluidized-bed combustion it is possible, for example, to dimension for a lower amount of re-circulating CO_2 than in the case of pulverized-coal combustion. This has a positive effect on the auxiliary power requirement and reduces the volumetric flow in the overall flue gas pass. Components can be built correspondingly smaller. Also, some progressive developments to reduce energy consumption in carbon capture are based on fluidized-bed combustion, which will be explained in greater detail in Section 11.3.4 on further developments in carbon capture.

To obtain the necessary purity of the CO_2 using the oxyfuel process, additional outlays are needed in the CO_2 treatment. The nuisance components are, above all, O_2 and H_2O, but other gaseous companion substances should be removed from the CO_2 flow as well. On the other hand, air pollutants that also have to be taken into account in principle in the transportation and storage of CO_2 are sufficiently contained, thanks to the usual flue-gas cleaning facilities.

Even where the oxyfuel process is optimally designed, O_2 cannot be avoided entirely owing to the oxygen excess needed for the combustion and owing to the leaks, while H_2O is still present in the CO_2 flow even after the lowering of the temperature and the condensing out in line with the temperature-dependent partial pressure. Additional process steps must be taken to separate O_2 and H_2O from the CO_2 flow. In earlier concepts, H_2O was bound by a hygroscopic solvent. Due to the patently rigorous requirements to be met by CO_2 purities, however, more recent concepts are based on costly distillation methods, so that both H_2O and O_2 as well as other gases are removed.

11.3.2.3 Importance for Chemical Industry

For the chemical industry, the oxyfuel process is unlikely to be of particular importance. It is a concept that was developed for the circumstances of a coal-fired power plant. Besides this application, the only possibly relevant interface to the chemical industry–as in the case of all carbon capture technologies–would involve CO_2 as feedstock.

11.3.3 ICGG with Carbon Capture

11.3.3.1 The Basic Idea of Carbon Capture

IGCC is short for 'integrated gasification combined cycle'. The name itself indicates that this is a completely different power plant technology. Here, the coal is not burned and used to generate steam within a steam turbine process as in the other two processes, but is first converted into a fuel gas. After a series of conversion and treatment steps, the fuel gas generates electricity in a combined-cycle gas turbine process.

In order to capture the CO_2 in the IGCC process, the nitrogen–much as in the oxyfuel process–is first removed at the very start of the process chain using air separation. The coal is then gasified with pure oxygen. A raw gas emerges with the main components CO and H_2. In a 'conventional' IGCC process without CO_2 capture, this raw gas would be directly burned in the gas turbine after undergoing various cleaning steps, and the CO share would form CO_2, which is emitted into the atmosphere. To avoid this CO_2 formation, the CO must be removed before the gas turbine process, though without losing its energy content. For this purpose, two process steps are supplemented. First, the CO is converted into CO_2 using water vapor. At the same time, hydrogen is produced, to which the CO's combustion energy is transferred. The chemical equation for this CO shift reaction is (Scheme 11.4):

$$CO + H_2O \rightleftharpoons CO_2 + H_2$$

Scheme 11.4 CO shift reaction.

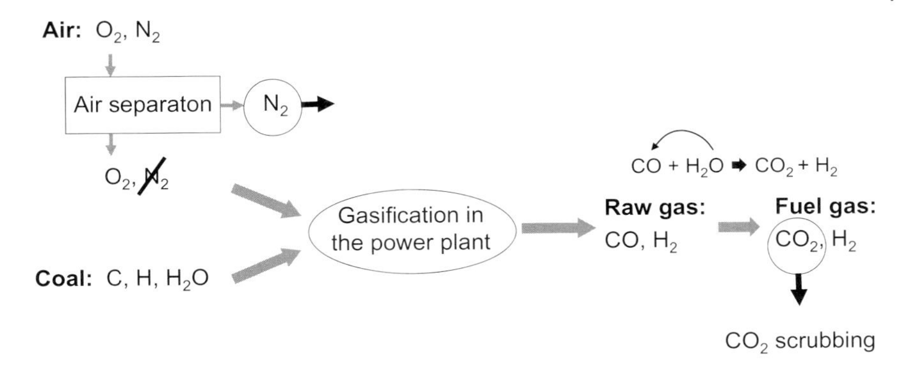

Figure 11.6 Basic idea of pre-combustion CO_2 capture based on the IGCC process.

The fuel gas mixture now consists of CO_2 and H_2, with the hydrogen stemming from gasification and the CO shift reaction. In a second process step, the CO_2 is now washed out of the fuel gas. Pure hydrogen finally reaches the gas turbine, so that the waste gas no longer contains CO_2.

Because the CO_2 is captured before (complete) combustion, this technology is also referred to as pre-combustion (Figure 11.6).

11.3.3.2 Technological Implementation

The IGCC process without CO_2 capture was implemented in the 1990s in four demonstration plants (Buggenum/Netherlands/250 MW, Wabash River/USA/250 MW, Polk/USA/250 MW, Puertolano/Spain/300 MW). After that time, a series of projects was started, only one of which was ultimately implemented. For these projects, most of which were launched in the USA, the IGCC process was an interesting option, because it mitigates air pollutant emissions. What is more, in the USA, unlike Europe, conventional steam power plant engineering was not so far developed, the hope in the USA being that efficiency advantages could be obtained with IGCC. However, IGCC is much more complex and has higher investment costs, so that it has not been able to gain traction to date, although the picture is changing thanks to the new carbon capture remit. Carbon capture in the IGCC involves lower additional outlays than in the case of the other two concepts, so that, proceeding from the higher outlays for the basic power plant process, IGCC with carbon capture overall is a competitive and promising proposition.

For the gasification of coal with oxygen, entrained-flow gasifiers are usually envisaged. The gasification temperature is about 1500 °C. For lignite, with its higher share of volatile matter compared to hard coal, too, a gasifier was developed on the basis of fluidized-bed technology which works at lower temperatures of below 1000 °C. Gasification is under pressure, with the pressure in the IGCC process being determined by the inlet pressure of the gas turbine. Depending on the gas turbine, the pressure can be in excess of 40 bar. Converting the coal into a fuel gas or synthesis gas in autothermal gasification is on the basis of numerous equilibrium reactions running in parallel. The most important are (Scheme 11.5):

$$\text{a) } C + CO_2 \rightleftharpoons 2\,CO$$
$$\text{b) } C + H_2O \rightleftharpoons CO + H_2$$
$$\text{c) } CO + H_2O \rightleftharpoons CO_2 + H_2$$

Scheme 11.5 Important equilibrium reactions in gasification; (a) Boudouard equilibrium; (b) Water gas reaction; (c) Shift reaction.

The energy necessary for the gasification process is partially added by combusting the coal.

Downstream of the gasifier, the raw gas must be cooled, so that the subsequent cleaning and conversion steps can be implemented. The greatest efficiency is achieved if the heat of the raw gas is used in a heat exchanger to generate steam which produces electricity in a turbine. Such a heat exchanger is prone to fouling, however, since the raw gas carries with it the ash of the coal which is sticky or liquid at high temperatures. One alternative for avoiding the fouling problems is quenching with water, although this involves losing much of the exergy in the raw gas. Quenching in the IGCC with carbon capture has one more merit, however: the CO shift reaction necessary for carbon capture needs water vapor as reaction partner, and this must be added to the raw gas. With the quenching action, this step is performed simultaneously.

Once the temperature is lowered, the process steps of raw gas treatment follow. These comprise desulfurization, CO shift reaction and CO_2 scrubbing. In gasification, H_2S emerges instead of the SO_2/SO_3 that forms during combustion in conventional steam power plants. For the sequence of the process steps, there are two alternatives, each having a different impact on energy requirements and on the efficiency of carbon capture. One factor is that H_2S and CO_2 scrubbing are marked by low temperatures, which – depending on the scrubbing solution deployed – can be well below 0 °C, whereas the CO shift reaction runs at a temperature of about 300 °C. If the CO shift reactor is located downstream of the H_2S scrubbing system (sweet shift), several temperature leaps must be performed, which entails exergy losses. However, there is also the option of locating the CO_2 shift reactor upstream of the H_2S scrubbing system (sour shift), so that the ups and downs of the temperature profile in the process are avoided. One drawback, however, is that the choice of catalyst is restricted for the CO shift where H_2S is present. Accordingly, converting CO into H_2 and CO_2 is less complete, so that some of the CO remains in the fuel gas and is emitted as CO_2 after combustion in the gas turbine. Consequently, the degree of CO_2 capture is lower.

The CO_2 in the IGCC is ultimately separated using a scrubber. In this respect, there are fundamental differences compared with post-combustion CO_2 scrubbing. In the IGCC, the fuel gas is under pressure. Moreover, the CO_2 concentration in the fuel gas is much higher than in the flue gas downstream of a conventional steam power plant. Under such marginal conditions, physical scrubbing processes can be used that get along with less energy. The most common scrubbing substances are methanol (Rectisol), dimethyl ether polyethylene glycol (Selexol) and *N*-methyl-2-pyrrolidone (Purisol).

A fuel gas finally reaches the gas turbine that consists of virtually pure hydrogen. Hydrogen has substantially different combustion properties than the natural gas which is usually input in gas turbines:

- hydrogen spreads flames eight times faster;
- the stochiometric combustion temperature is 150 °C higher;
- the volumetric flow is three times greater, with the same energy flow.

These properties necessitate various measures in the combustion area. Hydrogen is diluted with nitrogen, which is generated from air separation parallel to oxygen extraction. Diffusion burners are used, since the combustion process is not at present completely controlled with the modern pre-mix burners. However, diffusion burners produce too much NO_x, so that a denox system must be installed downstream of the gas turbine.

Basically, the principle of pre-combustion CO_2 capture can be used for natural gas as well. Here, the gas must be re-formed. However, the conversion into CO and H_2 leads to a high exergy loss. This being so, post-combustion CO_2 scrubbing is more suitable for gas-based power plants.

11.3.3.3 Importance for the Chemical Industry

For the chemical industry, the IGCC process with carbon capture can play a special role. The background is that coal gasification and synthesis gas treatment can be used not only for a downstream gas turbine process to produce power: synthesis gas is one of the basic raw materials in the chemical industry. Coal gasification with subsequent synthesis of, for example, methanol, naphtha, diesel and other products, is already a common method used especially in countries where – for various reasons – abundant coal resources will be used as a substitute for crude oil. The developments for the IGCC power plant with carbon capture and storage now combine coal gasification and synthesis gas production with CCS technology. This means that the use of coal as a basic raw material by the chemical industry is possible without increasing CO_2 emissions. In this respect, the development work for the IGCC power plant with carbon capture delivers the basis for how the CO_2 must be treated for subsequent transportation and storage below ground. In addition, the development work also supplies the necessary findings for the link-up of coal gasification/synthesis gas production with CO_2 transport and storage, which ultimately constitute an interdependent overall system. For a plant from which CO_2 is captured, carbon transport and storage represent a special marginal condition that must be taken into account in all operating conditions, like start-up and shut-down, load changes and failures.

Again, of course, the captured CO_2 is to be viewed as a possible feedstock for chemical processes. Since the CO_2 is ultimately separated using a scrubbing process, purity is especially high, as in the case of post-combustion CO_2 scrubbing.

11.3.4
Technologies to Reduce Energy Consumption for Carbon Capture

Capturing carbon from a process is energy-intensive. Depending on the technology employed, this can be due to various process steps. Besides the need to compress the CO_2 to the transport and storage pressure for all three methods of carbon capture, special mention must be made of:

- air separation in the oxyfuel process and in the IGCC to provide a nitrogen-free oxygen carrier;
- desorption in CO_2 scrubbing processes.

One future task will be to make CCS technology more efficient. The main objective here is to lower fuel consumption. Higher efficiency also reduces the expenditure on equipment needed (e.g., number of power plants). Another aspect is the additional lowering of CO_2 emissions.

The main causes of the energy consumption mentioned are found in the separation processes. When oxygen is produced, air is separated; in CO_2 scrubbing processes, components in the flue gas flow or the fuel gas flow are separated from one another. For these separation jobs, concept ideas exist that may be able to get along with much less energy than conventional cryogenic air separation or the CO_2 scrubbing processes:

- membranes;
- chemical looping combustion (CLC);
- carbonate looping.

For membranes, suitable materials must be developed that are able to perform the separation function with sufficient efficiency (selectivity) and limit the expenditure on equipment (throughput, space requirement), and are resistant to the disruptive substances in the gas flow. In addition, constructive solutions must be developed for the membrane modules, since mechanical stresses occur owing to the differences in pressure and temperature.

Chemical looping is used to provide pure oxygen for combustion. The oxygen is added to combustion by way of a metal oxide. The process involves a link-up between two fluidized-bed reactors. In the first fluidized bed, a metal absorbs the oxygen from the air. The metal oxide that emerges is circulated to the second fluidized bed where it releases the oxygen again directly in the combustion zone. The metal that is now available again is recirculated to the first fluidized bed where it absorbs more oxygen.

Carbonate looping is a post-combustion process for capturing carbon from a flue gas flow. For this purpose, the CO_2 is brought into contact with CaO; $CaCO_3$ forms. This process, too, involves two fluidized-bed reactors. The flue gas, which has a temperature of approx. 120 °C downstream of the power plant, is freed from CO_2 in the first fluidized bed (carbonator) in a reaction with CaO. The exothermal reaction to form $CaCO_3$ occurs in the fluidized-bed at a temperature of some 650 °C. At this temperature, the waste gas free of CO_2 is conducted to a heat exchanger to use its heat, while the $CaCO_3$ reaches the second fluidized bed (calciner). There, it is heated to a temperature of about 900 °C, which again drives out the CO_2 from the $CaCO_3$. In this way, CaO forms again and can bind CO_2 once more in the first fluidized bed. Also, the CO_2 is available in a separated form. Still, in the second fluidized bed, a considerable amount of heat must be added to the process. For this, coal is combusted in the fluidized bed. To ensure that the separated CO_2 does not mix again with inert substances (especially nitrogen), the coal is combusted together with oxygen from an air separation unit. So, at this point, an oxyfuel process is implemented. The coal fired in this oxyfuel section of the carbonate-looping process amounts to about 1/3 of the entire coal input in such a power plant. The main positive effect on energy consumption is due to the fact that:

- compared with post-combustion CO_2 scrubbing, no desorption heat is lost, since all heat expended can ultimately be used to generate steam at high temperatures;
- compared with the oxyfuel process, only 1/3 of the pure oxygen is needed, which significantly reduces the energy requirements for air separation.

The separation methods cited, besides their own direct development, also require a re-design and adaptation of the basic process, since their inclusion needs certain process parameters – above all high temperatures and increased pressures in places – which are usually not available at the points considered in the basic process.

11.4 CO_2 Transport

One essential module in the CCS process chain is the transportation of the captured carbon [dioxide] to the CO_2 storage site. Many discussions of CCS technology give CO_2 transportation low priority or even ignore it altogether. In most projects, however, there will be a greater or lesser distance between the capture and the sequestration sites. This being so, CO_2 transport plays a key role in the implementation of a CCS project. CO_2 transport crucially determines:

- the technical steps, as well as the expenditure on equipment and energy consumption for carbon capture and for the conditioning and preparation of the CO_2;
- the handling of all operating conditions and the associated technical outlays;
- costs and economic efficiency, that is, commercial feasibility;
- the efficiency of CO_2 reduction and, hence, effective CO_2 avoidance;
- the implementability of a CCS project in view of local circumstances.

For transporting CO_2, there are various possibilities:

- pipeline;
- ship;
- railway;
- truck.

The crucial issue in the choice of transport mode is where the carbon is captured and stored, that is, criteria like the distance between the two sites, onshore or offshore carbon storage, accessibility of a river, etc.

11.4.1 Pipeline

Transporting CO_2 by pipeline is the most efficient method. This is because, unlike the other three options, CO_2 is prepared for pipeline transport in the state that is also required for storage. This means that no further conversion of the aggregation state is necessary with the associated large energy consumption. Downstream of capture, the CO_2 is brought to supercritical pressure at the entry point to the pipeline. Also, pipeline transport is a continuous process. Immediately after capture and compression, the CO_2 is pipelined to the CO_2 depository for storage. In the other variants, ship, railway and truck, by contrast, transportation is discontinuous. This necessitates interim storage of the CO_2, both at the start and at the end of the transport chain. Depending on specific circumstances and the accessibility of the plant in which the carbon is captured, and on the CO_2 storage site, it might even be necessary to re-load the CO_2, entailing further interim storage.

Once the carbon [dioxide] is captured from the process, it is available in a virtually pure form under atmospheric pressure or slight overpressure. For transportation by pipeline, the CO_2 is compressed to a supercritical pressure in a range between about 100 and 200 bar. The lower threshold ensures that the pressure remains above the critical pressure of 74 bar along the entire transport route, even taking account of pressure losses. This rules out any phase change during transport. On long routes, however, it will become necessary to deal with pressure loss by having booster stations at certain intervals. An economic alternative is to choose such a high pressure at the entry point into the pipeline, that the pressure remains sufficiently high along the entire transport route even without further compression.

For the first CCS projects executed in the power plant sector and used to demonstrate CCS technology on a commercial scale with all the steps in the process chain, it is necessary to implement individual, low-capacity pipelines. For a typical plant size, the captured CO_2 to be transported here amounts to some 1 to 3 million tons per year. Compared with the CO_2 amounts that can occur and possibly be separated in processes in the chemical industry, however, these quantities are already very large. Such individual, small pipelines have very high specific costs, together with enormous planning and approval outlays. This being so, in any future commercial implementation and expansion of CCS technology, the aim must be to aggregate carbon dioxide transportation from regions with high total CO_2 and to channel the carbon by shared pipeline to the CO_2 storage sites. Such pipelines require far-sighted planning in sufficient capacities. Ideally, a comprehensive pipeline infrastructure will emerge which will connect all potential carbon-capture installations and the CO_2 depositories in a region or a country. In many cases, this would be the most efficient, flexible and least-cost concept for transporting CO_2.

For the chemical industry, with its relatively low amounts of CO_2 compared with the power plant sector, a link-up to other pipeline projects is of special significance, so that future CCS projects should aim at early inclusion in pipeline planning and, specifically, the planning of a pipeline infrastructure. For a specific location, there is also the job of connecting to a shared long-distance pipeline, that is, the planning and building of a spur line.

Pipelines for transporting CO_2 have been in operation since the 1970s. The USA runs a pipeline network with a total length of over 3000 km to provide CO_2 for enhanced oil recovery (EOR). Every year, approx. 35 million tons of CO_2 are transported in this way. Pipelines for transporting CO_2 are largely identical with those for natural gas transport, specifically as regards monitoring measures and security installations.

11.4.2 Ship

For certain constellations, shipping CO_2 can be an interesting or even necessary option. In transportation by ship, a distinction must be made between offshore transportation on oceans and onshore transportation by river.

Numerous potential carbon storage sites are located below the seabed. As a general rule, such CO_2 storage sites can also be reached by pipeline, although this may involve very high outlays, specifically for long distances, and will meet with technical constraints. In such a case, the option exists of transporting the CO_2 from the coast to the storage site using large tankers. In this respect, there is an analogy to transporting liquefied natural gas, because CO_2 must also be liquefied for shipping. Typical parameters for the liquefied state – depending on the design of the tanks – are, for example, 7 bar and −50 °C or 20 bar and −30 °C. CO_2 tankers with a capacity of some 10 000 tons today already ply the oceans. Even if much larger CO_2 ships are feasible in the future, it becomes clear that a very large number of ships would be necessary for the envisaged CO_2 quantities.

Transporting liquefied CO_2 onshore using, for example, barges on rivers is a practicable alternative only where marginal conditions are especially favorable. Shipping can be considered only if loading an inland ship is possible close to the site of carbon capture and only if unloading is at a favorable destination. The CO_2 quantities that can feasibly be transported by inland ships are altogether very limited, however.

The biggest drawback in shipping CO_2 is that the CO_2 must be liquefied for this purpose. Liquefaction means high additional energy consumption in the CCS process chain, since this step is being taken solely for the transportation stage. Before the CO_2 is injected into the depository, it must be compressed to a supercritical pressure, and reach normal temperatures by the addition of heat. The specific, electric energy requirement for liquefaction amounts to about 0.15 kWh per kg of CO_2. Furthermore, CO_2 liquefaction requires an additional plant with a capacity that can far exceed previous sizes, depending on the project. For discontinuous shipping of the carbon dioxide, the liquefied CO_2 must be placed in interim storage in large tanks. Liquefaction plant and tanks have considerable space requirements (e.g., in a magnitude of 1 ha for 1 million tons of CO_2 per year), which must be taken into account in addition to the space needs for capturing carbon.

11.4.3
Railway and Truck

Generally speaking, CO_2 can also be transported to the storage site by railway or truck. For technical CO_2, this is the standard method. However, for carbon captured from industrial plants and then transported to a storage site in correspondingly large amounts, railway and truck can play no more than a subordinate role. The feasible capacity is severely restricted by the railway cars needed and the available routes or by the truck units required and the burden on traffic.

One serious disadvantage, as in the case of shipping, is that the CO_2 must be liquefied, stored and re-loaded several times, which involves considerable expenditure on equipment and energy consumption.

11.5
CO_2 storage

11.5.1
Underground Storage of CO_2

At the core of the idea of CCS technology is the subterranean, permanent storage of CO_2 to prevent it entering the atmosphere. The CO_2 is injected below impermeable cap rock which prevents the CO_2 from escaping to the Earth's surface. This is modeled on natural gas which, in such conditions, has been fixed in subterranean deposits across geological periods. The subsoil, with its structure, offers various options that can serve as storage sites for CO_2.

Located below the impermeable cap rock are numerous porous rock layers with cavities filled with various media. The pores contain natural gas or oil, for example. One obvious possibility for storing CO_2, therefore, is to use depleted gas or oil fields. These two alternatives have the additional merit that the geology is very well investigated. Even gas or oil fields that are still in operation can be considered for storing carbon. Here, another benefit emerges, since the injection of CO_2 presses out more oil or gas from the deposit, thus increasing the yield. These techniques are called enhanced oil recovery (EOR) and enhanced gas recovery (EGR) resp. Here, CO_2 storage even generates economic added value. In the case of oil, this has been practiced in the USA ever since the 1970s. Today, the USA injects some 35 million tons of CO_2 for EOR annually into the underground. EGR, by contrast, is not yet state of the art. The problem here is a possible mixing of gas and CO_2 which drastically impairs the quality of the natural gas.

For many countries or plants in which carbon can be captured, the CO_2 storage capacities of the gas or oil fields within reach are so low, however, that they cannot make any significant contribution to a broad-based application of CCS. In such cases, recourse is had to another option for storing CO_2. In what are by far the largest areas, the porous rock beneath impermeable cap rock is filled with salt water. These areas are called saline formations or saline aquifers. The salt water makes it clear that there is no contact with the groundwater.

The salt water that fills the pores of the saline formations must be displaced when the CO_2 is injected. This can only be done to a very low degree, of course, since otherwise the pressure in the aquifer would be inadmissibly increased. Hence, only less than 1% of the pore volume is available for storing carbon. The precise value must be established for each particular storage site. All the same, the global capacities for storing CO_2 in saline formations are very large. It is these capacities that form the basis for CCS's crucial contribution to climate protection.

Storing CO_2 in saline aquifers, too, is already being practiced. In Norway, for example, CO_2 is separated, using amine scrubbing, from natural gas extracted from the Sleipner gas field. The carbon is then returned to the area of the gas field and injected into an adjacent saline aquifer. Every year, approx. 1 million tons of CO_2 has been stored in this way since 1996.

The suitability of a storage site for injecting and storing CO_2 depends on various factors. In addition to the intake capacity for CO_2, the speed with which the carbon dioxide can be injected into the storage site, too, plays a major role. To establish the pertinent, physical parameters, like porosity and permeability, injection tests ultimately have to be made. Only on this basis is a final characterization and assessment of a storage facility possible.

The CO_2 that is injected into a saline aquifer is prevented from escaping by several mechanisms. The impermeable cap rock under which the CO_2 accumu-

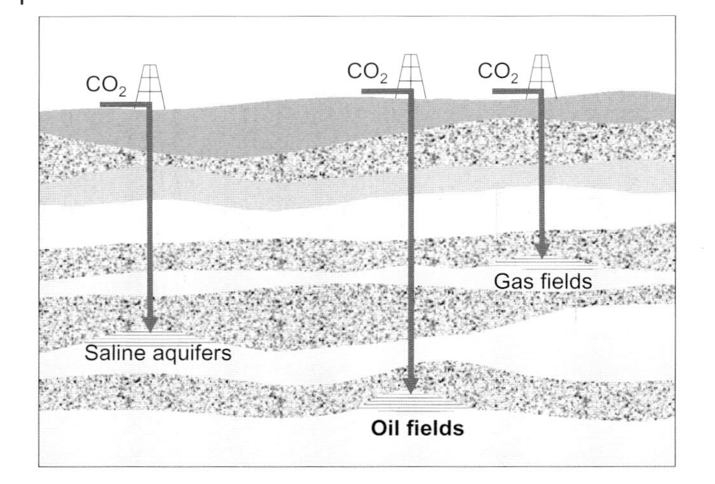

Figure 11.7 Three options for underground CO_2 storage in geological formations.

lates due to its lower density relative to the salt water is, of course, the main factor and the most important criterion for the storage site being considered. Beyond this, however, the carbon is trapped in a saline aquifer in several additional ways. Once the CO_2 has reached its position in the aquifer after injection, the surrounding salt water encloses the CO_2 in the pores. The capillary forces keep the water in the channels and prevent the CO_2 from displacing the water and from migrating. In a second mechanism, some of the CO_2 is dissolved in the salt water. Another part forms carbonates. These additional mechanisms ultimately mean that most of the CO_2 can no longer move freely. Besides the cap, which is actually already sufficiently leak-tight, this ensures that the CO_2 remains permanently in the depository. The different options for underground CO_2 storage can be viewed in Figure 11.7.

There are potential CO_2 storage sites both onshore and offshore below the ocean. Onshore storage sites have the advantage that they can be reached with much lower outlays, although the issue here tends to be one of public acceptance. Accessing offshore CO_2 storage sites usually requires immense material and energetic outlays.

The storage of CO_2 means that the CO_2 must fulfill certain purity requirements. Hence, when the CO_2 capture process is defined and designed, the possible impurities of the CO_2 must be precisely analyzed and, if necessary, removed in further treatment steps.

11.5.2 Carbonation

One alternative to subterranean CO_2 storage as a method for permanently trapping CO_2 is carbonation. In the present context, this means artificially imitating the weathering of rocks. Reaction partners for CO_2 are silicates like serpentine (magnesium silicate), and the process produces a carbonate like magnesite; CO_2 is permanently bound, the products are chemically stable, and can be deposited without any problem. Silicates are distributed around the world in such quantities that they would suffice for all known coal deposits or, rather, for the CO_2 quantities emerging from them, to form carbonate.

Despite its initially attractive properties, carbonation is only feasible at best for niche applications. The problems here lie in the logistics and in energy consumption. Each ton of CO_2 needs about 7 tons of rock as reaction partner. This means that, for each ton of coal, an additional 10 to 20 tons of silicate rock must be taken to the power plant and also hauled off again as carbonate rock, which is impossible to handle in logistical terms. In nature, the carbonation reaction is very slow, because of the low concentrations of CO_2 in the air, the small specific contact surface between silicate rock and CO_2 and the activation energy for the reaction. To obtain an acceptable reaction speed, various measures must be taken. Separating the CO_2 produces a high CO_2, concentration. In addition, the CO_2 is pressurized. The silicate rock is fine-grain milled, so that the reaction surface is greatly enlarged, and heat is added to the reactor to accelerate the reaction. Despite all these measures the carbonation reaction still takes hours, that is, it is not suitable for technical deployment involving large quantities. Moreover, the measures are so energy-intensive that the efficiency of a power station is halved.

This being so, carbonation is a process that can only be used where marginal conditions are extremely favorable, that is, where transport routes for the material are short and where the product can be marketed. It is no alternative to storing CO_2 below ground.

11.6 Efficiency and Economy Parameters of CCS

11.6.1 Efficiency Parameters of CCS

The efficiency of CCS as a measure for lowering CO_2 emissions is described using various parameters:

- rate of CO_2 capture;
- energy requirements for CO_2 capture, transport and storage;
- avoided CO_2.

First of all, in the process itself, the CO_2 capture rate is an important characteristic value and design parameter. It describes how much of the CO_2 that would otherwise be emitted into the atmosphere is captured from the process and stored. In the three methods developed for CO_2 capture from power plants, the capture rate is typically some 90%, possibly even 92%. A much higher capture rate can only be achieved with disproportionately large outlays. The CO_2 capture rate takes no account of the energy consumption for capture, transport and storage, and, hence, ignores the CO_2 emissions associated with such energy consumption.

In assessing the overall CCS process chain, it is of course important to quantify the energy requirements. Depending on the capture process, the chief consumers are: air separation, desorption of the CO_2 from the scrubbing solution, and any conversion steps. Irrespective of the capture method, the CO_2 must be compressed to a supercritical pressure (only transportation by pipeline as the standard case is considered here, ignoring the ship, railway and truck options as special cases with any given amount of complexity.) In this respect, energy is used both in the form of heat and in the form of electricity. In the power plant sector, the energy requirements can be covered by the power station process, so that optimal integration is possible. Any heating steam, for instance, is made available as turbine extraction steam, which makes this the most efficient concept. For power stations, therefore, the entire energy requirements for CCS can be combined and stated as a reduction in net electric output or efficiency. For the three capture processes, including transport and storage, the equivalent energy requirements per ton of captured CO_2 are in a range of roughly 0.25–0.3 MWh electricity, so that efficiency is lowered by approx. 10 percentage points. CO_2 compression, which is necessary for transport and storage irrespective of carbon capture, is responsible for a pro-rated efficiency fall of 3–4 percentage points.

To determine the effective energy requirements for CO_2 capture in other industrial plants, a precise examination of the provenance of heating steam and electricity is necessary.

For climate protection, finally, it is the amount of CO_2 avoided, that is, the effective reduction in CO_2 emissions, that is relevant. This is the result of the two parameters described above, the CO_2 capture rate and energy consumption with the associated additional CO_2 emissions. The basis for calculating the avoided CO_2 should be the target quantity of the product to be made. The CO_2 amount avoided is then obtained from the difference of the CO_2 emissions with and without CCS. In this respect, account must be taken of all components in the process chain, including the CO_2 emissions occurring in the provision of auxiliary energies. In plants where, as in power stations, all CCS components constitute an integrated, self-contained system for the energy supply, the specific CO_2 emissions of the product made can be established relatively simply (in power generation: ton of CO_2 per MWh of electricity). The specific CO_2 emissions for plants with and without CCS then very graphically express the efficiency of the CCS measure.

In summary, the following rough values may be estimated for a rough balance and assessment of CCS:

- CO_2 capture rate: 90%;
- fall in efficiency: 10 percentage points.

The effectively avoided CO_2 quantity depends on the efficiency of the basic power plant without CCS:

$$a_{\text{eff}} = 1 - \eta_{\text{CCS}} \cdot (1 - cr) \tag{11.1}$$

where

a_{eff}: effectively avoided CO_2 amount compared with the basic power plant
η_{basic}: efficiency of the basic power plant without CCS
η_{CCS}: efficiency of the power plant with CCS
cr: CO_2 capture rate

If the efficiency of the basic power plant is 50%, we obtain–with the rough values for the CO_2 capture rate and the fall in efficiency–effective CO_2 avoidance of 87.5% compared with the basic power plant. (NB: Of course, the CO_2 quantity avoided, relative to an existing power plant to be replaced, is obtained using a completely different calculation, and is often above the capture rate, since in many cases the new CCS power station will have much higher efficiency than the existing decades-old power plant.)

11.6.2
Assessing the Economic Efficiency of CCS

To assess and classify the economic efficiency of measures in reducing CO_2, the parameter 'CO_2 avoidance costs' has become established. This parameter describes how high the costs of avoiding 1 ton of CO_2 are in employing a process or measure. The CO_2 avoidance costs are obtained from the difference of the specific manufacturing costs for a product, divided by the difference of the specific CO_2 emissions:

$$a = (c_1 - c_2)/(e_2 - e_1) \tag{11.2}$$

where

a: CO_2 avoidance costs
c_1: specific manufacturing costs for the product according to process 1 (e.g., plant with CCS)
c_2: specific manufacturing costs for the product according to process 2 (e.g., conventional plant without CCS)
e_1: spec. CO_2 emissions in the manufacture of the product according to process 1
e_2: spec. CO_2 emissions in the manufacture of the product according to process 2

For power stations, the specific power generation costs 'c' are stated in €/MWh or $/MWh, and the specific CO_2 emissions 'e' in t CO_2/MWh. Accordingly, the CO_2 avoidance costs have the unit €/t CO_2 and $/t CO_2 resp.

Using the CO_2 avoidance costs, we can express in a simple manner how cost-efficient a measure for lowering CO_2 is. It can be used not only for CCS, but, ultimately, for all technologies or processes. Using this parameter, it is possible, therefore, to compare various processes with one another in order to identify the least-cost CO_2 reduction measure.

The CO_2 avoidance costs also permit the costs of CO_2 certificates to be compared. If the CO_2 avoidance costs are above the CO_2 certificate costs, it costs less, from a purely commercial angle, to apply the (conventional) process 2 and to buy CO_2 certificates for the higher CO_2 emissions. If, by contrast, the CO_2 avoidance costs are below the costs of CO_2 certificates, the CO_2-lowering process 1 is also advantageous from a commercial angle. The problem in this view of things, however, is assessing future trade in CO_2 certificates and the developments in certificate prices.

In applying this system, however, two points must be heeded. The CO_2 avoidance costs are used to assess a process 1. To this end, the difference between it and a process 2 is established. Specifically, if various processes 1 are compared with one another, attention must be paid to an appropriate selection of the basic process or also of the basic processes 2. The conclusiveness and interpretation of the parameter 'CO_2 avoidance costs' is very closely related to defining the basic process 2. Here, special care is needed.

The second point to be heeded is that the CO_2 avoidance costs do not express how much CO_2 is really saved by a process. Hence, use of the parameter is mainly meant for processes that lead to minimal CO_2 emissions, like CCS.

11.7 Upshot

CCS is a technology with a very great potential for slashing CO_2 emissions. This is especially true since fossil fuels and raw materials, for a long time to come, will form the backbone of a society's supply of electricity, heat, fuels and material products. In addition, for many industrial processes, CCS is the only way to obtain minimal CO_2 emissions. Accordingly, CCS or CO_2 capture as sub-technology can become an important module in CO_2 management for the chemical industry as well.

Before commercial maturity is reached, however, considerable development work must still be done and experience gained. The carbon capture processes, for instance, must be tested and optimized for each application. For processes in the chemical industry, the special deployment options and marginal conditions should be considered early on, so that suitable technologies and methods required for capturing carbon are developed. The concepts of power plant engineering can act as basis and incubator of ideas. In technical terms, besides carbon capture itself, the operation of a plant, too, constitutes a new challenge when viewed against the background of an interplay of all components in the CCS chain.

Depending on site conditions, and taking account of other CCS projects, concepts for transporting the CO_2 must be developed. In this respect, the chemical industry will not erect its own transport infrastructure; what matters for the chemical industry is a link-up to the infrastructure that is to be created for the power plant sector with its much larger CO_2 quantities. An early commitment here can help ensure that the concerns of the chemical industry are taken into account, for example, when it comes to pipeline routes and capacities. For CO_2 storage, too, as part of the CCS infrastructure, the chemical industry will follow on from the work of the power plant operators.

CCS or carbon capture can also become a component of a future alternative raw-material supply. Coal gasification in conjunction with CCS delivers a synthesis gas or basic raw materials, like methanol, naphtha and others without increasing CO_2 emissions. Captured carbon can serve as C1 module.

Besides the direct application of CCS technology, new business fields open up for the chemical industry. CCS requires new chemicals and materials for the various separation jobs.

What is important is that the chemical industry commits itself early on to resolving the technical issues in good time. All industries wishing to use CCS in future should ensure that their interests find their way into the underlying conditions to be created at the time when these are being defined, for example, in CCS legislation and approval law. Besides technical requirements, there are also issues of liability and security, that is, parameters that can directly decide economic efficiency. An early, visible and supportive commitment on behalf of CCS technology is also advisable, so that, in view of the public and political dimension, opportunities are improved for implementing this important option for climate protection.

Further Reading

1 El-Wakil, M.M. (2002) *Powerplant Technology*, McGraw-Hill.
2 Blum, R., Kjaer, S., and Bugge, J. (2009) USC 700 °C power technology – a European success story. *VGB PowerTech*, **89** (4), 26–32.
3 Federal Ministry of Economics and Technology (BMWi) (2008) COORETEC Lighthouse Concept: the path of fossil-fired power plants for the future. Research Report No. 566.
4 Intergovernmental Panel on Climate Change (IPCC), Metz, B., Davidson, O., de, Coninck, H., Loos, M., and Meyer, L. (Eds) (2005) *Carbon Capture and Storage – IPCC Special Report*, Cambridge University Press.
5 Imbus, S., Orr, F.M., Stanford U., Kheshgi, H., Bennaceur, K., Gupta, N., Rigg, A., Hovorka, S., Myer, L., and Benson, S. (2006) Critical issues in CO_2 capture and storage: findings of the SPE Advanced Technology Workshop (ATW) on carbon sequestration. SPE Annual Technical Conference and Exhibition, 24–27 September 2006, San Antonio, Texas, USA, Society of Petroleum Engineers.
6 Ogriseck, K., and Milles, U. (eds) (2006) Power plants with coal gasification, BINE projektinfo 09/06.
7 Tzimas, E., Cormos, C.-C., Starr, F., and Garcia-Cortes, C. (2009) The design of carbon capture IGCC-based plants with hydrogen coproduction. *Energy Procedia*, **1**, 591–598.
8 Gampe, U., Hellfritsch, S., and Gonschorek, S. (2006) Oxyfuel technology for fossil fuel-fired power plants with carbon sequestration – technical and economic feasibility. ENERDAY Workshop on Energy Economics and Technology, 21 April 2006, Dresden.
9 IEA Greenhouse Gas R&D Programme (2007) Capturing CO_2. IEA GHG Report.
10 Barker, D.J., Turner, S.A., Napier-Moore, P.A., Clark, M., and Davison, J.E. (2009) CO_2 capture in the cement industry. *Energy Procedia*, **1**, 87–94.
11 van, Straelen, J., Geuzebroek, F., Goodchild, N., Protopapas, G., and Mahony, L. (2009) CO_2 capture for

refineries, a practical approach. *Energy Procedia*, **1**, 179–185.

12 de, Mello, L.F., Pimenta, R.D.M., Moure, G.T., Pravia O.R.C., Loren Gearhart, L., Milios, P.B., and Melien, T. (2009) A technical and economical evaluation of CO_2 capture from FCC units. *Energy Procedia*, **1**, 117–124.

13 IEA Greenhouse Gas R&D Programme (2007) Storing CO_2 underground. IEA GHG Report.

14 Torp, T.A., and Gale J. (2004) Demonstrating storage of CO_2 in geological reservoirs: the Sleipner and SACS Projects. *Energy*, **29**, 1361–1369.

15 US Department of Energy, Office of Fossil Energy, National Energy Technology Laboratory (NETL) (2006) CO_2 EOR Technology.

16 BP Statistical Review of World Energy (2009) www.bp.com/statisticalreview (accessed 16 April 2010).

17 CO2-Capture Project (2008) CO_2-Capture Project, www.cplccp.net (accessed 16 April 2010).

18 European Technology Platform for Zero Emission Fossil Fuel Power (2010) CO_2 Capture and Storage, www.zeroemissionsplatform.eu/ (accessed 16 April 2010).

19 IEA GHG (2010) IEA Greenhouse Gas R&D Programme, www.ieagreen.org.uk/ (accessed 16 April 2010).

20 The Bellona Foundation (2010) Bellona CCS web, www.bellona.org/ccs/ (accessed 16 April 2010).

21 IEA Greenhouse Gas R&D Programme (2010) Sleipner project, http://www.co2captureandstorage.info/project_specific.php?project_id=26 (accessed 16 April 2010).

22 Statoil (2007) Sleipner Vest, http://www.statoil.com/en/TechnologyInnovation/ProtectingTheEnvironment/CarboncaptureAndStorage/Pages/CarbonDioxideInjectionSleipnerVest.aspx (accessed 16 April 2010).

23 Enhanced Oil Recovery Institute (2010) http://eori.gg.uwyo.edu/ (accessed 16 April 2010).

12
CO_2-Neutral Production – Fact or Fiction?

*Stefan Nordhoff, Thomas Tacke, Benjamin Brehmer, Yvonne Schiemann
Thomas Böhland, and Christos Lecou*

12.1
Introduction

Production systems that are CO_2-neutral are currently fiction. In 2004, the process industry emitted directly about 9.7 Gton CO_2, whereas the chemical and petrochemical industry accounted for around 1.5 Gton CO_2. On the basis of feedstock and feedstock including process energy, 7.9 and 12.7% of total global energy consumption is accounted for, respectively [1, 2]. These emissions are incurred at two stages along the process chain; firstly, during feedstock acquisition and secondly, within the utilities needed to provide process energy (see Chapter 10). More than half of the cumulative energy demand (CED) is for feedstock use, which cannot be reduced by attempts to increase energy efficiency, but only by feedstock and process change. Materials, fuels and energy derived from fossil sources will continue to dominate most process industries in the near future, despite the recent progress made in renewable energy systems and the growing use of bio-based materials and fuels. The fact is, however, that an increasing energy generation efficiency and sustainability (i.e., wind, solar, biomass, etc.), feedstock optimization and efficient production technologies, such as biotechnology, can contribute to reducing the CO_2 intensity of the chemical industry. With current best practice in commercial technologies, savings from 370 to 470 Mton CO_2 per year are already possible for the chemical industry [1]. However, renewable feedstocks may appear to be carbon neutral, but as will be explained, are not, due to energy intense agricultural and logistical aspects. It will remain a question of optimization between conventional fossil fuel feedstocks and biomass-based feedstock alternatives to reduce the overall CO_2 intensity of the chemical industry.

Managing CO_2 Emissions in the Chemical Industry. Edited by Leimkühler

ISBN: 978-3-527-32659-4

12.2 Renewable Feedstocks

12.2.1 Overview

Biomass resources, which could be used as renewable carbon sources in the chemical industry, are for the most part used for food and feed. Plant derived and refined starch, sugar, and oils/fats account for the biggest volumes and will therefore be considered in more detail. In terms of CO_2 reduction, however, a closer look has to be given to the use of plant residues and thus biomass as a whole.

In 2007, about 7% of all chemical sales were generated using bio-based feedstock, biotechnological production methods or a combination of the two. Products include pharmaceutical ingredients, enzymes, ethanol, oleochemicals and food and feed ingredients [3]. For the German chemical industry for example, fats and oils are the most important renewable feedstocks with about half of the total of 2.7 Mtons of renewable raw materials used, followed by starch (23%), cellulose (12%) and sugar (11%) [4]. More than a third of these triglycerides are used tenside production – not only to make cleaning detergents but also for the pharma-, cosmetic- and textile industry. Also lubricants, polymers and polymer additives as well as lacquers and colorants are important branches using these raw materials. Fats for the chemical processing are mainly derived from animals (e.g., milk or tallow), from plants (e.g., palm oils or seed oils), and from microorganisms (e.g., algae).

Sugar (esp. glucose) is mainly used for the production of ethanol. Complex carbohydrates such as starch and cellulose can also be used as a glucose source. The main sources of starch are maize (corn), potatoes, wheat, tapioca (cassava), rice, arrowroot and barley. Starch as well as sugar (here sucrose) is used mainly in the ethanol production, but also in the fermentation of organic acids (such as citric and lactic acids), amino acids, and antibiotics. Some of the other technical applications include biodegradable plastics (polylactic acid (PLA)) for the packaging and textile industries, surfactants, polyurethane, resins, binders, solvents, biopesticides and lubricants. On the other hand cellulose is mainly separated, along with lignin and proteins, from plants and biomass residues, such as wood or straw. In contrast to starch, the fermentation of cellulose with current enzymes is relatively slow and inefficient. Harsher environments as well as more complex and capital intensive enabling technologies are therefore necessary, leading to a relatively high cost constraint with current techniques.

Regarding CO_2 emissions from various biomass sources, Figure 12.1 compares total emission data from crop cultivation and acquisition for the entire biomass material. Best practice yield values (between 2005–2008) in large producer countries (see brackets) were selected. Palm oil, sorghum, cassava and sugar cane have a level of associated emission from transportation but their high yields result in relatively low emissions per material feedstock. On the basis of common expert opinion, the lowest relative feedstock-based CO_2 emission is held by the high-

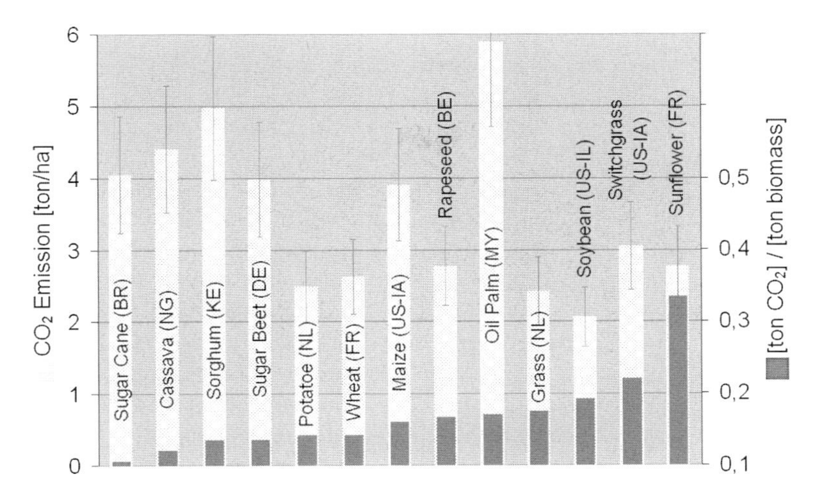

Figure 12.1 Carbon dioxide emission vs. biomass dry weight for regional half-products [5].

yielding Brazilian sugar cane, whereas sunflower with its low biomass yield results in the highest relative emissions.

12.2.2 Volumes, Trading and Pricing

12.2.2.1 Renewable Feedstock Trends

In recent years food prices have decreased generally in real terms. This trend seems to have an exponential tendency reaching its hypothetical base line in the near future (see Figure 12.2). However, because of growing world population, the increasing hunger for energy, and the demand for sustainable food and energy generation, future feedstock prices are expected to rise.

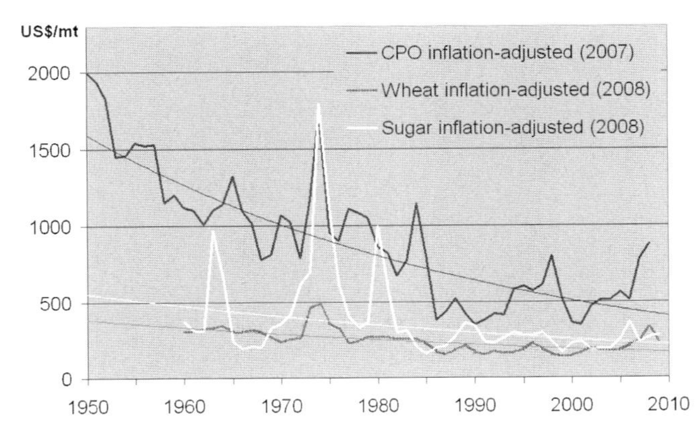

Figure 12.2 Development of prices of selected feedstocks [57], CPO: crude palm oil.

12.2.2.2 Sugar

Nowadays, world sugar production comes mainly from cane, the volume is growing by around 2.8% per annum; only 21% are made from beet with a flat growth rate. The major producing countries are Brazil and India accounting together for about a third in volume; Europe being around 10%. Total production was 150 Mtons in 2008/09 after 169 Mtons the previous year [6]. As a direct outcome of the EU regulation on the common organization of the markets in the sugar sector, several major plant closures have drastically reduced the production capacity. In the EU, sugar production peaked at 22.1 Mtons in the 2005/06 season and has reduced to 14.9 Mtons in 2008/09 [6]. The EU as well as the USA have become sugar importing countries whereas Brazil, India and Thailand are counted among the biggest exporting nations. China is the second largest producer in Asia, however, still importing substantial quantities. Asian consumption accounts for twice the world average, driven by both population and economic growth.

Seasonal variations in pricing are due to meteorological as well as to market reasons. A steady rise in prices in 2005/06, as can be seen in Figure 12.3, made the sugar volumes rise to top volumes during the next two seasons. However, growing demand for other crops like wheat and corn or cassava as starch sources are easily used for switching planting schedules in certain countries. Speculative activity contributed to the 2009 price hike as many investors moved from company shares into raw materials. Crude oil and commodity price spikes were also influenced by this development. Poor monsoon rainfall in India as well as floods in Brazil have led to all-time high sugar prices in 2009/10. The New York stock exchange listed raw sugar futures reached 379 €/ton in September 2009, the highest level recorded since 1981. London listed white sugar futures peaked at 420 €/ton, which was even above the new EU reference price of 404 €/ton valid from October 2009 [7].

Bioethanol derived from sugar can be used as a substitute for fossil fuels and when mixed with gasoline, it is suitable for use in motor engines. The mixture results in a fuel with a cleaner combustion. The Kyoto protocol requires signatory

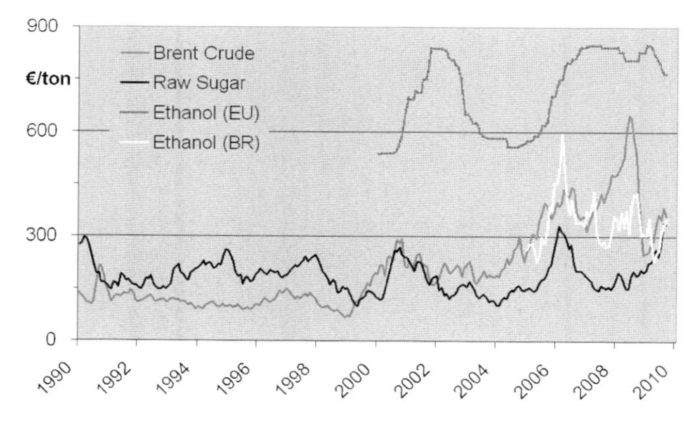

Figure 12.3 Nominal prices of crude oil, sugar and ethanol per ton [57].

nations to reduce greenhouse gas (GHG) emissions by at least 5% during the period 2008–2012. This has resulted in an unforeseen increase in the price of raw materials, which has doubled since 2005; and the trend is still going strong. In Brazil the relation between sugar and ethanol being produced in the same plants is readily influenced by the respective price relation in the fuel or sugar markets. About 58% of Brazilian sugar went into ethanol production in 2008 thereby reflecting the growing importance of this biofuel (see also Section 12.3.2.3). Many countries have implemented energy policies with the obligation to use certain shares of biofuels (ethanol, biodiesel) distorting the markets for the respective crops by coupling their prices to petrofuels.

12.2.2.3 **Starch**

The starch industry in Europe processes about 22.5 Mtons of agricultural crops, including 12.5 Mtons of cereals and 10 Mtons of potatoes and produces about 9 Mtons of starch and its derivatives. Starch is an ideal example of a commoditized market that has a wide array of industrial applications, which comprises paper and cardboard making, fermentation, biodegradable plastics and detergents, surfactants, resins, binders, solvents, biopesticides and lubricants. The annual turnover amounts to about 8 Bn$ and the industry invests approximately 210 M$ in R&D every year [7].

The food sector consumes the majority of the starch produced in Europe, about 4.5 Mtons or 50% of the starch production. In the non-food sector, about 1.3 Mtons of starch is used for paper and cardboard making, 1.1 Mtons in plastics and detergents (biodegradable, non-toxic and skin friendly detergents) and 1.2 Mtons in fermentation and other technical applications [8].

Average unit prices in the USA native and modified starch market have been steadily on the rise from 0.5 $/kg in 2004 up to 0.7 $/kg in 2007, where prices around one dollar are predicted for 2014 (see Figure 12.4) [8]. In Europe the prices

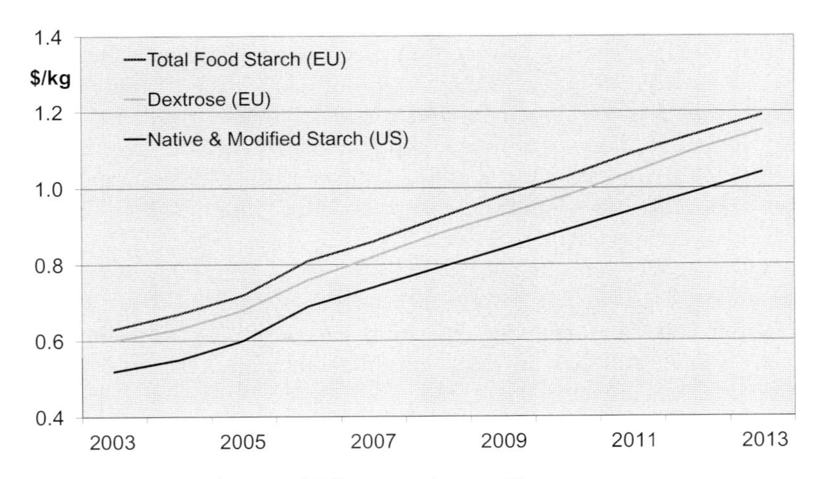

Figure 12.4 Nominal prices of different starch types [8].

developed from 0.63 \$/kg in 2003 to a predicted 1.19 \$/kg in 2013 [7]. The most pertinent aspect affecting the pricing of starch is the rising cost of raw materials. The raw material prices have been influenced by the increasing demand from the biofuels industry, and other issues similar to the sugar sector.

The type of raw material used also influences the pricing to a very large extent. In Europe, among wheat, maize and potato, wheat continues to be cheapest and potato the costliest. Crop yields affected by adverse weather can also play a vital role in increasing the raw material price. The other factors influencing the price of starch are increasing energy costs, supply demand equilibrium, product customization and the supply chain expenditures. The surging energy cost has had a twin effect by increasing the production costs as well as the transportation costs.

The global demand for bioethanol, which is also derived from starch-containing cereals such as maize, wheat etc., is increasing at a phenomenal rate, leading to the same issues as described in Section 12.2.2.2.

12.2.2.4 **Oils, Fats**

The total demand of fats and oils in 2008 is estimated to be around 165 Mton, whereof 83% stemmed from plant bases. The majority (125 Mton) has been used in food, only 9% was used by the chemical industry (e.g., coatings, additives, lubricants, pharmaceuticals, etc.) and 6% for feed. Already 9% arise from 3% in 2004 went into production of energy and biofuels [9]. This was mainly due to increased use of rapeseed oil for biodiesel production. Fats and oils are large-volume commodity products that are primarily consumed domestically. However, international markets have become increasingly important to producers. Geographic climates that are favorable to certain agricultural species often determine whether a region has a supply excess or shortage.

Overall, the fats and oils industry is growing because of an increase in population and growing industrial uses. Soybean oil became the world's predominant vegetable oil in less than two decades, as a result of the increasing demand for protein in feeding livestock and poultry. In the USA, soybean oil is the standard against which the majority of the other oils are priced. In the last 20 years, world production of both palm oil and canola oil has mounted rapidly. Southeast Asian countries continue to increase their acreage of palm trees and to replace older palm trees with higher-yielding tree varieties. In the northern climates, canola oil (low-erucic-acid, edible rapeseed oil) is being produced in greater quantities, especially in Canada, Europe and the USA.

Global fats and oils consumption is predicted to accelerate at an average annual rate of 4%, mainly due to growth in China and India. Also, demand for biofuels (mainly from rapeseed and palm oils) is expected to stimulate demand in Europe. This has an escalating effect on prices, as can be seen in Figure 12.5, where selected plant-based oil prices are compared with crude. Especially, since biofuels have a larger impact, prices are obviously coupled.

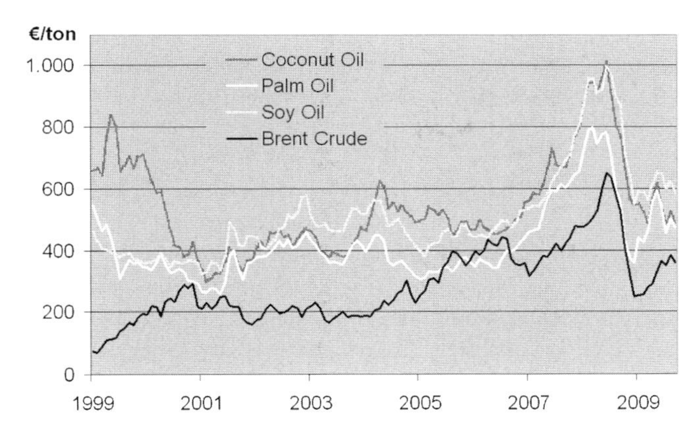

Figure 12.5 Nominal prices of tropic oils and crude oil per ton [57].

12.2.2.5 **Biomass and Residues**

Biomass can generally be subdivided into carbohydrates, lignin and others (fats, proteins, terpenes, etc.). They are produced by nature in a volume of 180 bn tons per year with carbohydrates (e.g. sugar, starch) playing a major role (75%) and lignin accounting for about a fifth of these renewable resources. Only 7 bn tons (as dry matter) of this biomass are being used for agricultural products, thereof the majority (95%) are consumed for nutritional purposes. These values might be compared with an overall 10 bn tons of fossil fuels mainly used in energy production [10].

Currently biomass use provides 13% of global final energy demand. World ethanol production for transport fuel tripled to more than 52 bn liters between 2000 and 2007, while biodiesel expanded eleven-fold to almost 11 bn liters. Policies have essentially triggered the development of biofuel demand by targets, blending quotas and mandates which have been enacted in more than 17 countries. Investment into biofuels production capacity exceeded $4 bn worldwide in 2007 and seems to be growing rapidly. Global use of biofuels is expected to double from 2007 to 2017. It is estimated that by then the share of biomass will increase to 14% of the current fossil energy use totaling 388 EJ in 2008 [11]. The major part should, however, be derived from organic waste and residues rather than from food-crops. This includes for example, biogas derived from manure and ethanol derived from agricultural and forest residues, as well as biodiesel from wood (biomass to liquid BtL, based on experimental plants). Plenty of reasons like higher GHG savings, better use of arable land and the mitigation of the food-vs.-fuel debate can be found. Advanced biofuels, such as cellulosic biofuels derived from timber processing residues, straw or corn stover, may be able to improve the resource efficiency of biofuels. A good example is cane-based bagasse as the residue from sugar production. In Brazil 73 Mtons of this residue fiber material from the processing of 5 bn tons of cane are effectively used for providing energy in the fermentation process or, increasingly, to produce electric energy for the grid. Apart from incineration, biodegradable waste can be treated by anaerobic digestion or by composting. Lignin is a widely named candidate as chemical feedstock but to date attractive large-scale uses as raw material are still scarce.

12.2.3 Competitiveness

12.2.3.1 Competition between Fossil and Renewable Feedstocks

There is a widespread belief that with rising fossil fuel prices, renewable raw materials will, in principle, become economically viable. But, as can be shown, price linkages to fossil-based raw materials do exist [12] (see also Section 12.2.2). Furthermore, 'Peak Oil' is often used as an argument to shift the focus of renewable feedstocks as the only 'natural' carbon source. It is true that, in the global chemical industry, 15–20% of the total fossil fuel energy resources are used as feedstocks and process energy (IEA, 2007 [1]). Yet, in the light of this aspect and by taking into account the high added value provided by the chemical industry compared with any other fossil fuel energy intense sector, it should be clear that the chemical industry will be one of the last sectors to 'run out of oil'. Furthermore, as opposed to feedstock use, the main driver of competition is as an energy source, since both fossil fuels and biomass can and are used for heat and power generation, as well as mobility. To understand the principles it is helpful to compare the mass-based and dry energy content of different raw materials (see Table 12.1):

For energy purposes, raw materials can only be compared if they are in the same aggregate state. This means that most of the renewable feedstocks must be compared against the price of coal, oils and fats and crude oil, as shown for example in Table 12.2.

Simply exchanging fossil fuels with biomass is hampered by some additional yet simple economic constraints. Logistics plays a key role since there are no

Table 12.1 Energy content of different raw materials.

	Fossil		**Renewable**	
	Energy content (lower calorific value) in MJ kg^{-1}			
Solid	Hard coal	30	Wood	15
	Hard coal briquette	31	Sugar	16
	Lignite (50–60% water)	9	Starch	15
	Lignite briquette	20	Fat	39
Liquid	Gasoline	43	Vegetable oil	37
	Diesel	43	Biodiesel	37
	Naphtha	44	Bioethanol	27
Gas	Natural gas	38	Biogas	27
	Ethylene	47		

Table 12.2 Mass-based price comparison, triggered by fossil-based prices.

	Fossil						Renewable					
	Price in $/ton and in $/ton C respectively; Average production in mio ton/a											
		2007		2009		Prod.		2007		2009		Prod.
		$/ton	$/ton C	$/ton	$/ton C	mio ton/a		$/ton	$/ton C	$/ton	$/ton C	mio ton/a
Solid	Hard coal	87	87	76	76	2950[a]	Sugar	218	519	402	957	150
Liquid	Naphtha	676	786	490	570	362	Rapeseed oil	943	1232	826	1079	20
							Bioethanol	509	976	467	896	68
Gas	Natural gas	370	493	216	288	2490[a]	Biogas	N.A.	N.A.	453	604	N.A.
	Ethylene	1240	1442	988	1149	130						

a) Oil equivalent tons.

concentrated 'oil wells' within the realm of bio-based feedstocks. On average 10 ton and at best 20 ton dry weight biomass can be harvested per hectare of arable land respectively. Thus harvesting and transport costs and the indirect CO_2 emissions (even for a single large-scale plant with a multimillion ton production capacity) must be taken into account for an overall balance.

One trigger for increasing the overall proportion of renewable raw materials in the chemical industry is the possibility of using simplified process parameters. High pressures and temperatures could be avoided, a decreased number of processing steps could be realized, etc. The overall production costs could be reduced despite of high raw materials prices. Investment costs for processes requiring extreme conditions, can be decreased analogously (see also Section 12.4.3). The other trigger is exclusivity, when a product with specific properties can only be produced via a biotechnological process. These basic economic considerations could be completely overruled, if legislation creates mandates for the incorporation of renewable raw materials (such as with existing biofuel policies). The CO_2 certificate prices could also be artificially elevated thereby emphasizing a greater role for the biomass-based processes, especially if one considers the general viewpoint that they are more sustainable, that is, the overall CO_2 emissions are substantially lower than a fossil-based process (see also Sections 12.3.2.3 and 12.5).

12.2.3.2 Yields and Efficiency of Chemical Processing

Due to the long history of the petrochemical industry, highly efficient process conditions, yields and selectivities are now commonly established. But, how do biotechnological processes work? In essence, the carbon source of biomass is innately associated with a high oxygen content due to the photosynthetic reaction.

As shown in Table 12.3, it is not surprising that significant quantities of CO_2 are emitted during the bacterial or yeast-based conversion of carbohydrate-derived renewable feedstocks, since the carbon is partially oxidized. An additional loss is consumed as an energy source for the metabolism of the cell. The overall stoichiometric yield is limited to around 50%. One of the most optimized and oldest large-scale biotechnological production processes worldwide–bioethanol–reveals a stoichiometric carbon efficiency of 51%. In reality 45–47% can be reached.

Therefore, to overcome this 'metabolic penalty' fermentation processes must be greatly simplified in comparison to existing petrochemical processes.

In order to use the synthetic potential of nature more efficiently both feedstock and product should be chemically closely related. Catalysis, traditional and in particular biocatalysis, can enable a faster and more efficient stepwise-based conversion compared with the fermentative processes. This and other options are described in detail in Section 12.3.2. Traditionally, in the petrochemical industry the desired product is directly purified after leaving the reactor. In the product stream, the desired product content is well above 10%–often higher than 95%. Aside from biocatalytic processes, fermentation processes operate at ambient temperatures and pressures in an aqueous medium. As a consequence, even if the metabolic carbon efficiency is high, the product concentration in the aqueous broth after fermentation is below 10%, only rarely somewhat above. An entirely different down-stream-processing is therefore required, which are associated with currently high specific investments and consumption of utilities.

Table 12.3 Carbon and oxygen content of different substances.

	Fossil			**Renewable**		
	Carbon and oxygen content in %					
		C%	**O%**		**C%**	**O%**
Solid	Bituminous coal	73	9	Wood	52	42
	Lignite	56	18	Sugar ($C_{12}H_{22}O_{11}$)	42	51
				Starch ($(C_6H_{10}O_5)_n$)	44	49
Liquid	Gasoline	87	<2.7	Biodiesel (RME)	77	13
	Diesel	86	~0	Bioethanol	52	35
	Naphtha (C_nH_{2n})	85	0			
Gas	Natural gas	69	~0	Biogas[a)]	75	0
	Ethylene	86	0			

a) Pure methane.

12.3 Industrial Biotechnological Processes

12.3.1 Market, Field of Application, and Currently Available Products

While biotechnology may give the impression of being a new and exciting branch of technological innovation, it has in fact encompassed continuous development throughout the ages. Few people, for example, would consider beer brewing or bread baking as marvels of biotechnology. Regardless, technological progress has recently achieved several key breakthroughs and with an increased frequency. This has presented numerous opportunities for the competitive and sustainable production of chemicals, materials and fuels derived from or synthesized with the help of biological systems. The total global biotechnology market in 2006 reached a value of approximately 125 bn€ and is expected to increase to 250 bn€ by 2011 [13]. As a field of study it has branched into individual niche categories. Depending on definition criteria, there are either 4 or 7 fields of biotechnology, which have been assigned a color label presented in Table 12.4.

Table 12.4 Overview of biotechnological fields (market size in billion €).

Colour	Field	Classical Products	Recent products or developments	Market Size
Red	Pharma	Penicillin, Aspirin	Stem cells, tissue engineering, recombinant mechanisms, etc.	65–80
Blue	Environmental	Microbial waste Water treatment	Extremophiles, bioremediation, waste material use, etc.	2–5
Green	Food production	Fertilizers, Pesticides	Functional ingredients, plant genomics, gene expression, etc.	10–20
Brown	Proteomics	Protease Enzymes	Cell signaling, biomarkers, targeted proteases, etc.	N/A
Yellow	Systems biology	Biochemical pathways	Regulatory and transcriptional networks, gene metabolism, etc.	0.1–0.5
Purple	Engineering	Bioreactors, Fermentation	Downstream processing, recovery, up-scaling, etc.	0.5–1
White	Industrial biotechnology	Alcohol, cheese, detergents, soap	Biomaterials, biocatalysis, biofuels, biorefineries, etc.	30–35

Data source: internal estimates.

White biotechnology is the branch devoted to the replacement of petroleum-based processes with bio-based processes that not only (partly) use renewable resources, but have a large-scale reduction potential in CO_2 emissions. Industrial biotechnology is already quite substantial with global sales estimated at 30–35 bn€ in 2009 and an expected annual growth rate of 5.1%. The largest growth potential, at 12.6% per year, is the subcategory of platform biochemicals; currently accounting for 1.9 bn€ global sales. Many industrial biotechnological chemical products are emerging as an economically and ecologically competitive option against the traditional products of the petrochemical industry (examples are shown in Section 12.3.2). In the realm of CO_2 mitigation potential, it is this white branch of biotechnology that presents the greatest potential.

It should be noted that while red biotechnology is approaching a market size of 100 bn€, considering the small quantity of active ingredients the market volume is constrained to the niche 'kton' range. White biotechnology on the other hand is geared towards the commodity market in the high volume 'Mton' range but at lower relative sales per product. The market volume is already approaching 100 Mton. Due to these product volume sizes and projected growth rates, the field of white biotechnology can significantly contribute to the CO_2-emissions savings.

12.3.2
Existing and Future Opportunities of Industrial Biotechnology

12.3.2.1 General Developments

High volume biofuels, such as bioethanol, and food and feed applications, such as amino acids and vitamins already exist. In the near future, biopolymers are expected with their substantial growth to be added to the list. Several companies have or intend to add a variety of products based on fermentable carbohydrates to their portfolio. Polylactic acid (PLA) and 1,3-propanediol (PDO) have already been commercialized by Cargill and DuPont, respectively. Bioethanol-based polyethylene (PE) is expected to be shortly commercialized by Braskem and Dow. Solvay has announced the commercialization of a partially bio-based PVC. Other announcements include succinic acid (DSM) and 3-hydroxypropionic acid (3-HP) (Cargill and Novozymes). Countless other derivatives are expected in the coming years as the economics prove competitive.

12.3.2.2 Amino Acids

Currently, almost 2 Mton of amino acids are globally allocated for feed applications. They are very effective in reducing the overall consumption of proteins and thereby indirectly reduce the environmental impact of animal production. For example, per kg weight gain, it is feasible, through selective feeding of amino acids, to reduce the nitrogen content of manure by 50%. This relates to a substantial reduction of nitrogen deposition in soil and lower atmospheric N_2O emissions. Another example highlights the impact of the feed additive L-lysine on the consumption of soybean flour. In 2005/2006 within the EU27 approx. 35.8 Mton of soybean flour were consumed. Without the ca. 300 kton per year of L-lysine currently supplied, an additional 10 Mton of soybean flour in the EU would be required. Amino acids are produced by fermentation, enzymatic synthesis, extraction from protein-hydrolyzates, and chemical synthesis. Due to the substantial progress in strain development over the last two decades, the fermentation industry has seen rapid growth and increased process efficiency. The feedstocks are the typical biomass streams sucrose, molasses, and starch hydrolyzates [14].

12.3.2.3 **Bioethanol and Bioethylene**

Biomass-based ethanol currently substitutes around 3% of the global gasoline consumption. It has a higher octane rate compared with gasoline (98 vs. 80), but has a lower volumetric energy content (67% of gasoline). Therefore, per kilometer driven, around 20% more ethanol is required [15]. Favorable political conditions for biofuels have stimulated a significant increased in global production. Currently, the major producers are Brazil (sugarcane) and the USA (corn). In Brazil, the integrated production of sugar and ethanol provides a certain degree of flexibility, depending on market demand. The production may be shifted from 55/45 sugar/ethanol to 45/55 sugar/ethanol. At the standard 50/50 ratio, roughly 67 kg sugar and 47 l ethanol is obtained per ton sugar cane. Most production plants are energy self-sufficient due to the use of the internal by-product bagasse. One ton of sugar cane yields between 240 and 280 kg bagasse (humidity ca. 50%) with an energy content that can replace 580 kWh by using a combined heat and power unit. The state-of-the-art integrated production allows a bagasse surplus of 7–15%, which might be sold on the fuel market [16, 17]. Aside from biofuels, ethanol can also be used as a chemical feedstock. A commercially proven system is the catalytic dehydration into ethylene; a bulk petrochemical. So-called 'bioethylene' could provide access to the C4-building blocks such as butadiene, 2-butene, etc. Through metathesis 2-butene and ethylene can be converted into propylene; the key C3-building block. Most advances are currently geared towards conversion of ethanol via catalytic dehydration and the subsequent polymerization into bio-based PE.

Braskem has announced the construction of a 200 kton bioethylene production site in Triunfo, Brazil with a planned completion for end 2010. The demand for such 'green' plastics could reach 2 Mton over the 2010–2019 period [18]. Nonetheless, this is still small compared with the global production of fossil-based plastics based on, for example PE and polypropylene (PP) which has reached the 245 Mton threshold [19].

Table 12.5 reveals the impact of CO_2 emissions by shifting from fossil-based to bio-based processing systems. Currently, due to the high energy intensity of cultivating wheat and other temperate carbohydrate-based crops, only the tropical sugar cane can lead to a reduction in CO_2 emissions in the case of PE production.

Table 12.5 Emissions for the production of chemicals/fuels and for their end use.

	Figures given in g CO_{2eq} MJ^{-1}			Figures given in kg CO_{2eq} kg^{-1} product		
	Gasoline (Crude oil)	**Bioethanol (Sugarcane)**	**Bioethanol (Wheat)**	**Polyethylene (Crude oil)**	**Bio PE (Sugarcane)**	**Bio PE (Wheat)**
Production of raw materials	4.5 (Extraction)	18.4 (Biomass)	40.3 (Biomass)	1.5 (Naphtha)	0.6 (Biomass)	2.2 (Biomass)
Processing	7.0	1.1	31.9	0.5	0.5	0.5
Distribution	1.0	2.3	1.5	0.2	0.2	0.2
Total emissions	12.5	21.4	73.7	2.2	1.3	2.9
Combustion	74.4	CO_2 cycle (71.4)	CO_2 cycle (71.4)	(Incorporated in product)	(Incorporated in product)	(Incorporated in product)
Total emissions (Burning)	86.9	21.4 (92.8)	73.7 (144.1)			

Source: Based on [20], [21], [22] and own estimations.

12.3.2.4 Building Blocks

Building blocks are chemical molecules with multiple functional groups that through transformation can be a source for a wide variety of useful chemicals and synthesis pathways. In 2004, a study from the Biomass Program of the U. S. Department of Energy (DoE) created the 'Top Value Added Chemicals from Biomass' in which twelve chemical building blocks originating from sugar were identified [23] – see Figure 12.6. This detailed report also gives a comprehensive overview regarding potential conversion pathways from these 12 bio-based building blocks to a number of high value added chemicals and/or materials.

For instance, the C3 chemical building block 3-HP has potential in both the commodity and specialty chemical sector. Many products including PDO, acrylic acid, methyl acrylate, acrylamide could be derived from 3-HP. Furthermore, the basic chemistry of 3-HP is not represented by the current petrochemical industry. Major technical hurdles for its success include the development of high yielding organisms with a potentially low-cost fermentation route [23].

12.3.2.5 Bioacrylic Acid

Currently, more than 3 Mton of acrylic acid as a monomer for super absorbent polymers (SAPs) and other applications are annually produced globally. The classical propylene process is based on the gas phase oxidation via acrolein to acrylic

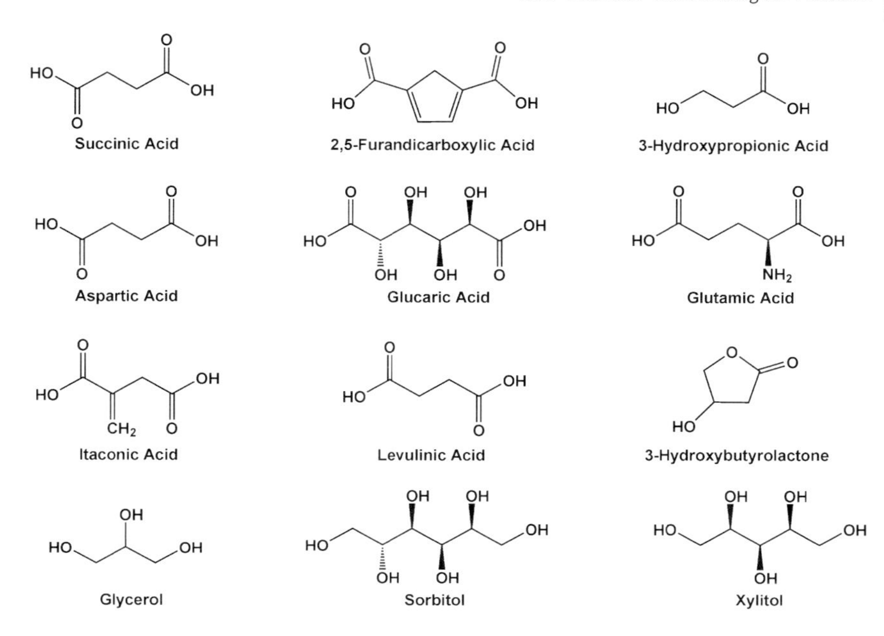

Figure 12.6 The 12 proposed sugar-based chemical building blocks [23].

acid. Several new bio-based production routes are becoming available. Carbohydrates could be converted via fermentation to lactic acid, 3-HP or to 3-hydroxypropionaldehyde (3-HPA) and finally into acrylic acid in consecutive catalytic steps. Lactic acid and 3-HP are catalytically dehydrated directly to acrylic acid; 3-HPA is catalytically dehydrated to acrolein and only then oxidized to acrylic acid. Glycerol as a by-product from biodiesel production could also be used in a two step catalytic process; first dehydrated to acrolein and then oxidized to acrylic acid.

Nexant and Novozymes predict the starch-based acrylic acid process to be competitive when crude oil prices exceed 65 \$/bbl. It could already be competitive using Brazilian sucrose at a crude oil price of 45 \$/bbl, which correspond to a sucrose price of approx. 9 ¢/lb [24]. On the technical side, further microbial development is required to obtain the desired metabolic pathway in a highly cost and energy efficient system.

12.3.2.6 **Oils for Chemicals**

Traditionally many soaps and detergents were originally produced from fatty acids and despite the gradual transition to petrochemical-based production routes, their organic oil basis still contributes to a major proportion to their production. This route can be classified as a sort white-based chemical, but due to its historical and continued use cannot be associated with any CO_2 reduction potential. Conversion of fatty acids to biolubricants could present significant savings. In the EU (2006), there were more than 450 different lubricants in excess of 4.6 Mton production. Each grade requires a unique compositional formulation for the specific

Table 12.6 Fatty acid to biolubricant production energy.

Category	Standard process		Recent improvement process[a]	
Crop	Seed-based	Fruit-based	Seed-based	Fruit-based
Overall conversion (ton esters/ton fatty acids)	0.898	0.903	0.940	0.946
Total energy (GJ ton^{-1})	7.71	13.89	4.95	7.00

a) Also produces PDO: 53.9 kg ton^{-1} for seed-based, 55.5 kg ton^{-1} for fruit-based.

application. In an integrated biorefinery system, bioethanol could be used as an alternative transesterification reactant. In this case, fatty acid ethyl ester (FAEE) would be produced instead of fatty acid methyl ester (FAME). Despite slightly different physical and chemical properties, FAEE could be applied as a base formulation for lubricants. Following the stoichiometric ratio, 0.142 g/g ethanol would be needed for the conversion which is more than the fossil fuel derived methanol which would be avoided. Despite the obvious renewable nature of bioethanol, in a full life cycle assessment (LCA (see Chapter 1)), the production energy and associate CO_2 emissions of the bioethanol route must be taken into account. Furthermore, according to a recent study, several process improvements associated with fatty acid processing can reduce the process energy requirements [25, 26]. Table 12.6 lists the conversion rate and process energy costs of FAEE for the main oil crop types.

Considering that the total process energy of FAEE is 5–7 GJ ton^{-1} and the standard production of lubricants is 50–65 GJ ton^{-1}, a high CO_2 savings potential is foreseeable.

As research in the field of white biotechnology continues to advance, more product examples of possible synthetic routes for fatty acids present themselves. While still in their infancy of development, long chain-based polymers and their relative monomers are being studied. Long chain lactams used to produce various polyamides (PA612, PA1010, PA12, etc.) present a grand potential in reducing the associated CO_2 production cost when based upon organic fatty acids [27, 28]. Biodiesel has a maximum potential to save 3.2 ton CO_{2eq} ton^{-1} whereby lactams range between 4–7 ton CO_{2eq} ton^{-1} [29].

12.3.2.7 Biocatalysis for the Production of Emollient Esters

The biocatalytic production of chemicals offers intrinsic properties. Commonly mentioned attributes are unusually high selectivity and operation under milder conditions. Recently the sustainability of biocatalytic processes for the production of cosmetic ingredients, in particular emollient esters has been quantified [30]. The frequently used emollient ester, myristyl myristate, was used as a model study.

An enzymatic production system using immobilized Lipase B from *C. antartica* under real industrial conditions (Evonik Goldschmidt site) including recycling of the enzyme until deactivation was compared with a state-of-the-art conventional process based on tin(II) oxalate as a catalyst at 240 °C for the synthesis of cosmetic esters [30]. The in-depth analysis revealed (based on a 5 ton production scale), that the global warming potential for the conventional system is 1.5 tonCO_{2eq}, whereas the enzymatic system enables a reduction of 62% down to 0.5 tonCO_{2eq}. While not mentioned this assessment gives an overall stance for the wide spread assumption that enzymatic catalyzed processes provide environmental benefits.

12.4 Expansion to Multiproduct Biorefineries

12.4.1 CO_2 Saving Limitations of Single Product-based Systems

In Figure 12.7 the cumulative energy demand (CED) of producing existing functional chemical families from the 'traditional petrochemical route' and the 'potential biorefinery route' is presented. CED and CO_{2eq} are interrelated and following the average stoichiometry reveals a conversion factor of 0.071 tonCO_{2eq}/GJ fossil energy. On the right hand side (crop image): regional agricultural intensity and the crops specific chemical composition yield an associated energy cost for the

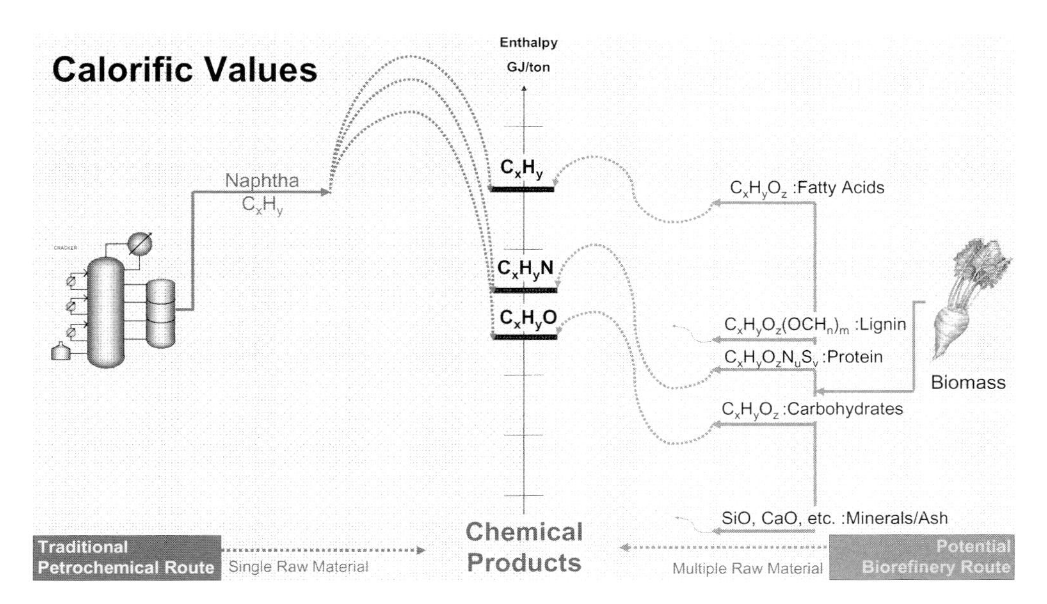

Figure 12.7 Cumulative energy demand of petrochemical and biochemical route. Adapted from Ref [5].

main biochemical group types (carbohydrates, protein, fatty acids, etc.) The right hand dotted lines connect the groups to the existing final chemical products and attempt to graphically represent the processing energy requirements to prepare, convert and purify these chemicals. On the left hand side (oil/naphtha cracker image): fossil fuel feedstock and the imbedded calorific value of the main derivate, naphtha. The left hand dotted lines connect the naphtha feedstock to the final chemical products also in term of process energy requirements. The resulting calorific values of the chemical products depend largely on the degree and type of functionality. Pure hydrocarbons (e.g., ethylene) have a higher calorific value than amines (e.g., caprolactam) which are in turn higher than the oxygenated chemicals (e.g., peroxide). This trend however, does not reflect the CED from the traditional petrochemical production route. Adding functionality requires several additional process stages which in turn demands more energy and indirectly linked CO_2 emissions. Biomass feedstock, however, may use the existing biochemical functionally and could be produced using milder and more direct process routes that should lead to less energy intensity. This is symbolized by the magnitude of the dotted lines.

12.4.2 Entire Biomass Use

12.4.2.1 Chemical Breakdown

Using a higher degree of the biomass feedstock and by upholding the biochemical structure eventually lead to significantly higher CO_2 emissions savings than the above-mentioned single product concepts. All plant-based biomass are composed of the same basic biochemical components as mentioned in Section 12.2.1 with only the concentration and yield quantity being different. In the single product system, biomass crops associated with a particularly high quantity or concentration of the desired component for conversion are selected. For example, corn is cultivated for starch and oil palm is cultivated for fatty acids. Typically the rest of the plant is combusted for thermal energy or used as low-value by-products. In the case of starch, more than 50% of the plant material and for the oil palm over 70% is not used as a main product. As any LCA will indicate, the industrial cultivation of biomass is highly energy intensive and not carbon neutral and could also come with a land use change carbon debt (see Section 12.5.2). The direct and indirect CO_2 emissions are weighted against the quantity of contributing products produced via the different biomass feedstock streams. The lower the amount of products produced, the higher the associated cultivation costs. As an explanatory guide, the starch contained in corn comprises of only 20–25% dry weight of the entire biomass material. This means that by focusing solely on bioethanol as the renewable product, the agricultural input energy of 2.5–5.0 GJ ton^{-1} becomes 10–30 GJ ton^{-1} in relation to the final product, should the by-products not contribute to a CO_2 saving system [25, 31]. Merely the acquisition of starch from corn thereby places a heavy burden on the overall energy balance of bioethanol. Using a higher portion of the biomass for the conversion to useable products will lower the associate agricultural cost and thereby contribute to a further reduction in CO_2; whereby even thermal combustion can partially contribute.

12.4.2.2 **Pure Syngas**

One of the methods in which the entire biomass material stream is used, or at least the vast majority (i.e., all organic chemical components), is the complete degradation of the material into volatile gases. Gasification reactors, such as the various pyrolysis designs, enable the production of hydrocarbon gaseous mixtures from biomass material [32]. These volatiles can be burned directly for the production of energy or further upgraded to syngas (hydrogen and carbon monoxide). Syngas is a valuable chemical product in the industry and has a higher associated CED in its traditional production method than energetic systems (approx. 20–30% higher). A yield exceeding 90% energy content can be easily attained using current technology [33]. Continuing with the corn cultivation analogy, at 90% useful product yield, the associated agricultural energy input now set at 2.8–5.6 GJ ton^{-1}. This allows a greater net energy and respective CO_2 emission savings potential, especially considering that the yielded products related to at least 14–15 GJ ton^{-1} biomass input material.

12.4.2.3 **Partial Syngas and Partial Biochar**

In complete contrast to the energy debate and assumed carbon emissions correlation, the product 'biochar' follows a distinctive carbon balance and is disconnected from the typical correlation. Biochar is produced by the slow-pyrolysis of biomass material containing a sufficiently high quantity of lignin [34]. Biochar has the strange attribute of being chemically and biologically inert and stable for thousands of years in soil. When applied to soils the carbon contained in the product is effectively removed from the ecosphere making it a sort of CO_2 sink. During the pyrolysis step, volatile hydrocarbons are emitted and could also be upgraded to syngas. Seeing that a proportion of the potentially combustible biomass material is locked within the biochar, a lower overall thermal output compared with Section 12.4.2.2 is present. The carbon sink attribute however leads to a significantly higher CO_2 savings potential. For every ton biochar committed to the soil, 3.67 ton of CO_2 is removed for the biosphere. Depending on the actions taken at post-Kyoto Protocol summits, biochar might receive double carbon credits as a carbon capture and soil storage (CCSS) mechanism.

12.4.2.4 **Chemical Structure Retention**

As mentioned, one way to improve the net energy balance, or CED, of white biotechnological processes was to use a larger portion of the biomass feedstock material (see Section 12.4.2.1). This addresses one option, another option is to retain the complex biochemical structure. Terrestrial plants are highly effective organisms in which efficiency for the photosynthesis step can reach well above 70%, but based on the resulting calorific value, the so-called *photosynthetically active radiation* (PAR), is usually below 1%. The reason for this huge discrepancy is the internal life functions and biochemical synthesis pathways to create a wide array of highly complex biochemical structures. Just as adding functionality costs process energy in the petrochemical industry, synthesizing complex biochemical structure also costs energy in the plant. Being able to use the complex chemical functionality, where plants had invested great lengths of energy, for existing chemicals of a similar functionality could by-pass many energy intensive process routes. A large part of the functionalized chemicals in the 3rd to 6th derivative classification are still considered commodities. Developing biotechnological process solutions to uphold the biomass complexes is needed to achieve large contributions to CO_{2eq} emission savings.

12.4.2.5 Proteins for Functionalized Chemicals

Protein is present in all plant matter from less than 5% to over 20% in some variants at a particular stage during the growth phase. Proteins are comprised of hundreds to thousands of interlinked amino acids. Figure 12.8 displays the traditional petrochemical route for a high-volume nitrogen-based functionalized chemical product vs. the potential biochemical route. The amino acids, arginine, which can be extracted from nearly any protein source, is used to highlight the potential of upholding the biochemical structure. Methane and propylene act as the carbon source with ammonia supplying chemical energy for nitrogen functionality in the existing industry. Classified as a bulk chemical commodity precursor, the production of 1,4-butanediamine is associated with a high CED of 114.7 GJ ton^{-1} and high emissions of 8.1 tonCO_{2eq} ton^{-1}. By employing enzymatic and a mild acidic/hydrolysis treatment, the protein-derived arginine can be directly synthesized into urea and 1,4-butanediamine. Urea is also a high volume nitrogen containing chemical with an associated CED of 50.3 GJ ton^{-1} and emissions of 3.6 CO_{2eq} ton^{-1}. In addition to the low process intensity involved in this biochemical route, as an additional portion of the biomass feedstock is being used meaning the agricultural energy intensity (as mentioned above) will also be reduced.

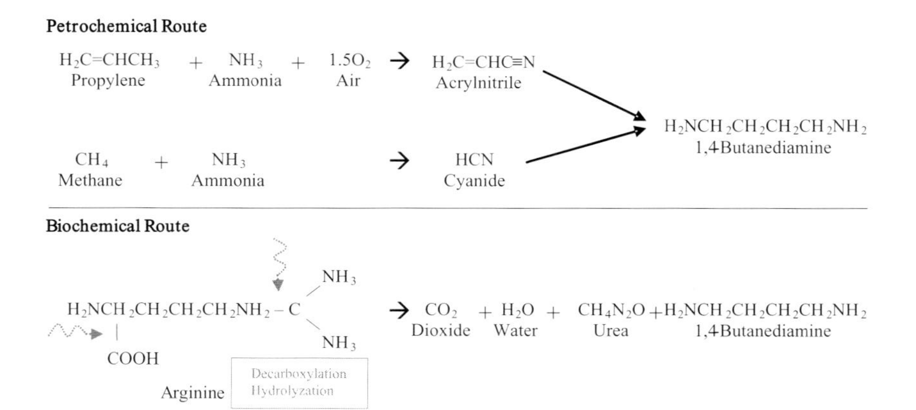

Figure 12.8 Chemical and biochemical process routes for nitrogen functionality. Adapted from Ref [35].

12.4.3
Technical Gaps and Future Development Considerations

For most biotechnological processes, milder temperatures and pressures are feasible. Even when taking into account the indirect production of enzymes or microbes, relatively low process energy and emission-based intensity are present. The big 'but' with these processes is water concentration. Biomass, biomass by-products and biotechnological processes generally contain or require vast quantities of water. The current separation, isolation and purification of intermediaries and final products might be energetically less intense but face a large economic burden. Several key separation technologies for bio-based material streams are simply missing on the industrial scale. To continue with the protein example above, isolation and purification of arginine from a protein-containing waste stream requires a cascade of separation steps; including expensive proteases, liquid chromatography and/or ion exchange membranes. While the emission savings potential is clear and valid, the economical and large-scale bioengineering processing equipment are vague or are completely lacking. Vast research and development in the purple field of biotechnology is required to enable the economic construction of biorefineries.

As progress in all fields of biotechnology advance, more products and more streams will become available for processing. In a multistream biorefinery, the main product will remain those based upon fermentable sugars, as these contribute to the vast majority of the biomass feedstock. These production systems will naturally evolve to become more efficient and more economically competitive. Enabling the technical use of the by-products for other useful production lines will further benefit the overall economic and ecological attractiveness. Unlike with a thermal oil cracker, finding the right temperature and catalyst will not suffice in making more streams available for processing. A biorefinery must incorporate new and tailored separation systems. The more useful products that emerge from the input biomass feedstock, the more CO_2 savings can be realized. Only when combined in a biorefinery can industrial biotechnological options fully realize their competitiveness against the well-established petrochemical industry. It is not satisfactory to rely on the end-of-oil strategies. In the climate debate, biotechnology already presents many potential advantages; the task will be to create economical advantages as well.

12.5
Determination of CO_2 Emissions in Processes of Chemical Industry

12.5.1
Data Generation

Companies that wish to assess the CO_2 emissions caused by the production of their products are facing the challenge of gaining detailed information about their processes. Moreover, a comparison of technologies used with other more sustainable alternatives (e.g., the shift from chemical to biotechnical processing) is necessary to evaluate the life cycle of products (see Chapter 1), showing the impact of possible process changes on CO_2 emissions. To assign a specific assessment with valid data, two complementary paths are conceivable. Firstly, second-hand data can be obtained and adjusted for the own issue. This kind of research is associated with lack of precision. In general, data has been generated for other issues involving different circumstances, for example an average case translates into mean values with broad assumptions of basic conditions and technologies. Furthermore, some studies offer data that is too aggregated to be exploitable for other situations. For example, in a study by the IFEU Institute the production of ethanol emits 42.0 gCO_2-eq./MJ in the agrarian stage and 62.6 gCO_2-eq./MJ in the industrial stage, leaving 0.4 gCO_2-eq./MJ for the distribution and the storage of fuel grade ethanol (see also Table 12.7) [36]. But if, for example, an ethanol producer uses more efficient production techniques and sources its feedstock from more sustainable producers, the emissions for the own case could be more favorable compared with the average branch. Secondly, individual research could be conducted to obtain more detailed and specific information. Particularly, small or medium-sized companies, however, rarely have the resources to conduct intensive screening. Therefore, a compromise between effort and accuracy is necessary to deliver a satisfactory proxy for actual conditions involved. Internally modeling the LCA is conducted in general specifically enough and efficiently, if second-hand data is used for parameters that cannot be influenced directly by the regarded company (e.g., the CO_2-emissions of average suppliers or for transport operations). Several institutions offer databases with common life cycle inventory (LCI) data which deliver standardized environmental and casually economical information for a specific region. For example, the average GHG emissions for the transport of goods per kilometer and weight in certain regions. A list of databases containing typical LCI is provided by the European Commission [22]. As location, capabilities, resource disposability and available technologies vary between companies and regions in general, these data need to be adjusted to fit the specific internal case. This leaves much room for interpretation upon conversion to real world conditions.

12.5.2
The Diversity in the Interpretation of Data

Numerous LCAs have been published to date, which tackle a wide variety of issues such as the comparisons of raw materials for certain products. However, different studies are not easily comparable, sometimes leading to contradicting conclusions. Table 12.7 shows an assortment of studies regarding CO_2 emissions for the production of ethanol as fuel alternative, derived from renewable feedstocks. In the case of bioethanol made from sugarcane the standard deviation of the pictured sources is around 30% for the base case. A study conducted by Fehrenbach *et al.*

Table 12.7 CO_2 Emissions for the production of bioethanol.

Biomass	Sugarcane	Sugarcane	Sugarcane	Sugarcane	Wheat	Wheat
Origin	Brazil	Brazil	Latin America	Brazil	North America	Europe
Source	Walter *et al.* [21]	Macedo *et al.* [20]	Fehrenbach *et al.* [36]	European Commission [22]	Walter *et al.* [22]	Fehrenbach *et al.* [36]
Cultivation	2.9	2.8	N/A	14.5	10.7	N/A
Other agricultural operations[a)]	17.3	15.6	N/A	0.9	29.6	N/A
Production of biomass	20.2	18.4	23.9	15.8	40.3	42.0
Industrial conversion	N/A	1.1	4.4	0.7	31.9	62.6
Distribution	1.6	2.3	1.8[b)]	0.4[c)]	1.5	0.4
Total emissions (Base case)	21.8	21.8	33.8	16.9	73.7	105.0
Credits for avoided emissions[d)]	N/A	9.4	N/A	10.3	31.8	N/A
Total emissions (With credits)	(21.8)	12.4	(33.8)	6.6	41.9	(105.0)
Direct LUC	N/A	N/A	180.1	N/A	N/A	47.8
Total emissions (Direct LUC)	(21.8)	(12.4)	213.9	(6.6)	(41.9)	152.8
Indirect LUC	N/A	N/A	48.5	N/A	N/A	69.5
Total emissions (Direct and indirect LUC)	(21.8)	(12.4)	262.4	(6.6)	(41.9)	222.3

All figures given in g CO_{2eq} MJ^{-1}.
a) For example, soil emissions, transport of biomass, etc.
b) Author's estimate. When transported to Europe 5.5 g CO_{2eq} MJ^{-1}.
c) If transported to Europe 7.7 g CO_{2eq} MJ^{-1}.
d) Byproducts (e.g., bagasse in the case of sugarcane) are used for energy generation.
Source: Based on [20], [21], [23], [36].

[36] compared the production of ethanol from sugarcane in Latin America and from wheat in Europe. In the base case Latin American bioethanol seems to be more sustainable in terms of CO_2 emissions than European bioethanol. But, if one considers the land use change (LUC) the conclusions could shift to the opposite.

Attempts to standardize LCAs have lead to the 'Code of Practice' [37] developed by the Society of Environmental Toxicology and Chemistry (SETAC) and the ISO 14040:2006 [38] standard established by the International Organization for Standardization (ISO). They provide principles and frameworks for the generation of LCAs. Regardless, a comparison between two separate studies from different institutions or individuals remains difficult (the studies shown in Table 12.7 follow these standards). The resulting climatic impact for the same issue can be positive or negative depending on the scope, the set boundaries, as well as different influencing drivers and different assumption settings. Moreover, a lack of transparency and the usage of aggregated data make a comparison tricky, as tradeoffs between diverse aspects occur regularly (e.g., costs vs. climate impact; see also Section 12.6). One of the topics most often assessed is that of the above-mentioned bioethanol for the use as fuel or as a fuel additive [21]. An overview of LCAs covering bioethanol for the use in automotive gasoline was performed by Hedegaard *et al.* and von Blottnitz and Curran [39, 40]. The main conclusion of these studies is that bioethanol leads in general to a net energy gain. Other environmental impact factors such as acidification, human toxicity and ecological toxicity impacts, however, tend to be more unfavorable for bioethanol [40]. In terms of reducing GHG emissions, recent doubts have been raised towards the application of biofuels. One example discussed was the conversion of land for industrial crop production, especially that of peat-land in Indonesia and Malaysia, since it is burdened with a carbon debt which may take several generations to approach break-even [41]. Other questions of biofuel impact have been posed, such as a disruption of food supply, increased water pollution, loss of biodiversity, and a lack of benefit of those directly affected by biofuels production [20].

12.6 The Three-Pillar Interpretation of Sustainability

12.6.1 Common Practice and Future Needs

Comprehensive sustainability is built upon three pillars which consider environmental, economical and societal aspects. These dimensions can be covered within the Sustainable Assessment (SustAss). It is based upon (i) the ecologically oriented LCA with the carbon footprint as an important element, (ii) the life cycle costing (LCC) and (iii) the social life cycle assessment (SLCA) [42]. Every assessment refers to a specific item (e.g., a product), typically evaluated from raw material acquisition to product disposal/recycling (cradle-to-grave). Ideally, the system boundaries of all three evaluation fields correspond with each other following consistent assumptions and are distinct to avoid double counting. Only under consideration of the entire life cycle, can problems referring to sustainability and their possible shifts to other issues when solved (e.g., the sourcing of renewable feedstocks could cause food shortage), be identified and anticipated [43]. Currently, the social dimension is rarely covered at the same level of accuracy as the economic and ecological dimension. This is due to the intrinsic difficulty in measuring it objectively. However, every dimension has a practical relevance, seeing that many multinational companies are shifting from ecological towards sustainability reporting. Analogously to the environmental LCA, guidelines for the SLCA have recently been published which cover the impact of a product towards different stakeholders (customers, workers, local community, etc.) [44]. A consideration must be made between many qualitative indicators (e.g., freedom of association and collective bargaining) and few quantifiable aspects (e.g., quantifying working hours per functional item). The SLCA does not lack indicators; over 200 social indicators are already available [45]. In fact, there is no consensus for how these should be incorporated in an SLCA.

12.6.2
Fuel vs. Food and other Misbalances

Ecological sustainable production and consumption often conflict with other ethical imperatives. Philosophers refer to such situations where two (or more) actions that are both morally imperative (or forbidden) cannot be realized simultaneously [46]. A paradigm of such a situation provides the use of biomass as a substitute for fossil resources. The required land area competes with agriculturally used land. The tortilla crisis gave some insight into this steeply growing potential for conflict – growing, as due to depletion the price of fossil resources rises, caused by a steep increase of world population. This growth, the increasing standard of living and the growing demand for energy supply based on renewable sources will lead to a bottleneck of land area in the future. To widen the area for cultivation, the use of marine-based feedstock is another option. For example, given the same solar irradiation algae can yield up to ten times more than typical land crops, since algae are able to use a greater proportion of the solar energy spectrum and is a much simpler organism, not wasting valuable energy on 'superfluous' things such as bark or flowers. Furthermore, algae reactors could be installed in areas which do not compete with food production. To date algae production is in its infancy. Metabolic optimization, nutrient recycling, and lower energy consumption during downstream processing need to be resolved. Competitive prices for algae below 500€/ton is unlikely to be reached in the near future.

Current research tries to reflect the above-mentioned regional differing impacts, where a market demand in one region could cause a disequilibrium in another region, with world models. For example, 'Model World 3' was developed in the 1970s to illustrate the mutual reaction between such factors as population, industrial growth, food production, exploitation of raw material reserves, land use and their influence on ecosystems [47, 48]. A later improved and updated version (1990s) added recent changes and innovations; for instance, more efficient technologies for energy generation and use, new materials, waste recycling and international conventions like the Kyoto protocol [49]. However, on the basis of these calculations, the conclusion was drawn that there are no borders for the innovativeness of mankind (which leads to a further efficiency) and the development of future generations (e.g., shrinking population in developed countries and growth in poorer regions) [50]. Other world models like the Bariloche model [2, 51], the DICE and RICE [52] or Stern's PAGE model [53] have been developed, representing the impact of various economic and ecological features on human well being, both on a global scale and over various decades. Up to now details on the energy mix, human behavior, religious and political influences still largely remain unconsidered, because of finite computational power, or it is simply not known as to how represent these features. Commonly these models equate human well-being humbly with economic welfare, the gross domestic product (GDP) being the key variable. At the RWTH Aachen a world model has been developed that aims to represent human well-being in a more comprehensive way [54], thereby building on Anand and Sen's capability approach that was put into practice by the Human Development Index (HDI), an alternative measure of human welfare extending the GDP and the Human Poverty Index [55, 56]. Since 1990 these indices are often used in world-development-reports.

The question as to how much invest in biomass as a substitute for fossil resources today, always invokes the key question: How valuable are the resources in 60 years time (or less) compared with the value of land area for people's immediate needs. Prices, just like any future gains and losses, are discounted when they appear in the future. The discounting rate varies from one analysis to another, which renders the comparison between various analysis even more opaque (see Section 12.5.2). This is a particular flaw in the economic analysis of global warming [52, 53]. Particularly when non-monetary losses are involved, arguments for or against some discounting rate always has to invoke not only economic, but moral arguments as well.

12.7
Outlook

If renewable feedstocks are used as a substitute for fossil raw materials, prices of all valuable alternatives start to correlate, as has been shown for the case of sugar, when it has been converted to biofuel. The argument that the prices of renewable feedstocks are relatively stable is only true, if they are not used for new applications, rising their demand. Therefore, new technologies should consider their impact on prices in the long run. Fossil-derived products lack the ecological and ethical dimension, since they are not CO_2-neutral and their exploitation limits the opportunities of future generations. Biomass derived products such as sugar and starch for biofuels also entail certain ethical dilemmas, seeing that they are one of the main sources of food. A possible way to avoid this conflict would be the use of waste streams and non-edible biomass such as wood. Progress made in modern biotechnology is gradually increasing the possibility of converting lignocelluloses (cellulose and hemicelluloses) into fermentable sugars, when the food vs. fuel (or chemistry) discussion would no longer apply, at least superficially. In this context, it must be noted that food as such is not the limiting factor but arable land itself. Non-edible crops which could be cultivated on marginal lands would provide that best augmentations. On the other hand, additional processes for pretreatment would be necessary for lignocellulosic biomass which increases investment and utility consumption. Only if all the dimensions of a issue are favorable, will the production of products in the chemical industry be sustainable. Should white biotechnology further develop, a significant quantity of biofuels, bioenergies and also biochemicals would be made readily available. This would help the chemical industry to diversify their feedstock supply, even though they will be in direct competition with other branches and applications. Mirrored to fossil fuels, the chemical industry will have a greater access to biomass feedstocks, due to the higher added value involved. Currently, CO_2-neutral production is fiction. Biotechnology and renewable resources will, however, contribute to future development towards CO_2-neutral production systems

References

1 International Energy Agency (2007) Tracking Industrial Energy Efficiency and CO_2 Emissions, IEA Publications, Paris.

2 Rotmans, J., and de, Vries, B. (1997) *Perspectives on Global Change: The Targets Approach*, CUP, Cambridge.

3 Denis, N., Meiser, A., Riese, J. and Weihe, U. (2008): Setting the right course in biobased bulk chemicals, in McKinsey on Chemicals, Number I, Autumn 2008, p.22–26.

4 Fachagentur Nachwachsende Rohstoffe (2009) Nutzung landwirtschaftlicher Rohstoffe durch die Industrie, http://www.nachwachsenderohstoffe.de/service/daten-und-fakten/industrielle-nutzung/rohstoffmengen.html (accessed 22 April 2010).

5 Brehmer, B., Boom, R.M., and Sanders, J. (2009) *Chemical Engineering Research and Design*, **87**, 1103–1119.

6 Licht, F.O. (2009) F.O. Licht's International Sugar & Sweetener Report, 141 (30).

7 Frost&Sullivan (2007) Strategic Analysis of the European Food Starch Markets, Frost & Sullivan.

8 Frost&Sullivan (2008) U.S. starch market. Report N107-88, Frost & Sullivan.

9 ISTA_Mielke_GmbH (2009) Oilworld, http://www.oilworld.biz/ (accessed 20 April 2010).

10 Arndt, J.-D., Freyer, S., Geier, R., Machhammer, O., Schwartze, J., Volland, M., and Diercks, R. (2007) *Chemie Ingenieur Technik*, **79**, 521–528.

11 Bringezu, S., Schütz, H., O'Brien, M., Kauppi, L., Howarth, R.W., and McNeely, J. (2009) Towards sustainable prouction and use of resources: assessing biofuels. United Nations Environment Programme, Nairobi, http://www.unep.fr/scp/rpanel/pdf/Assessing_Biofuels_Full_Report.pdf (accessed 23 April 2010).

12 Nordhoff, S., Höcker, H., and Gebhardt, H. (2007) *Biotechnology Journal*, **2**, 1505–1513.

13 Datamonitor (2007) global biotechnology market, www.datamonitor.com (accessed 23 April 2010).

14 Marx, A., Wendisch, V.F., Kelle, R., and Buchholz, S. (2006) *Towards Integration of Biorefinery and Microbial Amino Acid Production*, vol. (eds B. Kamm, P.R. Gruber, and M. Kamm), Wiley-VCH Verlag GmbH, Weinheim, Germany, pp. 201–216.

15 Goldemberg, J. (2008) *Biotechnology and Biofuels*, **1**, 1–7.

16 Chen, J.C.P., and Chou, C.C. (1993) *Cane Sugar Handbook*, John Wiley & Sons, Inc., New York, USA.

17 Lamonica, H. (2005) Geração de Energia Elétrica no Setor Sucro-Alcooleiro. Workshop Álcool de Bagaço de Cana-de Açúcar, 24th – 25th February. São Paulo.

18 BusinessNewsAmericas (2009) Demand for green plastics on the rise.

19 PlasticsEurope (2009) The Compelling Facts About Plastics 2009, PlasticsEurope.

20 Macedo, I.C., Seabra, J.E.A., and Silva, J.E.A.R. (2008) *Biomass and Bioenergy*, **32**, 582–595.

21 Walter, A., Dolzan, P., Quilodrán, O., Garcia, J., da, Silva, C., Piacente, F., Segerstedt, A., and Sustainability, A. (2008) Analysis of the Brazilian Ethanol, Campinas.

22 European Commission (2009) LCA Tools, Services and Data. http://lca.jrc.ec.europa.eu/lcainfohub/databaseList.vm (accessed 25 April 2010)

23 Werpy, T., and Petersen, G. (2004) Top Value Added Chemicals from Biomass – Results of Screening for Potential Candidates from Sugar and Synthesis Gas, US Department of Energy.

24 Schäfer, T., and Christensen, M.W. (2008) Industrial Biotechnology: Innovations within Petrochemicals, 3rd China Petrochemical Summit. Bejing.

25 Brehmer, B. (2008) Chemical Biorefinery Perspectives, PhD Thesis, Wageningen University

26 Hansen, S.B. (2007) *International Journal of Life Cycle Assessment*, **12**, 50–58.

27 Evonik Industries (2009) Evonik Launches Bio-Based Polyamide Vestamid Terra. Evonik Press Release 06/24/2009, Evonik Industries

28 Mutlu, H., and Meier, M.A.R. (2009) *Macromolecular Chemistry and Physics*, **210**, 1019–1025.

29 Ecoinvent (2009) database ecoinvent data v2.1 www.ecoinvent.org (accessed 25 April 2010).

30 Thum, O., and Oxenbøll, K.M. (2008) *SOFW Journal*, **134**, 44–47.

31 Brehmer, B., and Sanders, J. (2009) *International Journal of Green Energy*, **6**, 268–286.

32 Rajvanshi, A.K. (1986) Biomass Gasification, in *Alternative Energy in Agriculture*, vol. 2 (ed. D.Y. Goswami), CRC Press, Boca Raton USA, pp. 83–102.

33 Prins, M.J., Ptasinskia, K.J., and Janssen, F.J.J.G. (2007) *Energy*, **32**, 1248–1259.

34 Lehmann, J., and Joseph, S. (2009) *Biochar for Environmental Management*, Earthscan, London.

35 Sanders, J., Scott, E., Weusthuis, R., and Mooibroek, H. (2007) *Macromolecular Bioscience*, **7**, 105–117.

36 Fehrenbach, H., Fritsche, U.R., and Giegrich, J. (2008) Greenhouse Gas Balances for Biomass: Issues for Further Discussion, Institute of Applied Ecology (IFEU), Issue Paper for the informal workshop, Brussels.

37 Consoli, F., Allen, D., Boustead, I., Fava, J., Franklin, W., Jensen, A., de Oude, N., Parrish, R., Perriman, R., Postlethwaite, D., Quay, B., Seguin, J., and Vignon, B. (1993) Guidelines for Life-cycle Assessment: A 'Code of Practice', Society of Environmental Toxicology and Chemistry (SETAC), Brussels and Pensacola.

38 ISO (2006) 14040 *Environmental Management – Life Cycle Assessment – Principles and Framework*, European Committee for Standardization (CEN), International Organisation for Standardisation, Geneva.

39 Hedegaard, K., Thyø, K.A., and Wenzel, H. (2008) *Environmental Science & Technology*, **42**, 7992–7999.

40 von, Blottnitz, H., and Curran, M.A. (2007) *Journal of Cleaner Production*, **15**, 607–619.

41 Fargione, J., Hill, J., Tilman, D., Polasky, S., and Hawthorne, P. (2008) *Science*, **319**, 1235–1238.

42 Grießhammer, R., Benoît, C., Dreyer, L.C., Flysjö, A., Manhart, A., Mazijn, B., Méthot, A., and Weidema, B.P. (2006) Feasibility Study: Integration of Social Aspects into LCA. Discussion paper from UNEP-SETAC Task Force Integration of Social Aspects in LCA, Freiburg.

43 Klöpffer, W. (2003) *International Journal of Life Cycle Assessment*, **8**, 157–159.

44 Benoît, C., and Mazijn, B. (2009) Guidelines for Social Life Cycle Assessment of Products, UNEP (United Nations Environment Programme, Ghent.

45 Steen, B., Hunkeler, D., Schmidt, W.P., and Spindler, E. (2005) Integrating External Effects into LCC. SETAC Europe Working Group on LCC, draft paper.

46 Hillerbrand, R., and Ghil, M. (2008) *Physica D Nonlinear Phenomena*, **237**, 2132–2138.

47 Meadows, D., Meadows, D.L., Randers, J., and Behrens, W.W. (1972) *The Limits to Growth*, Universe Books, New York.

48 Mesarovic, M., and Pestel, E. (1974) Menschheit am Wendepunkt–2. Bericht an den Club of Rome zur Weltlage, Stuttgart.

49 Meadows, D., Meadows, D.L., and Randers, J. (2004) *Limits to Growth: The 30-Year Update*, Chelsea Green Publishing, Post Mills, VT, USA.

50 Meadows, D., Meadows, D.L., and Randers, J. (1992) *Beyond the Limits: Confronting Global Collapse, Envisioning a Sustainable Future*, Chelsea Green Publishing, Post Mills, VT, USA.

51 Herrera, A.O. (1976) *Catastrophe or New Society? A Latin American World Model*, International Development Research Centre, Ottawa.

52 Nordhaus, W. (2008) *A Question of Balance. Weighing the Options on Global Warming Policies*, Yale University Press.

53 Stern, N. (2006) *The Economics of Climate Change*, CUP, Cambridge.

54 Kopriwa, N., and Pfennig, A. (2008) *Chemie Ingenieur Technik*, **80**, 1393–1394.

55 Anand, S., and Sen, A. (1997) *Concepts of Human Development and Poverty: A Mulitdimensional Perspective* (ed. H.D.R. Office), UNDP (United Nations Development Programme), New York, pp. 1–9. Publisher:

56 Anand, S., and Sen, A. (1994) *Human Development Index: Methodology and Measurement*, UNDP (United Nations Development Programme), New York.

57 Thompson Reuters Datastream (2010) *Series Commodities*, available from Thompson Reuters Datastream.

Index

Managing CO_2 Emissions in the Chemical Industry. Edited by Leimkühler

ISBN: 978-3-527-32659-4

C